AF771437

Islam & Science

Series Editor

Shoaib Ahmed Malik, University of Edinburgh, Edinburgh, UK

The field of Islam and Science has been steadily developing since the 1980s, with authors writing about diverse topics like quantum mechanics, evolution, psychology, (scientific) miracles, methodological concerns, and the history of Islam and Science, to name a few. However, the field has been relatively slow in its development compared to the broader field of Science and Religion, which is mainly Christian-centric. One of the critical reasons for this is the lack of infrastructure for Islam and Science. With this gap in mind, this book series aims to be the first dedicated space exclusively for monographs and edited volumes on topics related to the field.

This book series not only recognizes the multidimensional nature of the discipline of Islam and Science but also aims to cater to its diverse dimensions for the benefit of the field's growth. This includes accommodating various methodological approaches (normative, descriptive, explorative, and constructive) and research foci (theology, philosophy, exegesis, jurisprudence, sociology, pedagogy, history, and more). Additionally, the series acknowledges the underrepresented voices within the Muslim world and seeks to create a platform for diverse representations of Muslims and their unique perspectives on the sciences.

Furthermore, the series acknowledges the global context of Islam and Science and the languages in which it has evolved, such as Arabic, French, Urdu, Turkish, Hausa, and Bahasa Indonesian, to name a few. Proposals for (complete or partial) translations of significant works into English with supporting critical essays explaining their significance are warmly welcomed.

For further information about the series, please contact Series Editor Shoaib Ahmed Malik at shoaib.malik@ed.ac.uk, or Palgrave Editor Amy Invernizzi at amy.invernizzi@palgrave-usa.com.

Aasim I. Padela
Editor

Bridging Bioscience and Islam

Engaging Big Questions about the Human Being

palgrave
macmillan

Editor
Aasim I. Padela
Department of Emergency Medicine
Medical College of Wisconsin
Milwaukee, WI, USA

ISSN 3004-913X ISSN 3004-9148 (electronic)
Islam & Science
ISBN 978-3-032-09995-2 ISBN 978-3-032-09996-9 (eBook)
https://doi.org/10.1007/978-3-032-09996-9

© The Editor(s) (if applicable) and The Author(s), under exclusive license to Springer Nature Switzerland AG 2026

This work is subject to copyright. All rights are solely and exclusively licensed by the Publisher, whether the whole or part of the material is concerned, specifically the rights of translation, reprinting, reuse of illustrations, recitation, broadcasting, reproduction on microfilms or in any other physical way, and transmission or information storage and retrieval, electronic adaptation, computer software, or by similar or dissimilar methodology now known or hereafter developed.
The use of general descriptive names, registered names, trademarks, service marks, etc. in this publication does not imply, even in the absence of a specific statement, that such names are exempt from the relevant protective laws and regulations and therefore free for general use.
The publisher, the authors and the editors are safe to assume that the advice and information in this book are believed to be true and accurate at the date of publication. Neither the publisher nor the authors or the editors give a warranty, expressed or implied, with respect to the material contained herein or for any errors or omissions that may have been made. The publisher remains neutral with regard to jurisdictional claims in published maps and institutional affiliations.

Cover illustration: © Pobytov via Getty Images

This Palgrave Macmillan imprint is published by the registered company Springer Nature Switzerland AG
The registered company address is: Gewerbestrasse 11, 6330 Cham, Switzerland

If disposing of this product, please recycle the paper.

Foreword I

The major theme of this book is the contemporary Muslim discourse on issues that lie at the interface of religion and science during the modern period. It is rather well known that the dominant shaper of this interface was not Islam but a West that had progressively distanced itself from traditional religion, thereby putting the relationship between religion and science on an entirely new epistemological foundation and toward a new kind of material civilization. The nature of the discourse is thus extremely important to the ongoing quest of the global Muslim *ummah* to become a competitive and ethically and morally based knowledge-driven community, which is itself a *raison d'être* of Islam, especially given Islam's scripturally based profound appreciation of religion and science as two of the major pillars of a praiseworthy and an exemplary human civilization.

Psychologically, the quest in question could be an emotive one for many Muslims since in their mind they are used to entertaining the idea that in Islam the harmony of religion and science is an axiomatic truth. Moreover, they have acquired some knowledge that their spiritual and intellectual ancestors have created historical precedents in the civilizational construction of a harmonious relationship between science and religion that keep reminding them of that harmony as a historical truth.

Islam's sympathy for the philosophical doctrine of harmony between religion and science is known to run deep into its very heart, which is the Qurʾanic revelation. As such, the tension and conflict that characterizes the greater part of the contemporary relationship between religion and science could only be a major source of intellectual disillusionment to many intelligent Muslim minds, if not anger. This book has captured some aspects of this tension and conflict, especially their epistemological dimension and their implications for the present and future Muslim scientific agenda and education. It is a great credit to the contributors of this book that, notwithstanding the immensity of the epistemological and ethical challenges posed by modern and contemporary science to Islam's philosophy and vision of science, they are courageous and ingenious enough to respond to the challenge by addressing selected topics that are rather instructive in displaying tensions between traditional Islamic thought and contemporary science that demand intellectually satisfactory outcomes.

The book's discourse on the tensions and conflicts between traditional Islamic thought and contemporary scientific thought is, however, confined to the discipline of bioscience. But this limitation in disciplinary scope in no way lessens the epistemological and civilizational significance of the discourse. In fact, it sharpens the significance of the issues under discussion. There is a good reason for this observation. The book discusses issues at the intersection of religion and science, focusing on what its editor, Aasim Padela, refers to as "Big Questions" about the human being. Now, the human being is generally accepted in Islam and in the pre-modern West as the microcosm of the universe, meaning that the human being mirrors the cosmos, or the human being is the cosmos in miniature. In the Islamic tradition, the human microcosm is termed *al-ʿālam al-ṣaghīr*, which literally means "the small universe." In the light of the Qurʾanic idea of creation, the human microcosm is understood as a hierarchically ordered reality. It may be further explained that the essential content of the cosmos is identical to that of the human being. Further, there is a one-to-one ontological correspondence between levels of reality in the human being and in the cosmos. This meaningful microcosmic understanding of the human being would imply that, in dealing with the Big Questions about the human being, the book is, in fact, addressing itself to similar questions about the cosmos of which our natural world (*al-ʿālam al-ṭabīʿī*) is an integral part.

Epistemologically speaking, bioscience easily stands as the most significant of the natural sciences, especially when viewed from the perspective of our attempt to fully comprehend the Big Questions about human beings. It is also within the discipline of bioscience that we could perhaps witness the most intense and the most problematic epistemological dispute between the Islamic theological-philosophical-scientific position and modern biological thought. Apparently, as is also made clear in this book, the core issue in this dispute pertains to the meaning of humanness. As such, the importance of the Islamic understanding of the human microcosm to the epistemological bridging agenda between Islamic scientific and philosophical tradition and modern bioscience could hardly be overemphasized.

Humanness emerges in this book as a major issue in the intersection between religion and science in general and between Islam and bioscience in particular. The idea of humanness discussed here underpins five Big Questions about the human being that the book has identified for its discourse, particularly from the contending modern bioscientific and Islamic perspectives. The editor describes the Big Questions as about the "origin, nature, fatedness, malleability, and uniqueness" of the human being. As far as ideas go, be these expressed in the form of questions or otherwise, bigness is to be understood in the conceptual or epistemological sense. The bigness of the idea is about its knowledge content or the size of its semantic field. Quite clearly, the ideas embodied in each of the five questions are neither of the same epistemological magnitude nor of the same degree of complexity insofar as their problematic nature is concerned. For example, the issue of the origin of human beings is widely acknowledged as more difficult to solve than the issue of fatedness or malleability. Notwithstanding all these considerations, in the final analysis, we contend that answering these five questions would require answering the more fundamental question of humanness, regardless of whether the answering attempt comes from bioscientific or Islamic perspectives.

The issue of the epistemological magnitude of each of the five questions may also be investigated and discussed from multidisciplinary and interdisciplinary perspectives. In Islamic tradition, each question may be approached from the viewpoint of several disciplines. The most well-established disciplines that engage these questions and their likes in some or other aspects are theology (*ʿilm al-kalām*), metaphysics, cosmology, psychology, biophysics, and biomedicine. Several chapters of this book address issues related to the five questions by means of an interdisciplinary approach. This attempt is praiseworthy since it is in conformity with the spirit of *tawḥīdic* epistemology that emphasizes both the horizontal and vertical interconnectivity of concepts and their unity. It has the real potential to generate a new interest in bridging between the traditional Islamic sciences, both *naqlī* (transmitted) and *ʿāqlī* (rational), and the modern biosciences.

The bridging initiative undertaken in this book and other initiatives, that we fervently hope will follow in the coming years, stand to benefit from new opportunities of our time, particularly in the form of fresh interpretations resulting from contemporary intersections between traditional Islamic thought and modern bioscience. These fresh interpretations pertain to both theoretical and practical issues of humanness. Significantly, the initiative is undertaken at this moment when the very "humanness" fabric of our global civilization is under constant assault with the help of increasingly destructive war technologies and steadily disintegrating. Indeed, even more scary is the specter of an emerging humanity that is still human only in name but that has lost sense of the true notion of humanness.

In one of my very early works, I argued that there is an urgent need for Muslim scientists and philosophers of science to revisit the traditional idea of "gradation of beings" that used to be applied as a supreme integrative principle in pre-modern biological sciences (Bakar 1987). This philosophical idea was known in traditional Islamic thought as *marātib al-mawjūdāt*, literally meaning hierarchy of existence. The corresponding idea in the Western tradition was the *great chain of being*. The gradation of beings refers primarily to a vertical hierarchy stretching from the lowest material form through humans to God. I emphasized in the cited work that "gradation in this sense is essentially qualitative and supra-temporal in nature. The temporal aspect of gradation is, however, not denied or ignored" (Bakar 1987: 5). Thus, in some classical Muslim biological works, we encounter arguments that the appearance of minerals after the elements, plants after minerals, animals after plants, and humans after animals, is temporal as well as in principio. The temporal or historical aspect of vertical gradation is of relevance to the discourse in this book since it is this area of inquiry that almost wholly interests modern bioscience.

From the perspective of Islamic epistemology, it is imperative that a comparative study be undertaken between the traditional theory of the vertical gradation of beings and the modern biological theory of evolution founded by Charles Darwin with the view of determining their convergence and divergence. Both theories seek to explain the phenomena of biodiversity and the natural order underlying them. We are also deeply interested in the two mega theories because each claims to have the distinction of being the supreme integrative principle that unites all the biological sciences. Both claims are, of course, debatable, but an enlightened discourse on them is worth pursuing since at stake is nothing less than the saving of the true meaning of humanness itself.

Integral to the role and function of an integrative principle in the sciences, particularly of the comprehensive type, is its ability and capacity to provide a conceptual scheme of interpreting, integrating, and synthesizing diverse sets of facts and theories. The epistemic status of the two theories in question as supreme integrative principles for the biosciences thus needs to be investigated. While all the five Big Questions are common to the rational speculations in bioscience and in Islam, significant parts of their respective answers are not. It is worth emphasizing here, though, that the greater part of the divergence in their answers stems not from the disparity between them in a wealth of scientific data in their possessions but rather from differences of interpretations, especially philosophical ones, of the same data. When it comes to big ideas like the ones that are addressed in this book, there are often multiple factors that may simultaneously influence the adoption of a scientific interpretation or a scientific theory.

Quite instructive in this respect is the issue of conflicting interpretations that have arisen from more recent research initiatives on the issue of humanness. In the era of religious revival in which we now live, humanness from the religious perspective has appeared to gain currency. However, providing a counter-narrative to this religious trend is the emergence of a new evolutionary perspective on humanness. Interestingly, however, a new challenge to the evolutionary humanness perspective emerges not from religion per se but from within the ranks of biologists. The core issue in the new contention is the humanness of the human genome. The generator of the new debate is the ENCODE (The

Encyclopaedia of DNA Elements) Project, which was sponsored by the US government through the National Institutes of Health and undertaken by the National Human Genome Research Institute (NHGRI). This mega research project on the landscape of the human genome has produced new insightful findings into the organization and regulation of human genes and genomes. These insights perhaps sum up the overall significance of the project.

What ENCODE revealed about the HG landscape mainly pertains to the active roles of the functional elements of the HG. The first phase of the HG project (1990–2003) made the revelation that 98 percent of HG was "junk, barren, non-functioning, and dark matter" and only 2 percent of it was active. But, in a big contrast, the second phase of the HG project (2003–2012) produced a startling new discovery. The essence of this discovery is that the so-called "junk DNA" is not junk after all, but at least 80 percent of it is found to be biochemically active and functional. It is not junk in the sense of being useless, but rather, it is essential DNA that contains all the crucial regulators that control gene expression. The vast stretches of seeming "junk DNA" are, in fact, the seat of crucial gene-controlling activity. It is pointed out by some of the researchers that the earlier so-called "junk DNA" is essential for successful embryonic development, and it contains the blueprint and tools required to create a baby.

There is an instructive lesson in philosophy of science, particularly on the issue of possible conflicting interpretations of research data. We witness here scientists clashing on ENCODE's genetic data generated by the research project. Frictions emerge over the use of the word "function" that may evoke contrasting meanings among evolutionists and non-evolutionists. Evolutionary geneticists are critical of the research findings because they view such terms as active, functional, gene-regulators, or controllers as undermining their evolutionism. Ideologically "neutral" geneticists welcome the findings on scientific grounds. Also welcoming the findings are the philosophically oriented scientists who see the life blueprint in HG as being irreducible to physics and chemistry.

These are some of the major issues that are epistemologically related to the five Big Questions that this book has raised about the human being as seen from the perspectives of the Islamic tradition and bioscience.

Editor Padela and the contributors of the book are to be congratulated for making available to readers a discourse in a single volume on multi-disciplinary and interdisciplinary approaches to the intersection between Islamic traditional thought and modern bioscience. *Wa bi'Llah al-tawfīk wa'l-hidāyah wa bihi nasta'in.*

Kuala Lumpur, Malaysia

Prof. Datuk Osman Bakar
Rector, International Islamic University Malaysia
Al-Ghazali Chair of Epistemology and Civilizational Studies and Renewal, International Institute of Islamic Thought and Civilization (ISTAC)
International Islamic University Malaysia
Emeritus Professor of Philosophy of Science, University of Malaya

Reference

Bakar, Osman. (ed.). 1987. *Critique of Evolutionary Theory: A Collection of Essays*. Kuala Lumpur: The Islamic Academy of Science.

Foreword II

This book, *Bridging Bioscience and Islam: Engaging "Big Questions" about the Human Being*, is the fruit of the authors' dedication to provide Muslim scientists, teachers, and students with an epistemological framework through which to navigate a myriad of critical discussions that cross religion and science. This noble effort should be applauded, and we pray that Allah accepts this work, makes it beneficial for its readers, and rewards the authors with an everlasting reward.

Seeking knowledge is obligatory for Muslims, but what kind of knowledge—conclusive, speculative, or both? The sources of true and conclusive knowledge are true reports (*al-khabar al-ṣādiq*), the intellect (*al-ʿaql*), and the five external senses. Islamic scholars (*ʿulamāʾ*) focus their efforts on acquiring conclusive knowledge as it pertains to beliefs, salvation, reward, and sin. The source of this conclusive knowledge is testimonial true reports (*al-khabar al-ṣādiq*) found in verbatim and non-verbatim revelation (*waḥy*), known as the Quʾran and Sunnah, respectively. Scientists tend to focus their efforts on gaining empirical knowledge through the five external senses. The invention of the microscope and the telescope accelerated the acquisition of sensory knowledge by informing human beings about the universe through enhanced observations of the universe itself.

The area where Islamic scholars and scientists clearly overlap is in knowledge gained from the intellect (*al-ʿaql*), which provides humans with rational knowledge. The human intellect can acquire knowledge within the limits of time and space. The Muslim must appreciate that acquiring knowledge that is beyond time and space requires revelatory sources that are supra-rational. For example, revelatory knowledge informs the human being about their existence before life in this world, their future existence in the grave, and their life in the hereafter. For the Muslim scientist, whose preoccupation should be to enhance their intellect, it is critical to appreciate that revelatory knowledge was revealed to the sound human intellect of Prophet Muhammad (pbuh).[1] What were the critical manifestations of the Prophet's (pbuh) sound intellect that existed prior to receiving revelation and supported his *qalb salīm* (content heart) that received revelation? How were the Prophet's (pbuh) *ʿaql* and five external senses enhanced by knowledge from revelation? What can the Muslim scientist learn from these manifestations that could unleash the creativity of their own *ʿaql* and empirical senses?

The critical attribute of the Prophet's (pbuh) sound ethical behavior is that he is *maʿṣūm*—protected from sin. The *ʿulamāʾ* identified this incredible feature of prophethood as the foundation for the Prophet's (pbuh) sound intellect. Even before receiving revelation, the Prophet's (pbuh) sound intellect restrained him from the greatest sin of *shirk*—worshiping idols or associating anyone or anything with the one true creator, Allah. The Prophet's (pbuh) ethical behavior was another clear manifestation of his sound intellect, as he was known among his people as as-Ṣādiq al-Amīn, the truthful and trustworthy one. Muslim scientists facing ethical challenges can find comfort in following the example of the Prophet (pbuh) by maintaining these fundamental ethics of truthfulness and trustworthiness in their current research and practice.

Based on the distinct categories of revelational, rational, and empirical knowledge, one may logically assume that developing and enhancing the ability to access them are also distinct and independent from one another. For example, a scientist may think that only scientific training will uniquely develop one's intellectual and empirical abilities to perform better science. However, knowledge from revelation enhanced the Prophet's empirical abilities as he "saw" the punishment of people

[1] Shorthand for may God's salutations, blessings and peace be upon him; Muslims are called to honor all Prophets of God by making this sort of prayer when mentioning a Prophet. Herein we have opted to use pbuh for shorthand to signal instead of the formal Arabic prayer. Moreover, in the main part of the text, we have omitted the honorific for ease of reading though we use the title of Prophet to signal the prayer is called for.

in their graves, and he "heard" the pens documenting destiny that exist beyond the heavens.

In the post-revelation period, Muslims excelled in their pursuit of scientific knowledge by accessing knowledge from revelation to inform and reform their intellect. The additional support for Muslim scientists along this path was their personal worship, prayers, *dhikr* (remembrance) of Allah, and following the Prophet's (pbuh) example of asking Allah for an increase in certain knowledge and perfection of behavior. Like their predecessors, today's Muslim scientists and students can also unleash the creativity of their *ʿaql* by not simply studying creation, but rather by connecting the knowledge of creation back to the Creator who is the source of all conclusive knowledge.

Shaykh Mohammed Amin Kholwadia
President, Darul Qasim College
Glendale Heights, Ilinois, USA

Acknowledgments

The idea for this book emerged during meetings for the *Fostering Connections between Science and Theology for Medical Track Muslim Students* project (greater details on this project are in Chapter 1). Our team of bioscientists, theologians, and educators felt that rather than authoring disparate academic and public articles, we should gather our work together in an edited volume for greater cohesion and impact. After many months dedicated to drafting the book proposal, soliciting publishing houses, conducting research, outlining, drafting, and revising chapters, and, of course, praying for blessings and acceptance, here we are. The accomplishment is no easy feat and many deserve appreciation.

To begin with, I express my sincere gratitude to Rashid Dar, formerly of the John Templeton Foundation (JTF), who shepherded the grant proposal and provided critical insights into the value propositions that would strengthen it. His assistance in widening my scholarly network by introducing me to kindred spirits in related fields was, and has been, invaluable. While Rashid facilitated the project getting funded, I would be remiss not to acknowledge Matthew Walhout, Vice President and Chief Programs Officer at JTF, who provided me with encouragement, advice, and support as I navigated the tumult of significant changes in my personal and professional life, as well as changes at JTF, during the grant period. Without his advocacy for and acceptance of changes in the project work plan and budget, this book would not have been possible.

The lion's share of appreciation is owed to the chapter authors: Omar Qureshi, Macksood Aftab, Mohd Yusof Mohamad, Abderezak Abdulahi Hashi, Bilal Ali Ansari, and Ahsan Arozullah, each of you has been with the project even prior to its inception and has put up with my task master alter ego who critiques much more than he encourages, may your patience be rewarded and your work be of immense benefit. Nazif Muhtaroğlu and Usama Javed Mirza joined our writing team much later on, and your pieces greatly improved our volume. I also appreciate Dr./Sh. Rafaqat Rashid's quick assistance and gracious understanding with the editorial process. Thank you most kindly. Most of all, however, it is Raudah Yunus who deserves collective gratitude. She single-handedly managed the multinational, multistakeholder, and multichallenge-riddled project with aplomb. Beyond this, she authored several chapters and dove in to assist each of the authors wherever she could. Your work has been exceptional, your dedication admirable, and your demeanor a comfort. May Allah (pbuh) reward you without measure. After Raudah's ministrations, Thérèse Wassily Saba, came in to perform the painstaking tasks of formatting and editing the chapters and bibliography; I thank you most kindly. Beyond these individuals, I also acknowledge the important contributions of Hamza Karamali, who contributed to and co-taught the epistemology-related seminar (see Chapter 2), and Shoaib Malik who not only co-taught the human origin-related seminar, but he also provided critical feedback on early versions of some chapters, guided us in our revisions, and helped see our book through to acceptance at Palgrave Macmillan in the *Islam and Science* series. I am also grateful to Amy Invernizzi, Senior Editor at Palgrave Macmillan, for her kind advice as we responded to various reviewer comments and critiques, and to her colleagues Steven Fassioms and Melvin Thomas in getting this to print. The anonymous reviewers of our grant and book proposal are also owed appreciation. I also thank all those scholars, friends, and teachers who have helped me in my academic and personal journey to bridge the worlds of Islam and bioscience, you know who you are, and I pray you are rewarded immensely.

Last but never in the least, I am ever-grateful to my wife, Maryam, who has supported each of my academic endeavors, sometimes at great cost to herself and our family. I greatly appreciate your sacrifices and support, particularly during the trying times that accompanied this project. I also acknowledge our children, Aaleeyah, Maaria, Ahmed Fateh, and

Faathima, who give me hope in the future of humanity, and sacrifice their time with me so that I may complete projects such as this.

Finally, all praise and gratitude are due to Allah for granting us with, and protecting, our religion, life, wealth, progeny, and intellect, for showing us His signs in the cosmos and within ourselves, and for sending the Prophet Muhammad with guidance and wisdom so that he may illuminate the path toward felicity in this life and the one to come. I pray that He makes this work a source of continuous reward for all those involved, and an immense benefit to its readers.

Aasim I. Padela, MD, MSc

Praise for *Bridging Bioscience and Islam*

"*Bridging Bioscience and Islam* edited by Aasim Padela, brings together Muslim theologians, scientists, and educators to examine fundamental questions about the human being at the intersection of Islamic thought and contemporary bioscience. Addressing themes such as human origins, free will, uniqueness, and emerging biomedical possibilities, the volume models a careful and multidisciplinary engagement that is intellectually serious and pedagogically attentive. Rather than assuming either a straightforward conflict or facile harmony, the contributors seek to clarify points of real tension and convergence but in a way that is accessible to students and valuable to scholars. This work will be of great interest to those concerned with religion and science, Islamic theology, bioethics, and the formation of Muslim students navigating modern scientific education."

—Safaruk Chowdhury, *Lecturer, Cambridge Muslim College, United Kingdom; author of* Islamic Theology and the Problem of Evil (2021)

"*Bridging Bioscience and Islam* is a compelling and urgently needed resource for educators, offering a rich intellectual framework for teaching science while engaging the profound questions students already carry about life, human origins, agency, and purpose. By challenging the assumed divide between Islam and bioscience, this volume empowers educators to move beyond compartmentalised curricula and model an integrated, confident approach to knowledge—one that nurtures coherent identities, ethical depth, and rigorous scientific inquiry."

—Nadeem Memon, *Associate Professor, Centre for Islamic Thought and Education, Adelaide University, Australia; editor of* Philosophies of Islamic Education: Historical Perspectives and Emerging Discourses (2016)

"*Bridging Bioscience and Islam* offers an intricate and deep dive into pre-modern Islamic theology, retrieving its rich intellectual traditions to serve as vital resources for contemporary bioethical inquiry. Bridging classical scholarship with modern bioscience allows authors to demonstrate how early frameworks of knowledge might provide meaningful answers to the most complex "Big Questions" facing humanity today."

—Ebrahim Moosa, *Mirza Family Professor in Islamic Thought & Muslim Societies, University of Notre Dame; author of* Muslim Theological Encounters with Science (2026)

"Asking what it means to be human is to stand at the crossroads of knowledge traditions. As bioscience probes origins, capacities and futures through empirical inquiry, Islam reflects on the human being through revelation, metaphysics and moral purpose. This book inhabits that crossroads, not to arbitrate between competing truths but to cultivate a disciplined conversation; one that bridges methods, honors epistemic limits and realigns the human being as both a biological reality and a theologically meaningful subject. In doing so, the book offers a timely and rigorous contribution to contemporary conversations at the intersection of Islam and science, specifically the life sciences."

—Shaikh Mohd Saifuddeen, *Institute of Islamic Understanding Malaysia, Kuala Lumpur, Malaysia*

Contents

Notes on Contributors

Macksood Aftab is a neuroradiologist and chair of radiology at Ascension Genesys Hospital. He is also clinical assistant professor at Michigan State, Central Michigan, and AT Still Universities. He is also founding editor of the *Journal of Islamic Philosophy*. Dr. Aftab completed his residency and fellowship training at Tufts and Harvard Medical School, respectively. He attended the AT Still University School of Osteopathic Medicine, where he earned his DO and MHA degrees. He holds a bachelor's in philosophy from the University of Michigan and a master's in history of science from Harvard University where his thesis title was "Historicizing Ghazali's Critique of Ibn Sina and Its Influence on Islamic Science."

Bilal Ali Ansari is an educator and research scholar with a background in secular and Islamic education. He earned a Bachelor of Arts degree with a focus on adult learning theory at National-Louis University in Lisle. He also graduated with a degree in Arabic and Islamic studies from *Jāmiᶜah Dār al-ᶜUlūm*, Karachi, and specialized in Hadith Studies at the *Jāmiᶜat al-ᶜUlūm al-Islāmiyya*. He is currently continuing his seminary studies in discursive theology, Islamic law, and Islamic philosophy at Darul Qasim College, where he is Head of the Department of Hadith Studies and lead academic advisor.

Ahsan Arozullah is Executive Vice-President, Research and Development, and Chief Medical Officer of TerSera Therapeutics. Dr. Arozullah completed his medical training at Northwestern University, followed by a fellowship at Harvard in clinical epidemiology, and has over 20 years of clinical research experience. Dr. Arozullah is a member of the Board of Directors at Darul Qasim College. Under the leadership of Shaykh Mohammed Amin Kholwadia (the founder of Darul Qasim College), Dr. Arozullah leads the bioethics cohort at the Institute.

Abdurezak Abdulahi Hashi obtained his bachelor's degree, his master's degree, and his PhD from the International Islamic University Malaysia (IIUM) in Theology and Comparative Religion. In 2014, he was awarded the Best Teacher award and appointed as the editor of the Kulliyyah of Science academic journal, *Revelation and Science*. Currently, Dr. Hashi is a professor at the Department of Biotechnology and the coordinator for the Multi-Disciplinary Academic Postgraduate Program.

Usama Javed Mirza has a Bachelor of Science from LUMS (Pakistan), where he studied physics and philosophy, and a Master's of Education from Columbia University, where he studied as a Fulbright Scholar. He is a New York state-trained emergency medical technician, an experienced educator, and a social entrepreneur. He is currently researching science, Islam, and education at the University of Cambridge as a PhD student in Education on the Gates Scholarship.

Mohd Yusof Mohamad is an assistant professor at the Kulliyyah of Allied Health Sciences, International Islamic University Malaysia (IIUM). He received his undergraduate and postgraduate education in the fields of Biomedical Sciences, Human Anatomy, and Tissue Engineering from IIUM, and a postgraduate Diploma in Islamic Studies from the same institution. Mohd Yusof currently teaches anatomy, physiology, and critical thinking at the Kulliyyah of Allied Health Sciences.

Nazif Muhtaroğlu is a research scholar at Brandeis University. He received his Bachelor of Arts in philosophy from Boğaziçi University (Istanbul), his PhD in philosophy from the University of Kentucky (2012), and completed postdoctoral studies at the Near Eastern Languages and Civilizations Department at Harvard University (2012–2014). He edited *Occasionalism Revisited* (KRM, 2017) and co-edited volumes in the Islamic Philosophy and Occidental Phenomenology in Dialogue Series (Springer, 2010 and 2014). His research concerns the history and philosophy of science, Islamic-Ottoman Intellectual history, modern philosophy, contemporary religious epistemology, philosophy of religion, philosophy of mind, and AI.

Aasim I. Padela is an emergency medicine physician, health researcher, and bioethicist whose multidisciplinary scholarship bridges the Islamic tradition and biomedicine using empirical, sociological, health services, bioethical, and discursive methods. He holds an MD from Weill Cornell Medical College and an MSc in Healthcare Research from the University of Michigan. He completed emergency medicine training at the University of Rochester, a research fellowship at the University of Michigan, a clinical medical ethics fellowship at the MacLean Center for Clinical Medical Ethics, and visiting fellowships at the Oxford Centre for Islamic Studies and the International Institute of Islamic Thought. He has authored over 170 peer-reviewed articles and book chapters and is an author/editor/co-editor of five books: *Medicine and Shariah: A Dialogue in Islamic Bioethics* (Notre Dame University Press 2021), *Islam and Biomedicine* (Springer 2022), *Organ Donation in Islam: The Interplay of Jurisprudence, Ethics and Society* (Lexington Books 2022), *Islam, Muslims, and COVID-19: The Intersection of Ethics, Health, and Social Life in the Diaspora* (Brill 2024), and *Maqasid al Shariah and Biomedicine: Bridging Moral, Ethical and Policy Discourses* (IIIT Press 2024).

Omar Qureshi is Director of Integrated Curriculum at Yaqeen Institute. He earned his Bachelor of Arts in Microbiology and a Master's in Curriculum and Instruction—Science Education, both from the University of Missouri-Columbia. He completed his PhD in Cultural and Educational Policy Studies with a specialization in the Philosophy of Education from Loyola University. Previously, he was the provost and assistant professor of Liberal and Islamic Studies at Zaytuna College, and lecturer in Western and Islamic philosophy at Cambridge Muslim College. Dr. Qureshi has also held the position of principal and dean of academics at a parochial school in Chicago.

Raudah M. Yunus is a researcher, writer, and social activist who recently completed a postdoctoral fellowship with Dr. Padela in Muslim community health research and bioethics at the Medical College of Wisconsin. Dr. Yunus obtained her medical degree (MBBCH) from Alexandria University (Egypt) and completed a Master of Public Health (MPH) and a Doctor of Public Health (DrPH) at the University of Malaya thereafter. She has previously taught at the Faculty of Medicine of Universiti Teknologi MARA in Malaysia. Her research areas span Islamic bioethics, behavioral gerontology, epidemiology of aging, long-term care, human trafficking, and the health of marginalized populations.

List of Figures

List of Tables

PART I

Preliminaries on the Engagement of Islam and Bioscience

CHAPTER 1

Introduction: Islam, Bioscience, and the Human Being

Aasim I. Padela

سَنُرِيهِمْ آيَاتِنَا فِي الْآفَاقِ وَفِي أَنفُسِهِمْ حَتَّىٰ يَتَبَيَّنَ لَهُمْ أَنَّهُ الْحَقُّ ۗ أَوَلَمْ يَكْفِ بِرَبِّكَ أَنَّهُ عَلَىٰ كُلِّ شَيْءٍ شَهِيدٌ

> We will show them Our signs in the universe and within themselves until it becomes clear to them that this [Qur'an] is the truth. Is it not enough that your Lord is a Witness over all things? (Q.41:53)

There are many Qur'anic verses that call on believers to reflect upon the natural world, to ponder its purposes, and, in doing so, bear witness to the majesty and might of the One who created the cosmos. Among these verses, the verse quoted above links the cosmos and human beings together as signposts signaling the existence of a Creator and notes that by reflecting on these signs, humankind will come to realize the veracity of the Qur'an.[1]

[1] There are multiple commentaries on this verse. Among them, some interpret the clause regarding signs within themselves as referring to the (miraculous) military and political victories achieved by the early Muslim community (see Tafsir ibn Kathir n.d.).

A. I. Padela (✉)
Medical College of Wisconsin, Milwaukee, WI, USA
e-mail: apadela@mcw.edu; draasim@medicineandislam.org

The Initiative on Islam and Medicine, Glendale Heights, IL, USA

© The Author(s), under exclusive license to Springer Nature Switzerland AG 2026
A. Padela (ed.), *Bridging Bioscience and Islam*, Islam & Science,
https://doi.org/10.1007/978-3-032-09996-9_1

To some extent, this verse inspired this book which uses scriptural and scientific data to better understand the human being.

This edited volume is several things: a discursive intervention, an educational resource, and a research deliverable. Relatedly, it seeks to serve multiple audiences, including academics, educators, and students of knowledge. Similarly, the volume gathers essays from diverse disciplinary experts—theologians, bioscientists, academics, and clinicians (some authors wearing multiple hats)—to address disparate topics about the human being. Despite this multiplicity, however, the book is unified: its scholarly focus is on addressing fundamental questions about the human being; it is wedded to a Sunnī perspective as it tackles these questions; and the chapters emerge from a single overarching project that designed, implemented, and evaluated a novel curriculum drawing on insights from Islam and bioscience to address fundamental questions about the human being. In order to aid the reader in navigating this "unity in diversity," in what follows, I describe the origin of this volume and the contexts it addresses, and then provide an overview of the chapters and how they relate to each other.

Why this Book? The Context for the Text

Three observations motivate this volume: (i) contemporary religion and science discourse often assumes a conflictual relationship within which Muslim engagement is uneven; (ii) biomedicine increasingly shapes our understandings of the human being, yet both premedical and medical education rarely provide curricular space to deliberate over fundamental questions regarding humankind; and (iii) curricula that bridge Islam and the biosciences could equip current and future Muslim allied health professionals with beneficial knowledge, skill sets, and capacities.

Religion and Science Discourse

One needs to look no further than their local newspaper or personal social media newsfeed to find headlines framing religion and science as being at odds. Illustratively, during the dark days of rapidly spreading SARS-CoV-2 virus infections, stories that featured religious leaders flouting social distancing regulations due to their belief in protection from God or religious adherents forgoing vaccines due to their beliefs were commonplace (Khurana 2020; Evans 2021; Reeves 2021). Similarly, articles portray religious scholars and scientists clashing over the ethical implications of artificial intelligence (McDonald 2017). This clash-based framing also plays out in bioethical and policy discourses where religious views are perceived as constraining the rights of the public and forestalling biomedical progress. Debates around abortion exemplify this problem as religious views on what constitutes a human being are seen as restricting access to reproductive healthcare, constraining liberty, and throttling the development of novel therapeutics to treat disease. This so-called conflict narrative reflects, or alternatively shapes, public perspectives, as a recent national survey of Americans found a majority to believe that religion and science are often in conflict (Funk 2015).

But how did we get here? There are many versions of an "origin" story. Importantly, notions of conflict are linked to developments in Europe during the sixteenth, seventeenth, and eighteenth centuries in eras termed the Scientific Revolution, the Age of Enlightenment, and thereafter. Several factors contributed to tensions between religion and science, including public reactions toward an overbearing Catholic Church that governed public and private lives, the development of philosophies that privileged rational inquiry to discern the workings of the natural world, and scientific discoveries that appeared to counter deeply held religious views about human beings and the universe. The thesis was formally developed, however, by historians John William Draper and Andrew Dickson White in the nineteenth century, which maintains that there is an intrinsic intellectual conflict between religion and science that inevitably leads to hostility (Draper 1881; White 1876).[2] Of particular relevance to this project, however, is that the Church, in the face of the rising power of science in the sixteenth and seventeenth centuries, came to define religion "as a set of propositions to which believers gave assent, and which could therefore be judged and compared between different religions and as against natural science" (Stevenson 2011: 96). This reduction of religion from a worldview and accompanying life practices into a set of beliefs was intended to bolster religious authority. Instead, it opened up an epistemic space for confronting religious "beliefs" with scientific "proofs."

In the present times, multiple factors support the persistence of a conflict narrative. Demographic factors, such as the decline in religious affiliation (Pew Research Center 2022), political factors, such as the push to

[2] Their reading of history, however, has been heavily critiqued by modern historians; see, for example, Lawrence M. Principe (2006: 7) and Gary Ferngren (2002: ix).

relegate all religions, and minority religions in particular, to the private sphere, and educational factors such as the absence of dedicated teaching on civil discourse, philosophy of science, and epistemology within secondary schools and colleges before students differentiate into disciplinary silos, contribute to the oppositional framing. Mass media, with its focus on capturing attention using evocative means, may also exacerbate the problem by preferring to run stories that highlight the tension between views. Notably, the American political context where the First Amendment to the Constitution contains a free exercise clause that aims for the public to have unfettered capacity to religious practice, as well as an anti-establishment clause that constrains the government from privileging one faith over another by providing fiscal and legal support, makes US society a particularly ripe setting for vigorous debate. Indeed, perceptions of overreach by religious groups, or on the part of the government, can easily slide into religion versus science debates.

Contrastingly, discussions over scientific and religious evidences were not marked by conflict in the medieval Islamic world (Iqbal 2002; Bakar 2008). Instead, scientific methods were seen as complementary tools by which to probe the mysteries of the universe and thus generate a holistic understanding of God's creation. To be sure, there were debates about the interpretation of Qur'anic verses that spoke about the natural world[3] and contentions over the ontological and epistemic authority of reason (Emon 2010). But, for the most part, scientific activities in the Muslim world took place in the context of a comprehensive paradigm that sourced conflicting religious and scientific truth claims to the inadequacy of the human intellect to grasp reality fully. The organically developed Islamic scientific tradition, fine-tuned a hierarchy of sciences that generated frameworks of inquiry, containing both the metaphysical and physical realms (see Iqbal 2002: Ch. 3). As such, competing truth claims were the anomaly rather than the norm. Over time, and for multiple reasons, this tradition declined, and the great cosmopolitan centers of learning in Baghdad, Cordoba, Istanbul, and Isfahan hollowed out.

The arrival of European colonial powers with advanced armaments generated by scientific and technological advancements was a watershed moment for the Muslim world. They destroyed the remaining vestiges of holistic Muslim academies and installed "Western" forms of education

[3] For an example of such debates, see Chapter 6's discussion about seven earths.

that provided tangible pathways for social and economic advancement within a reorganized public sphere. Consequently, an inferiority complex set in within which political leaders and public intellectuals attributed the loss of Muslim political and economic sovereignty to intellectual stagnation and an unwillingness to embrace modern science and technologies. Hence, many saw their "liberation" tied to replicating Western models of education and societal organization, and thus, a narrative of science versus Islam gained purchase among the intelligentsia. The newly imported educational systems separated the academies of religion and science, thus generating intellectual gaps between students and scholars of science and those in religious studies. Additionally, as scientific innovations wholly transformed Muslims' vocational, social, and leisure activities, science became reified in the public imagination, and public and private funds flowed into institutions that promised progress in helping Muslims "catch-up" to the West. Centuries on, the "catching up syndrome,"[4] particularly in science education, remains in place as the West's political, economic, technological, and cultural might is relatively undiminished.

Consequently, in Muslim lands where once there was, more or less, a unity between the knowledge systems of science and scripture, in the aftermath of colonialism, there came to be a growing chasm. This chasm bolstered perceptions that the worlds of science and Islam are on opposite sides of the spectrum with respect to reality and, in a certain way, subtly fostered the conflict narrative's growth in Muslim communities in both Muslim-majority and Muslim-minority contexts, for it is this narrative that shapes/d the educational systems that educate the Muslim intelligentsia in both geographies.

The story I share above is not without its detractors and admittedly lacks some nuance, but having laid out a history of the discourse in broad strokes, let me now turn to Muslim scholarly engagement with contemporary religion and science debates. My treatment is not intended to be exhaustive nor a critical appraisal; rather, I orient the reader to four models of engagement that demonstrate that Islam and science discourse has been uneven.[5] The first type involves scientific exegesis, *tafsīr ʿilmī*,

[4] This term is borrowed from Muzaffar Iqbal.

[5] Certainly, there are more than four types of engagement between Islam and modern science. I have restricted myself to these four because each of these has a significant amount of academic literature devoted to them, they are mentioned in major review works on Islam and science, and they illustrate distinct patterns of engagement.

which generated vast literature in the late eighteenth and early nineteenth centuries. In light of the rapid changes in Muslim educational systems and the power of science in the public sphere, prominent academics sought to fortify the Qur'an, and thus Islam, from being deemed irrelevant and inaccurate. They did so by attempting to demonstrate that the Qur'an contains theories and facts that science has only recently uncovered. Thus, an entire genre of scientific exegesis emerged, accompanied by popular writings that tied the miraculous nature of the Qur'an, *'ijāz*, to its scientific claims. Though prominent scholars such as former Grand Mufti of Egypt, Muḥammad 'Abduh (d. 1905), and his Syrian/Lebanese seminary-trained student Muḥammad Rashīd Riḍā (d. 1935), as well as public intellectuals such as the French physician Maurice Bucaille (d. 1998),[6] championed this manner of engagement with modern science, they were met with critical reception in Islamic scholarly circles. Indeed, for most Islamic scholars, using correlations with modern scientific data as a litmus test for the veracity of the Qur'an was absurd, and the approach seemed inflected with the inferiority complex noted above. While such apologetics gained traction in both public and academic spaces within the Muslim world, due to the development of alternative models of discourse and the lack of sustained institutional support, analogous efforts have largely dissipated.

A second type of engagement is termed the "Islamization" approach (al-Faruqi 1982; Safi 1993; Hussain 2015; Hashim and Rossidy 2000). Taking aim at the separation of knowledge into religious and secular sciences, scholars in this camp sought to unify knowledge streams under a single epistemic framework and instill a comprehensive worldview. Said another way, the approach aims to integrate Islamic values with modern scientific and academic disciplines by developing a robust Islamic epistemology and ancillary disciplinary methodologies. This approach was the brainchild of Ismā'īl al-Fārūqī (d. 1986), a Palestinian-born Islamic scholar dually trained in the Western academy and the Muslim seminary.[7] Al-Fārūqī recognized that the project depended on curriculum development and subsequent educational implementation, and thus co-founded

[6] An exemplar work in this genre that has been popularized the world over is Maurice Bucaille's *The Qur'an & Modern Science* (1980).

[7] Some scholars note that "Islamization" began with Syed Naquib al-Attas's (1969) *Preliminary Statement on a General Theory of the Islamization of the Malay-Indonesian Archipelago*.

a think tank, the International Institute of Islamic Thought, devoted to research and development in the area and assisted in the founding of International Islamic Universities around the globe to implement and iterate the curricular endeavors. The Islamization approach also had its critics, who pointed out that the natural sciences were almost wholly left out of the Islamization project.[8] This conspicuous absence signaled to critics that the philosophical and epistemic underpinnings of modern science were overlooked unintentionally or, worse, seen as unproblematic within an Islamic worldview. While the idea of Islamization continues to captivate scholars and educators, the development of integrated curricula lags far behind as fiscal support for continued research and iteration ebbs and flows (Maulana 2017).

The Islamization approach focuses on epistemology and education, while the sacred science approach seeks to foreground the metaphysical roots of scientific endeavors (Nasr 1993). Scholars in this camp posit that modern science is disconnected from its spiritual roots and thus leads to a fragmented understanding of the world. They call for a reintegration of science and spirituality, where scientific inquiry is conducted within a metaphysical framework that acknowledges the sacredness of nature. For them, the cosmos' observed order indicates that it is teleological and has a Divine Creator. Accordingly, data from the natural sciences must be interpreted through a metaphysical lens. Such metaphysical frameworks are found within perennial philosophy. However, the fountainhead of truth is the Qur'an and Sunnah, and scientific evidence, correctly interpreted, cannot conflict with these sources. Proponents of this view include Sayyed Hossein Nasr (b. 1933), an Iranian philosopher and professor of Islamic studies at George Washington University. The sacred science approach is not eisegetical like the *tafsīr 'ilmī*, nor is it primarily focused on epistemology like the Islamization model; instead, it aspires to be broader in its critique of modern science and more foundational in subsuming all rational inquiry underneath a framework of metaphysical truths that encompass all of reality (Nasr 1981; Iqbal 2002).

[8] Partially in response to the Islamization approach's blind spot, a group of scholars, collectively calling themselves the Ijmalis and led by Ziauddin Sardar (b. 1951), a British-Pakistani public intellectual, emphasized the social aspect of scientific endeavors. They held that science was instrumental in nature and that its main goal was not to uncover truth but to assist humankind in solving problems ethically.

A more recent approach to Islam and science discourse attaches itself to the classical Islamic discipline of ***kalām***, speculative or dialectical theology 2002 (Altaie 2016; Setia 2012; Waters and Fosl 2015; Guessoum 2011). The objective here is to recover a theology of nature[9] that can serve as the foundation for a contemporary philosophy of science. Such a philosophy would be intrinsically grounded to creedal and ontological commitments, yet it would allow for resolving thorny issues at the heart of religion and science debates. As expounded by Mohammad Basil Altaie (b. 1952), an Iraqi physicist, some fundamental theological commitments include holding the world to be created and continually re-created from moment to moment by God and to be indeterminate and contingent (Altaie 2017). These views maintain the omnipotence of God, acknowledge the possibility of miracles, and allow for Newtonian physics' apparent determinism and the apparent randomness and probabilistic nature of events at the quantum level to both be true. This novel approach aims to open up avenues for research and dialogue that yield a nuanced yet comprehensive understanding of the natural world.

Recent scholarship has expanded the scope of ***kalām***-based engagement with science to include topics such as evolution, quantum mechanics, the problem of evil, extraterrestrial life, and intelligent design (Chowdhury 2021; Malik 2021; Karamali 2024: 25; Jalajel 2009; Malik et al., 2022). Of particular relevance to the questions probed in this volume are the work of Malik (2021) and Jalajel (2009) on human evolution, which explores how Islamic theological anthropology can accommodate evolutionary biology without compromising core doctrinal beliefs. Scholarship in this area is nascent and, to date, has been predominantly funded by the John Templeton Foundation, with a focus on a few institutions in the Muslim world (Kalam Media, n.d.).[10]

[9] A theology of nature is differentiated from natural theology based on their grounding assumptions and goals of inquiry. Theology of nature seeks to understand and interpret the natural world in light of religious beliefs and creedal commitments. Natural theology, on the other hand, attempts to infer knowledge of God solely from reason and the observation of nature, without relying on specific religious doctrines. These terms do not fully map onto the inquiries of Islamic theologians but are helpful in navigating Islam and science discourses.

[10] One such institution is the now quiescent Kalam Research and Media headquartered in the United Arab Emirates, whose landmark work on Islamic Analytical Theology was sponsored by the John Templeton Foundation.

Nonetheless, taken together, this growing body of work underscores that *kalām* can serve as a robust intellectual resource for engaging modern science in a way that preserves theological integrity while addressing scientific developments.[11]

While these approaches have merits and shortfalls, the main challenges each faces relate to sustainability and implementation. At the discursive level, platforms for bringing Muslim scientists and Islamic scholars together are rare, and those that cross theological lines to invite religion and science experts from other religions into dialogue are even rarer. Hence, opportunities to sharpen arguments, redress critiques, and deeply engage contemporary science are nearly nonexistent. Relatedly, at the level of educational systems, while scholars in each camp have authored numerous academic works that focus on theoretical issues and concepts, program implementation models and curricular resources are scant. Similarly, iterative and cyclical implementation to fine-tune theory and teaching practices, as well as determine curricular effectiveness and long-term outcomes, has, generally speaking, not been systematically performed. Indeed, an entire ecosystem of thought leaders, theoreticians, researchers, pedagogical experts, program implementers, evaluators, and institutions is needed for platforming, implementing, and testing programs. While scholars dedicated to the intersection of Islam and science exist, and pieces of the institutional aspect of the ecosystem for specific models of engagement are in place, much work remains to be done to link scholars and platforms and to tie theory to practice.

Biomedical Education and Conceptions of the Human Being

Biomedicine signifies a theoretical framework and a body of multidisciplinary knowledge used to identify and address human maladies.[12] Biomedicine's epistemic and philosophical commitments are inherited from European Enlightenment thought (Ali 2018). Consequently, three features mark biomedicine's approach to human maladies. First, diseases are tied to biological, chemical, and physical phenomena. Second, biomedicine privileges knowledge garnered through laboratory and clinical

[11] In Chapter 2, Dr. Omar Qureshi will discuss a *kalām* epistemological framework.

[12] One could argue that biomedicine is also applied to the care of animals (veterinary medicine) and would be applied to organisms discovered elsewhere in the universe (exobiology). In this book, our concern is with the application of biomedicine to humans.

experimentation. Third, biomedicine commits itself to, both philosophically and methodologically, reductionism. Thus, biomedicine holds that the human being is best understood through examining its subsidiary parts, and, in concert, the phenomenon of human illness is primarily explained by disorder, malformation, maladaptation, and dysfunction of these constituent parts (Krieger 2011; Valles 2020). These features are mutually reinforcing; a naturalistic account of disease requires methodological naturalism to be employed in research, which lends explanatory power to reductionism. Extending the line a bit further, the biomedical worldview generates a research paradigm that delivers a body of knowledge that furnishes the grounds for a specific type of healthcare practice. It follows that, in Aristotelian terms, biomedicine (and healthcare systems operating out of this model) primarily focuses on material and efficient causes of the human condition. It eschews final causes, and its relevance to deciphering formal causes is debated.[13]

The biomedical approach to human maladies is hegemonic; all other systems of assessing and treating maladies must submit to the epistemological framework of biomedicine or risk being dismissed as quackery. This dominant position results from the authentic explanatory power and resulting therapeutic proficiency of biomedical frameworks, and from cultural power dynamics. For example, many traditional systems of identifying, assessing, and treating diseases were undermined or directly attacked by colonial governments. The power dynamics and cultural imperialism have continued long after the era of colonialism ended, as biomedical education and practices continue to be exported from Western countries into low- and middle-income countries and supplant traditional and folk practices (Valles 2020).

Biomedicine has become globally dominant in defining health and disease, evaluating their presence or absence, and providing remedies to social ills through professional channels. The globalization of biomedicine is promulgated through formal medical education systems that are exported from, or mimic, Western medical academies. The curricula taught by these academies produce physicians steeped in biomedicine's technologies, practices, ways of thinking and knowing, and understandings of the human being. Since the physician's daily work of

[13] For more analysis related to the practice of medicine and the four causes, see Bishop (2011).

identifying, diagnosing, and treating patients is wedded to the naturalistic, reductionist, and mechanistic understandings of the human being, medical education follows suit and prepares trainees to carry out these practices. Accordingly, medical students, more or less, come to see the human being as a "machine in motion," and dysfunction or disorder in this machine's parts is held to be the root of all disease. This "indoctrination" is witnessed at the outset, with the first encounter medical students have in training being the dissection of the human body to uncover its mechanistic connections and pathological deformities. Thereafter, students are introduced to the "hard" sciences of biochemistry and pharmacology, which demonstrate the power of chemical processes in creating or removing disease. Medical education next proceeds to cover specific organ systems and their pathophysiology, and trainees are enveloped in healthcare systems where most of their preceptors have been certified to care for disorders of body parts or address specialized disease types.[14] All of this education reinforces the naturalism and reductionism of biomedicine, rendering the human being into (dys)functional parts. To be sure, the biopsychosocial model of medical education does not neglect the psychological and social parts of the human being (Engel 1977, 1980); rather, addressing the psychological and social dimensions of human maladies is often left to non-clinicians.

Medical education typically emphasizes the naturalistic and reductionistic approach, breaking down complex human experiences into simpler biological components. This approach, while valuable for understanding specific physiological mechanisms, not only overlooks the holistic nature of human beings, but it also leaves off addressing fundamental existential questions about the human being and human experience. The formal curriculum does not routinely discuss questions such as its nature, essence, purpose, and future. Informally, in the "hidden curriculum," however, some answers are supplied with evolutionary mechanisms being discussed in relation to human anatomy and concerning bacterial resistance to antibiotics. Similarly, in discussing neuroanatomy, mental health,

[14] I have sketched out the prototypical progression of medical education based on the Flexner report (1910). Obviously, each medical school has a unique curriculum and adapts it based on its own vision of how students should develop into doctors. For example, many incorporate medical humanities teaching to restore a more balanced understanding of human experiences with illness.

and topics such as "brain death," implicit connections between the brain and notions of the self and personhood are made. In this way, a biological account of humankind is imparted.

Just as existential questions about the human condition are left unattended, so are concepts such as illness and healing. The biomedical model can define disease and measure cure, but human experiences of illness and philosophical and theological conceptions of healing are subjective and varied. Hence, these topics are discussed at the margins of the formal curriculum and in a limited fashion. Medical education may be excused for focusing on the concrete problems of disease and the hard sciences, if premedical education required courses on these "softer" topics. Yet, premedical education requires, at least in the United States, coursework in the physical and biological sciences, mathematics, and English; a robust foundation in the humanities is not needed. Many medical schools have developed elective-based medical humanities tracks to redress these deficiencies. While such interventions are welcome and sorely needed, these additions are small in scope and have a relatively minor presence within medical education.

The Promise of Education at the Intersection of Bioscience and Islam

As hinted above, fundamental questions at the heart of contemporary science and religion discourses revolve around the human being. Advances in cosmology, physics, neurosciences, and biology have provided data that make it possible to furnish answers to questions that were traditionally within the remit of religion, such as: What is life? How did it begin? Do we have free will? Given the possibility of differing answers, academic and public discourses on the relationships between science and religion focus on conflict and commensurability. As the truth claims of one knowledge domain are measured against claims from the other, research scholars, professional scientists, religious leaders, students, and the informed laity seek to clarify areas of contention, zones of overlap, and spaces for alignment.

Despite the overarching assumption of inherent conflict, this renewed interest generates the opportunity for Muslims to authentically engage in cross-disciplinary dialogue on "Big Questions" about human beings. Beyond the benefit a renewed discourse could have for academic and public stakeholders, integrating Islam and biosciences in educational curricula offers a profound opportunity for present and future Muslim allied health professionals.

Since medical education largely operates out of a naturalistic and reductionistic worldview, religiously inclined students may experience tension and stress when religious perspectives on the nature, origin, and other aspects of the human being appear to conflict with the biomedical views they are learning. In response, trainees may compartmentalize their identities by relegating Islam to the personal sphere, experience psychological stress as they struggle to affirm their religious views, or avoid the matter altogether by leaving the profession. These difficulties may be more profound for Muslim clinicians and trainees in non-majority contexts who are often subject to religious discrimination from their peers and patients, do not have their religious practices accommodated by institutional leaders, and experience mental health distress from Islamophobic workplaces (Padela et al. 2016; Padela et al. 2023; Baqai et al. 2023; Murrar et al. 2022, 2023a, 2023b; Azam et al. 2024).[15] Since their religious identity already engenders negativity, sequestering religious views further away from their professional life may be easier than trying to address potentially conflicting narratives about the human being. Moreover, given the expectation of conflict between religion and science in these societies, Muslims may be expected to check their religion at the workplace door and thus may not find meaningful ways to engage these issues at work.

Hence, a curriculum that integrates Islam and biosciences can help in forging a stable and coherent professional identity. By understanding how Islamic perspectives align with scientific discoveries, professionals can approach their work holistically, honoring their religious beliefs as well as their commitment to bioscientific inquiry. Moreover, by gaining skills in identifying areas of potential conflict and learning to analyze truth claims from both scripture and science critically, such curricula

[15] I led two national surveys of Muslim clinicians in the United States, and one in the United Kingdom, examining their experiences with religious discrimination and accommodation, as well as the personal and professional impact such experiences have upon their careers. Our work demonstrates a rise in the prevalence of perceived religious discrimination for Muslim clinicians in the United States from 2013 to 2021, with a greater proportion noting such experiences (24–53%) as well as greater job turnover due to discrimination. We also found those who were more religious faced greater challenges (see Initiative on Islam and Medicine 2013–2021).

would enable them to contribute meaningfully to religious and scientific discourses. Muslim allied health professionals educated through such curricula would be uniquely positioned to address misconceptions by clarifying religious views on bioscientific issues and promoting a more nuanced and informed dialogue. Such exposure may also instill within them the desire to pursue further study and research at the intersection. For example, equipped with a greater understanding of the intersection of Islam with bioscience, they could initiate research projects that explore the ethical, spiritual, and practical dimensions of contemporary bioscientific issues from an Islamic perspective. Alternatively, they could shape the design of biotechnologies that align with Islamic values, thus ensuring broader acceptance within Muslim communities.

At the level of practice, such exposure, when coupled with further training, may facilitate the development of integrative medical practices. For example, integrating Islamic perspectives on treating maladies of the soul, mind, and body with therapies based on neuroscience and psychology may generate treatment plans that offer a holistic approach to health and wellness.[16] Additionally, such learning could foster greater awareness of religious sensitivities within healthcare and thus improve communication and care delivery, particularly in diverse and multicultural settings. Professionals trained in this way could serve as bridges between healthcare systems and religious communities, advocating for culturally appropriate policies and practices.

Integrating Islam and biosciences in educational curricula tailored for Muslim allied health students has the potential to create well-rounded, ethically conscious professionals. These professionals would be equipped with the knowledge, skills, and capacities to navigate their careers with confidence and coherence and hold the potential to make significant contributions to religion and science discourses and the practice of healthcare in ways that reflect their religion. Our project represents a first step in such an enterprise with a modest curricular intervention tailored to Muslim students interested in medical careers (see details below and in Chapter 10).

[16] Such integrative models are presently being developed. For example, the Khalil Center in the United States healthcare delivery model is based on Traditional Islamically Integrated Psychotherapy (TIIP) (see Khalil Center n.d.).

How Did this Volume Come about? And Who Is it for? An Overview of the Larger Project and its Activities

In light of the problem carved out by the first two observations described above, and the potential solution offered by the third observation, I convened a multidisciplinary team of theologians, scientists, and educators[17] to design, implement, and evaluate a seminar-based curriculum tailored to Muslim medical track students and focused on addressing fundamental questions about the human being from bioscientific and Islamic perspectives. Topically, we homed in on five "Big Questions" about the human being: its origin, its nature, its fatedness, its malleability,[18] and its uniqueness. These questions allowed for juxtaposing theological and bioscientific understandings and thus made the instruction relevant to Muslims pursuing biomedical careers.

After successfully securing a grant from the John Templeton Foundation,[19] we targeted two populations of learners: 11th- and 12th-grade students at Muslim high schools in the United States and first- and second-year students completing foundational studies at the International Islamic University of Malaysia.[20] The selection of these cohorts was purposeful; their differences in educational level and geography allowed for comparing curricular efficacy across Muslim students at Islamic schools in a Muslim-majority and a Muslim-minority context and in a secondary or collegiate setting. Their similarities, in terms of interest in medical careers, Islamic studies background, and English proficiency, on the other hand, made the curriculum sufficiently relevant. With respect to curricular goals, drawing upon Kirkpatrick's model for

[17] The project team of bioscientists and theologians are detailed in Chapter 10. With respect to educators, our team included pedagogical experts from High Quality Educational Consultants led by Habeeb Quadri and we collaborated with the Islamic Schools League of America led by Shaza Khan.

[18] In this context, we also address the future of humankind.

[19] The project title is "Fostering Connections between Science and Theology for Medical Track Muslim Students," with Project Leaders Aasim I. Padela and Abdurezak Hashi; for more details, see John Templeton Foundation (2022–2025); and Initiative on Islam and Medicine (n.d.).

[20] We further tested the curriculum with a third population of learners, Muslim medical students in the United States. These results of this third pilot mirror the findings presented here, but were not included in the volume due to time constraints.

learning evaluation and the Theory of Planned Behavior (Ajzen 1985, 1991; Kirkpatrick 1994), we set out to increase student knowledge related to bioscience and Islam, as well as student interest, intention, and preparedness to pursue further study at the intersection of bioscience and Islam. Finally, we aimed to promote attitudinal change away from perceived inherent conflict toward the possibility of harmony between Islamic and bioscientific truth claims. Importantly, given the learner stage, the learning objectives were modest: we expected students to gain awareness of religion and science as domains of knowledge and an appreciation for, as well as some literacy in engaging with, critical inquiry concerning truth claims from both domains. However, mastery in resolving potential conflicts or in integrating data from the two domains of knowledge was not aimed at. Greater details about the project team, seminar topics, curriculum design, and educational evaluation are found in Chapter 10 of this volume. Herein, however, I would like to comment on several critical features of the project work plan. Each of the topical seminars was co-designed and co-taught by a bioscientist and a theologian.[21] The dyad jointly researched the varied bioscientific and religious vantage points on the question at hand. This learning, which involved both canvassing the academic literature and original research into the scientific and scriptural evidence, supported their development of a lesson plan for the seminar. The draft lessons were modified by pedagogical experts, tailoring them to the appropriate learner level and revising them based on mock sessions delivered by the dyad. During the implementation of the seminars, a 360-degree teaching assessment was conducted whereby each instructor wrote down five areas of improvement, a team of experienced educators provided focused feedback on teaching skills based on their observation of the seminar, students shared their perspectives on seminar delivery and content both informally in open discussion sessions and formally via a survey, and pre-post student survey data was based on curricula aims (knowledge, behavioral intention, and attitudinal change as noted above). These data streams informed the further

[21] The five seminars related to the five "Big Questions" were co-taught in this manner. The introductory seminar on epistemology was taught by two theologians, Omar Qureshi and Hamza Karamali, each with science backgrounds, while the last seminar, which involved a review of the seminars and moderated discussions, was co-taught by two bioscientists, Aasim I. Padela and Raudah Yunus, each with religious studies backgrounds.

tweaking of lesson plans,[22] as necessary, in preparation for the seminar implementation in Malaysia after having completed the sessions in the United States (see further details below).

After implementing and quantitatively and qualitatively evaluating the curriculum across the two settings, the team collaborated with educational consultants[23] to design teaching modules for educators in high school, college, and medical school settings. The modules contain lesson plans, templated student activities, video recordings of seminar instruction, and teaching scripts. The modules are designed in such a way as to be adaptable for teaching over a semester/term or in an intensive seminar series and for teachers to cover the Islamic and bioscientific content themselves or to utilize videos of our team delivering this part of the lesson. We also collaborated with educational institutions to host professional development workshops to support educators in implementing and adapting the curriculum for their settings.[24]

This book was motivated by both the desire to provide supplemental reference material for educators and academics embarking on teaching Islam and bioscience and to meaningfully contribute to the academic literature. Each core chapter is thus authored by an individual(s), who has researched and taught the related seminar topic and draws upon the research they conducted in preparing for the seminars. Thus, the volume offers a strategic approach to teaching these controversial topics; the chapters set out the evaluative logic and evidence streams so that educators can absorb the material and use it, as appropriate, in their courses or for public/professional dialogue. However, the chapters probe the Big Questions more deeply than the teaching exercises did, for this book was also motivated by the authors' desire to engage with the extant scholarly

[22] In order to maintain scientific rigor of the efficacy trial of our educational intervention, the tweaking was restricted to changing the order of didactics and group activities, modifying slide sets to make the concepts clearer and remove extraneous material, and revising the way in which instructors engaged students and fielded questions. The learning objectives of each session, as well as the core content delivered, remained the same. Consequently, the teaching modules maintain fidelity to the curriculum as delivered.

[23] This work was primarily performed by Habeeb Quadri's team at High Quality Educational Consulting. The modules are available on the Initiative on Islam and Medicine website, http://medicineandislam.org.

[24] For example, an online workshop was hosted by the Islamic School League of America, while in-person workshops were hosted at the Kuantan campus of the International Islamic University of Malaysia and the Riphah University in Lahore, Pakistan.

discourse. In this way, the chapters go beyond the seminar teaching's focus on identifying the truth claims, the source of knowledge the claim is based upon, and critically appraising the degree of certainty the evidence supports.[25] Rather, each chapter provides provisional normative stances on the Big Question it addresses and material for scholarly inquiry. Consequently, the volume will benefit multiple scholarly and professional communities working in fields that straddle religion and science. Since it fills in scholarly lacunae related to Islam, bioscience, and the medical humanities, the volume can benefit Islamic studies experts, bioethicists and moral philosophers, scientists and intellectual historians, and researchers working on dialectics between tradition and modernity and science and religion. It models a scholarly approach to navigating religion and science discourses and delivers Islamic perspectives on fundamental questions on the human being. Similarly, social scientists and religious studies experts who examine how religious communities respond to contemporary bioscientific and technological advancements will find the work of value. Bioscience is altering the ways in which we conceive humanity and fueling expectations of healthcare delivery; this volume chronicles how Islamic scholars draw upon their tradition to contend with the age-old questions that have taken on new dimensions in light of biomedicine. Finally, the volume can benefit Muslim allied health professionals. As hinted above, the volume may make them more aware of the connections between bioscience and Islam, spur their interest in further study at this intersection, and provide them with ways by which to resolve perceived tensions between the accepted beliefs of their profession and the doctrinal views of their faith.

What Is in this Book? An Overview of the Content

This volume seizes upon the opportunity to foster humble discourse on fundamental questions about the human being by bridging bioscientific and Islamic understandings. Using contemporary scientific discussions on evolution, artificial intelligence, determinism, transhumanism, and

[25] This framework is detailed in Chapter 2. Importantly, the seminar teaching did not aim at providing definitive answers and a complete resolution of the possible tensions between religious and bioscientific claims. Rather, given the learner stage, we focused on identifying truth claims and grading supporting evidence, so as to enable critical inquiry. The chapters dive into the authors' perspectives on the topics more fully.

extraterrestrial intelligence as the platform for discussion, the volume features Muslim scientists and Islamic theologians addressing complex questions related to the origin of the human being, its nature, its fatedness, its malleability, and its uniqueness. In addition to scholars focusing on the *what* of these critical topics, the volume also speaks to the *how* of teaching Islam and bioscience. Indeed, each core chapter provides an approach to discussing these topics through the lens of epistemology in a course dedicated to Islam and bioscience.

Authors' Positionality

Before detailing how the book unfolds, a few comments on our normative perspectives are warranted. We take the view that Islam is a religious tradition marked by a single revelatory event comprising two scriptural parts: the Qur'an and the Sunnah of the Prophet Muhammad.[26] Normativity is rooted in this scripture and linked to a scholarly class of interpreters, *'ulamā'*, who explicate scriptural values and tenets and coalesce this body of knowledge into a coherent worldview accompanied by ontologic, epistemic, moral, social, and other frameworks that allow Muslims to live out the tradition. That said, the tradition is orthopraxical rather than governed by a clerical orthodoxy[27] and inherently plural. There are but a few scriptural texts that are univocal and classed as certain knowledge,[28] and thus, a relatively small number of creedal commitments.[29] On the other hand, the tradition contains a vibrant multiplicity of approaches for extending scriptural values onto contemporary realities. This plural space denotes the incorporation of a human element in constructing the tradition and generates virtue in exerting one's intellect to discern "truth" while, at the same time, facilitating plasticity and flexibility across time and space.

[26] In the Islamic tradition, the Qur'an and Prophetic Sunnah are considered to be two parts of a single revelatory event. The former, termed the recited revelation, *waḥy matlū*, and the latter, termed the revelation that is not recited, *waḥy ghayr matlū*, represent the Islamic scripture and mutually reinforce each other.

[27] While this holds true for Sunnī Muslims, the Shī'ī sect has a more rigid clerical structure which is vested with moral and political authority.

[28] Chapter 2 discusses the notion of certain knowledge.

[29] A classic treatise on the matter was authored by the polymath jurist-theologian Imam Abū Ḥāmid al-Ghazālī (d. 1111), *Fayṣal al-tafriqa bayna al-Islām wa al-zandaqa*. The work is translated into English by Prof. Sherman Jackson (see Jackson 2002).

All of the authors are affiliated with the Sunnī[30] legal tradition and, accordingly, the Māturīdī and Ashʿarī schools of theology. As such, our approach to deciphering the scriptural evidence utilizes Sunnī hermeneutics, and our views on God, creation, and the role of human reason draw from the teachings of the aforementioned theological schools. Moreover, we accord interpretative and labeling authority[31] to those with traditional licenses granted by seminarians and to classical scholars widely recognized as authorities within the tradition. Hence, we do our best to connect our analyses to viewpoints propounded by such scholars, although we recognize that the tradition is a dynamic intellectual conversation between diverse scholars across time and space.

Concerning conflicts between religious and scientific claims, our preconceived notion is that there is harmony to be had. Said another way, we held the view that scientific and scriptural claims, or in different parlance, rational and transmitted proofs cannot be contradictory. Islam upholds that knowledge has one ultimate source, God, and the natural world, as well as revelation, are indicants pointing toward Him. Accordingly, there can be no true conflict between scriptural and scientific evidence; limited human understandings of each generate perceived conflict.[32] Relatedly, much of the public perceptions of conflict between religion and science, in our view, stem from unfounded metaphysical assumptions that are smuggled along with scientific claims, that is, that reality can only be known through empirical investigation or that all that exists is the observable universe, or from faulty reasoning applied to scientific and scriptural evidences. Nonetheless, an apparent conflict may present itself, and in these situations, a robust epistemic framework through which to evaluate and resolve tensions among truth claims is particularly valuable—Chapter 2 details such a framework.

Finally, as I have described elsewhere, we all recognize the need for multidisciplinary inquiry at the intersection of Islam and bioscience (al-Akiti and Padela, 2022; Padela 2021, 2022). Reality is

[30] Namely, the Mālikī, Ḥanafī, Shāfiʿī and Ḥanbalī schools of Islamic law.

[31] By labeling authority, I refer to the idea of pronouncing something as consonant with the religious tradition; determining something is "Islamic."

[32] As expounded on in Chapter 2, the principle of non-contradiction also buttresses the claim that conclusive evidence arising from these two domains of knowledge cannot be in actual conflict.

multidimensional, and single-discipline views only disclose a part of this reality. Moreover, methodological blind spots and biases temper the claims that can be made about reality based on a discipline-specific line of inquiry. Thus, multiple vantage points and disciplinary experts are needed to map out what we know and, perhaps more critically, what we don't know. This perspective informed our teaching exercises and chapter writing; we conversed as a team over each lesson plan and book chapter and sought to open up discussions and dialogue as opposed to providing a singular "orthodox" Islamic view on the fundamental questions. Hence, this volume presents a multidisciplinary engagement with pressing questions that foregrounds epistemic humility.

An Overview of the Chapters

The volume is organized as follows. There are three parts; Part I covers preliminary discussions on the engagement of Islam and bioscience and includes this introduction and a chapter detailing an epistemic framework through which to engage Islam and science dialogue. Part II represents the core of the book with six chapters addressing fundamental "Big Questions" about the human being in consideration of the Islamic and bioscientific evidence. Generally speaking, these chapters highlight key concepts related to the religion and science debate over a specific fundamental question, critically appraise the religious and scientific evidence that undergird the disciplinary truth claims, and weigh in on the potentially competing accounts. Each core chapter, thus, offers a normative perspective on a specific "Big Question."[33] Importantly, these chapters illuminate different facets of humankind and thus build upon one another. Following these, the final part, Part III, covers practical field perspectives in two chapters. The first discusses the historical backdrop shaping contemporary Muslim science education and the need for holistic curricula that introduce students to epistemology, and the other presents the results of an empirical evaluation of our novel Bioscience and Islam curriculum.

Accordingly, the chapters unfold in the following way. Chapter 2, "First Principles, Certainty, and Uncertainty: Unburdening Ourselves

[33] As an exception, there are two chapters related to human malleability and future, each attending to a different topic.

of the Science–Religion Conflict" completes Part I of the book. This chapter is authored by Prof. Omar Qureshi, a seminary-trained Islamic scholar and a Western academy-trained educator. The chapter begins with an account of knowledge developed by logicians and metaphysicians of the Sunnī tradition. It uses that foundation to map out an epistemological framework, including conceptions of certainty and a classification scheme of propositions, that can be used to evaluate perceived conflicts between religious and scientific truth claims. As noted above, the chapter holds that there can be no actual conflict between the two domains of knowledge, for certainly true knowledge has a single source, that is, God, and the principle of non-contradiction rules out the two propositions being valid in the same sense and at the same time. Instead, the chapter contends that the apparent conflict will recede when the evidentiary bases for the claims are critically evaluated, for such examinations will lead to finding greater certainty on one or the other side. Hence, the author advocates for "unburdening ourselves from the conflict thesis."

Within the proposed epistemic framework, the author incorporates scripture-based claims, or revelatory knowledge, as part of the larger category of testimonial knowledge and transmitted proofs, thus allowing for dialogue across disciplines based on rational principles and agreed-upon means and sources of knowledge, namely, the intellect (*al-ʿaql*) which provides humans with rational knowledge; the five external senses which provide sensory knowledge; and true reports (*al-khabar al-ṣādiq*) which are the basis for testimonial knowledge. After identifying these sources of knowledge, Chapter 2 discusses rational, empirical, and Divinely stipulated judgments and a scheme of propositions based on their source, evidentiary nature, and logical modal status as immediate/inferred, necessary, possible, and impossible. It also explains degrees of certainty surrounding judgments and truth claims. These discussions lay out an epistemic framework for critical inquiry. The framework, albeit in a much simpler form, was used to structure didactics and activities within the aforementioned Bioscience and Islam curriculum. The chapter details four scenarios where apparent conflict may arise between rational (i.e., scientific) and transmitted (i.e., scriptural) claims and describes a stepwise process to evaluating and resolving the apparent conflict. Since the author holds that conflict is the exception and is sourced in the incomplete and/or imprecise interrogation of the certainty surrounding a truth claim, the chapter does not detail processes for harmonizing/integrating truth claims, nor does it outline when one should suspend judgment (*tawaqquf*).[34] Nonetheless, Chapter 2 provides educators, theologians, and academics with the cognitive tools and processes to evaluate knowledge from religious and scientific sources.

However, each of the subsequent core chapters refers to the concepts and terms described in Chapter 2 and therefore evaluates knowledge based on the schema provided, each author follows a variable approach in resolving perceived tension. In other words, as a group while we took a uniform, epistemically oriented approach to evaluating truth claims by identifying evidentiary sources and classifying degrees of certainty surrounding these claims, but we did not hold each other to a specific method for integrating, harmonizing, or resolving the tension between the sources. This diversity is purposeful as we desire readers to appreciate that there are multiple ways to bring the differing viewpoints together. Indeed, there are various approaches to resolving conflicts between Islam and science within the academic literature.

Illustratively, my approach to resolving perceived tensions between religious and scientific claims proceeds as follows. As noted in Prof. Qureshi's chapter, the first step in evaluating truth claims is to specify the claim based on which type of evidence is being used to furnish the claim and then to grade the evidence according to its epistemic value. Such a grading may resolve the perceived tension, as it could be that one side asserts a claim based on certain knowledge, and the other is grounded in probabilistic knowledge. Therefore, priority must be given to the side that has more certain evidence. When the evidence is of equal grade, one may opt for suspension of judgment, *tawaqquf*. Between these two poles of suspension of judgment and prioritization are the possibilities of harmonization or integration.[35] In my view, harmonization follows a similar line of analysis where the evidence from each domain is interrogated, and an attempt is made to hold both claims as true insofar as possible. Such an approach may involve multiple analytic, interpretive, and hermeneutical methods, including metaphorical interpretation, *ta'wīl*, of revelatory

[34] *Tawaqquf* refers to the pausing on making a determinative statement, or in other words, a specific truth claim. The notion has been much debated in recent Islam and evolution discourse, and Chapter 2 explains the concept contextually.

[35] Harmonization is related to the ideas of *jamʿ* and *taṭbīq* within the Islamic ethico-legal sciences. In a general sense, the core idea is to develop a hermeneutical approach that resolves the tension. There are other avenues to resolve conflicts as well including *naskh* (abrogation).

texts. Critically, metaphorical interpretation is only available as a method in specific circumstances, such as when the scriptural evidence is probabilistic in its semantic import, *ẓannī al-dalāla*, and the scientific evidence is more certain.

Part II of the book comprises the core chapters and begins with Chapter 3, "Scientific Advances in Understanding Human Nature: An Islamic Occasionalist Perspective," co-authored by an Islamic philosopher, Prof. Nazif Muhtaroğlu, and a physician-scientist, Dr. Macksood Aftab. The volume's inquiry into the "Big Questions" begins with the authors defining human beings as "the hereditary descendants of Adam and Eve… defined by the union of body and soul… In modern scientific discourse, this distinction aligns with 'anatomical Homo sapiens'" (Chapter 3: xx). Taking this foundational understanding as a starting point, the chapter weighs in on human nature. It highlights how neuroscientific views that adopt a naturalist and physicalist view of the mind and link that to human nature appear to conflict with Islamic theological stances about the nature of the human being, which assert the existence of a soul. This perceived conflict arises, in the authors' collective view, from the overstep represented by reifying a wholly naturalistic interpretation of neuroscientific findings regarding the link between the human brain and mind, as well as explanations for the existence and experiences of consciousness. Chapter 3, thus, discusses the mind–body problem, its origin, and contemporary bioscientific attempts to resolve the problem by recourse to various types of physicalism. The authors find these attempts wanting, noting that the scientific claims are based on presumptive knowledge and that explanatory gaps persist. Though also speculative, the authors assert that two models for the human soul within the Islamic tradition—occasionalist property dualism within physicalism and occasionalist substantial dualism—better accommodate contemporary neuroscientific data and offer a more plausible interpretation of the link between the human brain and mind (soul). In closing, the chapter comments on the implications of these models and the possibility of artificial intelligence achieving human (or human-like) consciousness.

In recent years, there has been an explosion of literature discussing the compatibility of the neo-Darwinian evolutionary account of the origin of humankind and Islam (Malik 2021, 2023, 2024; Khan 2023; Jalajel 2009; Iqbal 2010, 2023; Moran 2021; Qadhi 2018; Zaman 2022). In Chapter 4, "Human Origins and Evolution in Light of Revealed Truths: Theological Anthropology, Scholarly Consensus, and

Similarity," Prof. Qureshi adds to that discourse through the lens of Islamic scholarly consensus *ijmāʿ*. He describes the critically important role *ijmāʿ* plays in Islamic epistemology by bounding the interpretation of scripture and meting out creedal positions. He then asserts that the aforementioned contemporary literature on the topic appears to invoke scholarly consensus on some aspects of the human origin question but overlooks it in others. Using his epistemic framework to navigate scripture-based truth claims, Prof. Qureshi contends that the Prophet Adam was the first human and the father of all humankind. He further asserts that the possibility of pre-Adamic humans is precluded by scriptural interpretation supported by scholarly consensus. Indeed, he notes that these creedal positions are accorded the epistemic grade of certainty (*qaṭʿī*) based on the nature of transmitted knowledge behind the views, namely, scholarly consensus grounded in scripture. Having weighed in on the origin of the human being from a theological perspective, he next turns to bioscientific claims where the neo-Darwinian account asserts common ancestry on the basis of morphological similarity. He destabilizes the inductive logic behind this truth claim by noting that the observance of similarity does not logically necessitate common origin. Indeed, he cites both contemporary evolutionary biologists and classical Muslim scholars who agree on the observation of similarity between humans and other species but do not "jump" to assert common ancestry between them. In his view, the observations are not definitive proofs, and the accompanying rational argument is not conclusive, rather, the assertion of common ancestry between humankind and other natural kinds must be accorded a reduced epistemic grade. Subtlety, the author thus appears to argue that the Islamic doctrinal account of human origin with the Prophet Adam is to be privileged due to its more certain nature.

Chapter 5, "Are We Programmed or Do We Have Free Will? Discussions on Fate and Genetic Determinism in Islam and Bioscience," addresses the "Big Question" of whether human actions are predetermined. Co-authored by an academic bioscientist, Prof. Mohamad, and an academic theologian, Prof. Hashi, the dyad considers theological and religious perspectives on determinism. The chapter begins by describing the ethical problem-space as one related to moral accountability and introduces the reader to Islamic theological conceptions of free will, determinism, and moral accountability. In doing so, the authors heavily reference scriptural texts that offer nuance in affirming that all events are within God's knowledge and dominion, yet some aspects of human life are predetermined, and others are subject to human choice. Next, the authors describe the theory and science behind genetic determinism and the extent to which certain human behaviors are based on genetic predispositions. Utilizing a case study related to moral culpability for criminal behavior, the authors bring both domains of knowledge into conversation and find that there is no inherent conflict between religious and bioscientific claims. Rather, the state of the science is that human behaviors are not fully determined, and thus, there is moral agency to act and moral culpability for wrong action. The scriptural evidence also assigns human beings with moral agency despite God's dominion over, and foreknowledge of, human action. Accordingly, on their view, humankind is not fated, and human actions are not predetermined.

Chapter 6 also represents a collaboration between a theologian and a bioscientist. In "Assessing Evidence for Extraterrestrial Intelligent Life and Implications for Human Uniqueness," the team of Bilal Ali Ansari, a seminarian, and Dr. Ahsan Arozullah, a bioscientist, concern themselves with the following two interrelated questions: Are humans the sole bearers of intelligence in the universe? And if extraterrestrial intelligent life exists, what would that mean for Islamic and scientific notions of human uniqueness? The chapter leans on the aforementioned epistemological framework to address these questions by considering the relative degrees of certainty associated with claims about intelligent life based on revelatory, rational, or empirical sources.[36]

As background, Chapter 6 describes Western philosophical foundations for the search for extraterrestrial intelligence (SETI), including the contrasting views of Aristotle and Epicurus regarding human uniqueness and human accountability. It further details the Drake Equation, which enumerates the possibility of there being detectable extraterrestrial intelligent life and organizes technological efforts toward its search. Given that the empirical evidence for such life is unavailable, the chapter engages scriptural evidence regarding the possibility of such life in

[36] Of note, the authors slightly deviate from the categories presented in Chapter 2 by not using the terms transmitted and rational proofs and instead opt to separate the categories of knowledge into revelatory, rational, and empirical sources. Similarly, they offer a slightly modified approach to resolving perceived conflicts between revelatory and empirical or rational evidence involving literal or figurative interpretation of the revelation based on either side's epistemic strength/certainty. These slight modifications notwithstanding, Chapter 6 aligns with Chapter 2's framework and approach to evaluating truth claims.

greater depth. Several Qur'anic verses and *ḥadīth* addressing the existence of extraterrestrial life are examined with respect to their meaning, epistemic status, and semantic import. Specifically, commentaries and reports regarding Qur'anic verse Q.65:12, which can be read as indicating seven earths, and Q.42:29, which mentions creatures existing through the earth and skies, are analyzed. These texts are also used to engage in theological discussions regarding prophethood and the uniqueness of humankind. Bringing the knowledge domains together, the authors assert that empirical, rational, and revelatory evidence indicate the possibility of extraterrestrial life.[37] Significantly, confirming such existence would not disrupt creedal statements that hold humankind to be exceptional/unique among all of creation and that the Prophet Muhammed is the final prophet. Indeed, the authors seek to motivate theologically informed scientific inquiry, saying that "the scientific pursuit of empirical evidence for the existence or non-existence of intelligent extraterrestrials is potentially beneficial for theological understanding" (Chapter 6: xx).

The following two chapters address "Big Questions" about the malleability and future of humanity vis-à-vis biomedical interventions, and, as such, tie into preceding discussions of human origin and nature. In my chapter, Chapter 7, "Is Human Height Changing, and Should Medicine Intervene? Transhumanism, Biomedicine, and Islamic Bioethics," I take up three interrelated questions: Is humankind hurtling toward a trans- or post-human future? Is contemporary healthcare, wittingly or unwittingly, facilitating this end? And if we are, should we be? Using human height as a case study for potential natural and induced changes in human features, the chapter engages religious and bioscientific knowledge to address the *are we* question and initiates a multidisciplinary and multilevel Islamic bioethical discourse over the role of medicine to tackle the *should we* question. I begin with an account of transhumanism and its relationship to biomedicine and contemporary healthcare. This discussion highlights the shared philosophical and epistemic commitments of transhumanism and biomedicine, as well as distinctions between therapy, enhancement, and alteration, bioethical concepts that are used to differentiate ethical

[37] If one includes angels in the category of extraterrestrial intelligent life, then scriptural evidence conclusively establishes their existence.

medical practice from activities aligned with transhumanism. I next introduce the "disease" of idiopathic short stature to underscore the sociocultural construction of disease states and allow for analyzing the role of medicine in inducing bodily changes. Considering human height as a feature of the human being, I bring scientific and scriptural evidence for the changing of height over time together and evaluate each side's truth claims using the epistemic framework. Notably, I juxtapose scriptural viewpoints on the height of the Prophet Adam and subsequent human beings alongside archaeological evidence and rational claims. In Chapter 7, I offer several approaches to resolving the perceived tension between scriptural texts that suggest the Prophet Adam's height was 60 cubits and that human beings have been decreasing in height since then and bioscientific evidence that shows a relatively stable human height across time with modest increases over the last two centuries. Specifically, given the degrees of certainty surrounding the evidence, one could prioritize the bioscientific claims regarding human height in recent millennia, and a harmonization approach is also viable where the scriptural texts are deemed to be referring to specific epochs. In the end, the two sources need not be seen as conflicting accounts.

Turning to the Big Question about human malleability and the future, both scientific and scriptural sources evidence that human physical features such as height have changed over time. While height is not central to human nature and essence, radical changes to human biology could portend the move toward a trans- and eventually post-human future. Hence, greater ethical deliberation about the future of humanity is sorely needed. Accordingly, the final section of Chapter 7 attends to the moral boundaries of medical practice by initiating a multidisciplinary and multi-tier Islamic bioethical deliberation over medicine's role in inducing bodily changes. In this part of the chapter, I bring together different genres of Islamic bioethical literature and introduce the reader to ethical guideposts such as the theological notion of changing the creation of God, which is used to proscribe human enhancement and alteration activities. In concluding the chapter, I outline an intervention that links various Islamic moral sciences to theoretical, practical, and applied bioethical deliberation processes.

The final core chapter, Chapter 8, "Radical Life Extension: Intersecting Bioscientific Possibilities and Religious Sensibilities," is authored by Dr. Raudah Yunus, a clinician and public health expert who completed a postdoctoral fellowship in Muslim health disparities and

Islamic bioethics.[38] Building upon the preceding chapter's discussion of transhumanism and the role of medicine, Chapter 8 analyzes Radical Life Extension (RLE) through the bioscientific and Islamic lenses. The chapter details the biomedical science and interventions that support the possibility of RLE and subsequently explicates an Islamic perspective on human life, aging, and longevity using Qur'anic verses and *ḥadīth* reports. Thereafter, scriptural evidence relevant to the possibility of life extension, specifically those that discuss the lifespan of the Prophet Nūḥ and the lifespan of the nation of the Prophet Muhammad, is measured against empirical evidence derived from contemporary demographic data and anti-aging research. In applying the epistemological framework to appraise the revelatory and empirical proofs regarding the possibility of radical life extension, Dr. Yunus asserts that no actual conflicts exist because both the scriptural and bioscientific proofs indicating such a possibility are probabilistic/presumptive (*ẓannī*). Although humankind may have radically extended lifespans in the future, whether due to environmental factors or induced biomedical interventions, she suggests that such ventures may not align with Islamic morality. As to whether humankind is changing, she asserts that "we are, in fact, changing," in features but not necessarily in essence. "What is unclear is to what extent" (Chapter 8: xx–xx). In her view, natural changes are unlikely "to transform humans into a different species with extraordinary capabilities, as depicted in science fiction or imagined in the evolutionary account" (Chapter 8: xx–xx).

The last part of the book, Part III, contains two chapters that move directly into the educational domain. Chapter 9, "The Importance of Teaching Muslim Students about Scientific and Religious Constructions of the Human Being," is authored by Usama Javed Mirza, a Gates scholar and doctoral candidate at Wolfson College at the University of Cambridge. Taking a "notes from the field" approach, this developing scholar discusses the need for redesigning science education for Muslim students and educators in multicultural and faith-based schooling contexts, both in the West and in Muslim-majority countries. He argues that contemporary scientific education is inflected with particularistic worldviews and is often taught as a universal monoculture. The "Monoculture

[38] The postdoctoral program concluded in September 2025.

of Modern Science" (Santos 2016)[39] holds that there is only one valid way to produce scientific knowledge and is linked to US–Western European historical and cultural traditions.[40] He notes that Muslim students receiving such education encounter numerous challenges related to epistemic authority and religious identity when scientific claims are perceived to conflict with religious beliefs, values, and doctrines. This type of education thus engenders a sense of cultural inferiority and may diminish students' religiosity. It also may breed a professional and personal identity bifurcation.[41] Consequently, the author argues that science curricula in Muslim contexts need to cover epistemology and the history of science and to engender critical thinking. Having glossed this background, the author discusses pedagogical approaches to developing curricular materials that are more expansive than the monocultural scientific approach within which Islamic theological, philosophical, and epistemological perspectives are incorporated. He concludes the chapter by highlighting several seminal initiatives in this space and issuing a call for more theoretical and practical research into bridging Islam and science within education.

The last chapter in the volume, Chapter 10, is "Addressing 'Big Questions' about the Human Being through Bioscience and Islam: Findings from an Educational Intervention across Two Countries." This piece shifts the reader's attention away from theory and onto practice and empirical validation. Having described the educational project that seeded this volume in broad strokes above, this chapter dives into the practical aspects of its implementation and presents results from our mixed-methods evaluation. Authored by Dr. Yunus and me, the chapter describes our novel seminar-based enrichment program that sought to build Muslim students' knowledge, behavioral intent, and positive views regarding the intersection of bioscience and Islam. Given that the teaching exercises represented an acceptability, feasibility, and efficacy trial, the chapter lays out the design of the study, including the sociological

[39] Boaventura de Sousa Santos, a Portuguese sociologist and legal scholar, coined this term.

[40] Cross-reference my discussion regarding biomedicine above.

[41] Indeed, we see this in healthcare, and our curricular intervention hoped to foster more integrated identities.

contexts and theoretical frames that informed our curriculum design, and the pedagogical elements and strategies utilized during the seminars. Relatedly, the chapter details the evaluation methods used, including the outcome measures, tools, and analytic strategies employed. Notably, our assessment involved both quantitative, that is, pre- and post-participant surveys, and qualitative, that is, participant reflective essay data analysis. Concerning results, Chapter 10 concludes that the educational intervention was acceptable in content based on the reactions of school administrators and students. It was also feasible to conduct based on the instructors' perceptions and the relatively low attrition rate among students. Finally, with respect to the curriculum, it was efficacious in that students' knowledge gain was substantial, and levels of interest, intention, and preparedness to learn more, engage with, and pursue further opportunities at the intersection of Islam and bioscience increased significantly.

In summary, this book offers fresh multidisciplinary discussions on epistemology, theological anthropology, ontology, moral theology, and pedagogy through essays at the nexus of Islam and bioscience. Consequently, the chapters provide helpful insights to a wide variety of scholars, educators, and students and intend to spur further research and dialogue at this intersection. The three parts of the book complement each other, taking the reader on a journey that first covers key concepts and paradigms related to Islam and bioscience discourse, and next addresses five "Big Questions" about human beings by using an epistemic framework to bring together bioscientific and scriptural claims about humans and resolve competing accounts as necessary. Finally, it discusses the merits of curricular approaches to (Islam and) science education for Muslim students. In closing, I must underscore that this book does not offer a final, definitive answer to the "Big Questions," nor does it claim that the epistemological approaches offered herein are the only legitimate methods by which to understand and analyze potential conflicts at the bioscience–Islam interface, and neither does it declare the Bioscience and Islam curriculum described herein to be the most effective or best approach to educating students at this intersection. Instead, this book adds to the ongoing discourse by offering new insights and novel approaches to spur further research and educational efforts.

References

Ajzen, Icek. 1985. "From intentions to actions: A theory of planned behavior." In *Action control: From cognition to behavior*, ed. Julius Kuhl and Jürgen Beckmann, 11–39. Berlin: Springer-Verlag.

Ajzen, Icek. 1991. "The theory of planned behavior." *Organizational Behavior and Human Decision Processes* 50 (2): 179–211. https://doi.org/10.1016/0749-5978(91)90020-T.

Al-Akiti, Afifi, and Aasim I. Padela. 2022. "Taking on the Ghazālian challenge of integrating religion and science in Islam and biomedicine." In *Islam and biomedicine*, ed. Afifi al-Akiti and Aasim I. Padela, 1–14. Cham: Springer International Publishing.

Ali, Muna. 2018. "The 'bio' in biomedicine: Evolution, assumptions, and ethical implications." In *Islamic perspectives on the principles of biomedical ethics: Muslim religious scholars and biomedical scientists in face-to-face dialogue with Western bioethicists*, ed. Mohammed Ghaly, 41–67. Qatar: World Scientific/Imperial College Press.

Altaie, Basil. 2016. *God, nature, and the cause: Essays on Islam and science*. Kalam Research and Media LLC-Fz.

Altaie, Mohammad Basil. 2017. "Daqīq al-Kalām: A basis for an Islamic philosophy of science." *Cambridge Muslim College Papers* 4: https://cambridgemuslimcollege.ac.uk/wp-content/uploads/2020/04/CMCPapers4-Daqiq-al-Kalam.pdf.

Al-Attas, Syed Muhammad Naquib. 1969. *Preliminary statement on a general theory of the Islamization of the Malay-Indonesian archipelago*. Kuala Lumpur: Dewan Bahasa dan Pustaka.

Azam, Laila, Sohad Murrar, Usman Maravia, Omar Davila, and Aasim I. Padela. 2024. "Religious identity-based discrimination in the physician workforce: Findings from a survey of Muslim physicians in the UK." *BMJ Leader*. https://doi.org/10.1136/leader-2024-001004.

Bakar, Osman. 2008. *Tawhid and science: Islamic perspectives on religion and science: Shah Alam*. Malaysia: Arah Publications.

Baqai, Benish, Laila Azam, Omar Davila, Sohad Murrar, and Aasim I. Padela. 2023. "Religious identity discrimination in the physician workforce: Insights from two National Studies of Muslim clinicians in the US." *Journal of General Internal Medicine* 38 (5): 1167–1174. https://doi.org/10.1007/s11606-022-07923-5.

Bishop, Jeffrey P. 2011. *The anticipatory corpse: Medicine, power, and the Care of the Dying*. Notre Dame, IN: University of Notre Dame Press.

Bucaille, Maurice. 1980. *The Qur'an & modern science*. Peace Vision.

Chowdhury, Safaruk. 2021. *Islamic theology and the problem of evil*. Cairo: American University in Cairo Press.

Draper, John William. 1881. *History of the conflict between religion and science.* Washington, DC: D. Appleton and Co.

Emon, Anver M. 2010. *Islamic natural law theories.* Oxford: Oxford University Press.

Engel, G. L. 1977. "The need for a new medical model: A challenge for biomedicine." *Science* 196 (4286): 129–136. https://doi.org/10.1126/science.847460.

Engel, G. L. 1980. "The clinical application of the biopsychosocial model." *American Journal of Psychiatry* 137 (5): 535–544. https://doi.org/10.1176/ajp.137.5.535.

Evans. John H. 2021. "Insights on vaccine hesitancy from religious people's view of science." Berkley Center for Religion, Peace and World Affairs. Berkley Forum. https://berkleycenter.georgetown.edu/responses/insights-on-vaccine-hesitancy-from-religious-people-s-view-of-science. Accessed February 16, 2021.

Al-Faruqi, Ismail Raji. 1982. *The Islamization of knowledge: General principles and work plan.* Washington, DC: IIIT Press.

Ferngren, Gary, ed. 2002. *Science & Religion: A historical introduction.* Baltimore, MD: Johns Hopkins University Press.

Flexner, Abraham. 1910. *Medical education in the United States and Canada: A report to the Carnegie Foundation for the Advancement of Teaching.* New York: Carnegie Foundation.

Funk, Cary. 2015. "Perception of conflict between science and religion." *Pew Research Center.* https://www.pewresearch.org/science/2015/10/22/perception-of-conflict-between-science-and-religion/#fn-94004-2.

Guessoum, Nidhal. 2011. *Kalam's necessary engagement with modern science.* Dubai: Kalam Research & Media. https://islam-science.net/wp-content/uploads/2014/09/Kalam's-Necessary-Engagement-with-Modern-Science-Nidhal-Guessoum.pdf.

Hashim, Rosnani, and Imron Rossidy. 2000. "Islamization of knowledge: A comparative analysis of the conceptions of Al-Attas and Al-Fārūqī." *Intellectual Discourse* 8 (1): 19–44. http://irep.iium.edu.my/id/eprint/25955.

Hussain, Yasir. 2015. "The works and contributions of Ismaiʿil Raji al-Faruqi in Islamization of knowledge." *Journal of Islamic Thought and Civilization* 5 (1): 25–45.

Initiative on Islam and Medicine. n.d. "Fostering Connections Between Science and Theology for Medical Track Muslim Students." https://www.medicineandislam.org/fostering-connections-between-science-and-theology-2/.

Initiative on Islam and Medicine. 2013–2021. "Advancing Equity for Muslim Physicians in the Healthcare Workforce." https://www.medicineandislam.org/advancing-equity-for-muslim-physicians-in-the-healthcare-workforce/.

Iqbal, Muzaffar. 2010. "Darwin's shadow: Evolution in an Islamic mirror." *Islam & Science* 8 (1):

Iqbal, Muzaffar. 2002. *Islam and science*. London: Routledge.

Iqbal, Muzaffar. 2023. "Engaging new Muslim evolutionists." *Islamic Sciences* 16 (1): 3–7.

Jackson, Sherman A. 2002. *On the boundaries of theological tolerance in Islam: Abū Ḥāmid al-Ghazālī's Fayṣal al-Tafriqa*. Karachi: Oxford University Press.

Jalajel, David Solomon. 2009. "Islam and Biological Evolution: Exploring Classical Sources and Methodologies." MA thesis, University of the Western Cape, Bellville.

John Templeton Foundation. 2022–2025. "Fostering Connections Between Science and Theology for Medical Track Muslim Students." Project Leaders: Aasim I. Padela and Abdurezak Hashi, July 2022–June 2025, ID# 61863. For greater details visit, https://www.templeton.org/grant/fostering-connections-between-science-and-theology-for-medical-track-muslim-students.

Kalam Media. n.d. "The Importance of Kalam and the Revival of Theological Discourse." Kalam Research & Media. https://kalamresearch.com/the-importance-of-kalam-and-the-revival-of-theological-discourse/.

Karamali, Hamza. 2024. "Theological information on the existence of intelligent life outside our solar system: Metaphysics, scripture, and science." In *Islamic theology and extraterrestrial life: New Frontiers in science and religion*, ed. Shoaib Ahmed Malik and Jörg Matthias Determann, 25–41. London: I.B. Tauris.

Khalil Center. n.d. "TIIP Certification." https://khalilcenter.com/school-of-ip/tiip-certification.

Khan, Tahseen N. 2023. *The provenance of man: A Sunni apologetic of the original creation of Ādam*. Chicago, IL: Philasufical Publications.

Khurana, Mansee. 2020. "The Right to Worship: Church and State Clash over Religious Services in the Coronavirus Era." NBC News. May 9. https://www.nbcnews.com/politics/politics-news/right-worship-church-state-clash-over-religious-services-coronavirus-era-n1201626.

Kirkpatrick, Donald L. 1994. *Evaluating training programs: The four levels*. San Francisco, CA: Berrett-Koehler.

Krieger, Nancy. 2011. *Epidemiology and the people's health: Theory and context*. New York: Oxford University Press.

McDonald, Glenn. 2017. "Church that worships AI god may be the way of the future." *Live Science*. https://www.livescience.com/60728-church-that-worships-ai-god.html.

Malik, Shoaib Ahmad. 2021. *Islam and evolution: Al-Ghazālī and the modern evolutionary paradigm*. London: Routledge.

Malik, Shoaib Ahmed. 2023. "Defending 'Islam and evolution: Al-Ghazālī and the modern evolutionary paradigm': Abrahamic dialogues and interdisciplinary insights." *Theology and Science* 21 (4): 745–780.

Malik, Shoaib Ahmed. 2024. "Adam, eve, and human evolution: Is there a conflict?" In *Islamic philosophy of religion: Essays from analytic perspectives*, ed. Mohammad Saleh Zarepour, 261–281. Abingdon: Routledge.

Malik, Shoaib Ahmed, Hamza Karamali, and Moamer Yahia Ali Khalayleh. 2022. "Does criticizing intelligent design (ID) undermine design discourse in the Qur'ān? A Kalāmic response." *Zygon*® 57 (2): 490–513. https://doi.org/10.1111/zygo.12773.

Maulana, Mohammad Firman. 2017. "The Islamization of knowledge (IOK) an experience of International Islamic University Malaysia." *Journal of Creative Writing* 3 (2): 23–41.

Moran, Glen. 2021. "The final domino: Yasir Qadhi, YouTube, and evolution." *Zygon: Journal of Religion and Science* 56 (1): 34–53. https://doi.org/10.1111/zygo.12666.

Murrar, Sohad, Laila Azam, Benish Baqai, Omar Davila, and Aasim I. Padela. 2023a. "Relationships between religious commitment, workplace experiences, and professional and psychological outcomes among US Muslim physicians." *Academic Medicine*. https://doi.org/10.1097/ACM.0000000000005686.

Murrar, Sohad, Laila Azam, and Aasim I. Padela. 2023b. "A religious minority tax in healthcare? Insights from Muslim American physicians." *SSM-Qualitative Research in Health* 4: 100316. https://doi.org/10.1016/j.ssmqr.2023.100316.

Murrar, Sohad, Laila Azam, Omar Davila, Benish Baqai, and Aasim I. Padela. 2022. "The impact of practicing both medicine and religion: Muslim identity as a predictor of discrimination, accommodation, and career outcomes in academic medicine." *Academic Medicine* 97 (11S): S169.

Nasr, Seyyed Hossein. 1981. *Knowledge and the sacred*. Edinburgh: Edinburgh University Press.

Nasr, Seyyed Hossein. 1993. *The need for a sacred science*. Albany, NY: State University of New York Press.

Padela, Aasim I. 2021. "Jurists, physicians, and other experts in dialogue: A multidisciplinary vision for Islamic bioethical deliberation." In *Medicine and shariah: A dialogue in Islamic bioethics*, ed. Aasim I. Padela, 227–234. Notre Dame, IN: University of Notre Dame Press.

Padela, Aasim I. 2022. "Integrating science and scripture to produce moral knowledge: Assessing *Maṣlaḥa* and *Ḍarūra* in Islamic bioethics and the case of organ donation." In *Islam and biomedicine*, ed. Afifi al-Akiti and Aasim I. Padela, 295–316. Cham: Springer International Publishing.

Padela, Aasim I., Huda Adam, Maha Ahmad, Zahra Hosseinian, and Farr Curlin. 2016. "Religious identity and workplace discrimination: A National Survey of American Muslim physicians." *AJOB Empirical Bioethics* 7 (3): 149–159. https://doi.org/10.1080/23294515.2015.1111271.

Padela, Aasim I., Laila Azam, Sohad Murrar, and Benish Baqai. 2023. "Muslim American physicians' experiences with, and views on, religious discrimination and accommodation in academic medicine." *Health Services Research* 58 (3): 733–743. https://doi.org/10.1111/1475-6773.14146.

Pew Research Center. 2022. "Key findings from the global religious futures project." December 21. https://www.pewresearch.org/religion/2022/12/21/key-findings-from-the-global-religious-futures-project/.

Principe, Lawrence M. 2006. *Science and religion.* Chantilly, VA: The Teaching Company.

Qadhi, Yasir. 2018. "Human origins—Part 1: Theological conclusions and empirical limitations." August 31; updated. https://yaqeeninstitute.org/read/paper/human-origins-part-1-theological-conclusions-and-empirical-limitations. Accessed July 22, 2024.

Reeves, Jay. 2021. "Many Bible Belt preachers silent on shots as COVID-19 surges." *AP.* (August 16): https://apnews.com/article/health-religion-coronavirus-pandemic-7a03fed8e7548797487f2b7a20161e4e.

Safi, Louay. 1993. "The quest for an Islamic methodology: The Islamization of knowledge project in its second decade." *American Journal of Islamic Social Sciences* 10 (1): 23–48. https://doi.org/10.35632/ajis.v10i1.2522.

Santos, Boaventura de Sousa. 2016. *Epistemologies of the south: Justice against Epistemicide.* London: Routledge.

Setia, Adi. 2012. "Kalām Jadīd, Islamization, and the worldview of Islam: Applying the neo-Ghazālian, Attasian vision." *Islam & Science* 10 (2): 25–73.

Stevenson, Lisa. 2011. "Praying along: Interfaith chaplaincy and the politics of translation." In *Shattering culture: American medicine responds to cultural diversity*, ed. Mary-Jo Delvecchio Good, Sarah S. Willen, Seth Donal Hannah, Ken Vickery, and Lawrence Taeseng Park, 94–111. New York: Russell Sage Foundation.

Tafsir ibn Kathir. n.d. "Tafseer Ibn-e-Katheer." Word of Allah. https://www.wordofallah.com/tafseer.

Valles, Sean. 2020. "Philosophy of Biomedicine." *Stanford Encyclopedia of Philosophy*, summer ed. Edited by Edward N. Zalta. Metaphysics Research Lab, Stanford University. https://plato.stanford.edu/entries/biomedicine/.

Waters, Benjamin Victor, and Peter Stanley Fosl. 2015. "Toward a new *Kalām* cosmological argument." *Cogent Arts & Humanities* 2 (1). https://doi.org/10.1080/23311983.2015.1062461.

White, Andrew Dickson. 1876. *The warfare of science.* New York: D. Appleton.

Zaman, Muntasir. 2022. *The height of prophet Adam: At the crossroads of science and scripture.* Oldham: Beacon Books.

CHAPTER 2

First Principles, Certainty, and Uncertainty: Unburdening Ourselves of the Science–Religion Conflict

Omar Qureshi

Introduction

This chapter provides an account and analysis of several salient epistemological principles, concepts, and modes of reasoning related to scientific and theological inquiry drawing from the works of Sunnī *kalām* theologians and logicians in the Islamic tradition. Not merely an excavation of a past an intellectual tradition, herein I develop an epistemological framework that serves as a foundation for scientists, science educators, theologians, and religious educators to employ when engaging competing Islamic and scientific truth claims about reality. In so doing, the aim is to reframe the dialogue from the conflict more toward more fruitful engagements. Understanding the Islamic tradition in all its diversity as a

O. Qureshi (✉)
Integrated Curriculum, Yaqeen Institute for Islamic Research, Carrollton, TX, USA
e-mail: omar.qureshi@yaqeeninstitute.org

Logic and Theology, Qalam Seminary, Carrollton, TX, USA

© The Author(s), under exclusive license to Springer Nature Switzerland AG 2026

A. I. Padela (ed.), *Bridging Bioscience and Islam*, Islam & Science,
https://doi.org/10.1007/978-3-032-09996-9_2

living tradition adhered to by a majority of the 1.9 billion Muslims on this planet, it is hoped that this framework will be utilized to understand revealed and rational truths in light of each other. Given the projected expansion of scholarly literature on Islam, science, and education (Mustafa et al. 2021), this study contributes to the discourse by critically engaging with the epistemological frameworks of Sunnī *kalām* and modern science. By moving beyond thin, and often reductive, interpretations of Islam's theological and scientific traditions that focus more on civilizational achievements, this approach fosters a more comprehensive and intellectually richer understanding. Consequently, it offers a perspective that resonates more deeply with members of the faith community.

The following account on epistemology stems from the Sunnī *kalām* tradition, focusing mainly on the Ashʿarī school of theology, but not excepting the other two theological school of the Sunnī *kalām* tradition are the schools of Abū Manṣūr al-Māturīdī (d. 944) and Aḥmad ibn Ḥanbal (d. 855). For our purposes, in the area of epistemology, the three schools are mainly in agreement on the most salient issues. For this reason and due to being a dominant school within the Sunnī *kalām* tradition, Ashʿarī theological works will serve as the source for the account with references made to the other theological schools. The account will start with a definition of knowledge, which leads into concepts of certainty, and an examination of the various sources of human knowledge. This will be followed by an inquiry on which propositions are conclusively certain. Next, the kinds of propositions and their relationship to each other will be investigated. All of this will lead to the final section of the chapter, which addresses principles, along with their epistemological justifications, to employ when there is a perceived conflict between rational and transmitted proofs. This last section is pertinent to the "religion and science conflict" thesis and will demonstrate that such a conflict, as conceived in Western discourse, is difficult to read into and impose onto the Islamic tradition. Aside from the important question of whether scientific findings can yield certainty, the main argument of this chapter is that, looked from within the Sunnī *kalām* tradition and based on their conception of the unity of truth, there can only be an apparent conflict between conclusively certain and presumptive proofs, whether rational or transmitted, however, this is not understood as a conflict between science and religion. In fact, such a conflict between conclusive rational and transmitted proofs is rationally inconceivable and rejected outright as a possibility by Sunnī *kalām* theologians. The approach adopted instead is to accept whichever proposition has greater epistemological strength and yet not to discard propositions of lesser strength as human knowledge grows cumulatively in all sciences. Consequently, this forces the religious scholarly community and the scientific community to discuss their truth claims and the deliverables of each discipline in areas where a conflict is apparent.

Knowledge, Certainty, and Its Sources

One encounters a rich, nuanced discourse in the Islamic tradition on the nature of knowledge, its kinds, and sources. Different definitions of knowledge exist among different theological and philosophical schools in the Islamic intellectual tradition and within schools themselves (see Abrahamov 1993; and Benevich 2022). In this section I provide an overview of various definitions of knowledge as found in advanced works of Ashʿarī theologians in order to set the foundations for pursuant discussions on the sources and various kinds of knowledge.

When treating the topic of knowledge, the first question of inquiry, prior to its definition, is into the conception of knowledge for what is being defined in a definition of the mental concept of the object. Specifically, the inquiry is on whether the conception of knowledge is necessary (*ḍarūrī*) or acquired (*naẓarī*). Ashʿarī theologians have generally three views on this question. While Fakhr al-Dīn al-Rāzī (d. 1210)[1] maintained the conception of knowledge to be necessary, al-Juwaynī (d. 1085)[2] and al-Ghazālī (d. 1111)[3] both held its conception to be acquired and therefore requiring human reasoning (al-Jurjānī 1893: 1:41, 44).

[1] Fakhr al-Dīn al-Rāzī was a pre-eminent Ashʿarī theologian of the post-classical time period of the Islamic intellectual tradition. For more on his life and works, see Adamson and Benevich (2023).

[2] ʿAbd al-Malik ibn ʿAbdallāh al-Juwaynī, also known as Imām al-Ḥaramayn was one of the foremost of the early theologians of the Ashʿarī school. His theological work *Kitāb al-Irshād ilā Qawāṭiʿ al-Adilla fī Uṣūl al-Iʿtiqād* had significant impact in the school and received numerous commentaries from scholars in the Islamic regions of North Africa. His scholarship also extended to law and legal theory. He composed *Nihāyat al-Maṭlab fī Dirāyat al-Madhhab*, a work in the Shāfiʿī legal school, and *Al-Burhān fī Uṣūl al-Fiqh*, in Shāfiʿī legal theory, both works have had significant impact in their respective disciplines.

[3] Abū Ḥāmid Muḥammad ibn Muḥammad al-Ghazālī, one of al-Juwaynī's top students. Born in Ṭūs, a city in Khurāsān (present-day Iran), he had an illustrious scholarly career. He taught at several prestigious institutions during his lifetime, the most notable being Niẓāmiyya of Baghdad and the Niẓāmiyya of Nishāpūr. While he is known for his scholarly contributions in law, legal theory, logic, and philosophy, he is a master in the science of

Accordingly, al-Rāzī holds that, due to its clarity, knowledge cannot be defined, as any definition offered would be less clear than the conception of knowledge. Al-Juwaynī and al-Ghazālī maintain that the concept of knowledge is acquired, yet it is obscure, and consequently cannot be defined with a real definition (*al-ḥadd al-ḥaqīqī*),[4] that delineates its essential attributes. Thus, most definitions of knowledge suffer from critical deficiencies. These definitions are provided by most theologians and Peripatetic philosophers, constituting the third approach, holding that the conception of knowledge is acquired yet able to be defined.

Of this third approach is Ashʿarī's own definition of knowledge as "that which necessitates the entity which [knowledge] subsists in to be a knower" (*alladhī yūjibu kawn man qāma bihi ʿāliman*) (al-Jurjānī 1893: 1:47). This definition suffers from circularity in that it uses the term "knower," which is derived from the very word intending to be defined. Another definition provided by Ashʿarī is that knowledge "is the perception of the object of knowledge as it is with respect to the thing itself" (*idrāk al-maʿlūm ʿalā mā huwa bihi*) (al-Jurjānī 1893: 1:47). This definition uses figurative language when using perception (*idrāk*) in place of knowledge and is circular because it contains the term "object of knowledge" (*al-maʿlūm*) when defining knowledge. Some Muʿtazilī theologians defined knowledge as "the belief of a thing according to what it is based on necessary [knowledge] or proof" (al-Jurjānī 1893: 1:45). This definition, Ashʿarī theologians say, is too inclusive, due to it extending to probable knowledge, and too exclusive, due to it excluding knowledge of what is impossible. The Peripatetics offer a definition of knowledge as being "a simulacrum (*tamthīl*) of the essence of the perceived object in the soul of the perceiver" (al-Jurjānī 1893: 1:50). This definition also suffers from being too inclusive as it extends to compound ignorance, doubt, and probable knowledge.

ʿAḍud al-Dīn al-Ījī (d. 1355)[5] provides us with his preferred definition of knowledge which, according to him, fulfills all the conditions of a real

Sufism as well. Two of his well-known works in Sufism are: The Revival of the Religious Sciences (*Iḥyāʾ ʿUlūm al-Dīn*) and The Niche of Lights (*Mishkāt al-Anwār*).

[4] A real definition is one composed of the genus and differentia. It is also referred to as a scientific definition or a logical definition.

[5] ʿAḍud al-Dīn al-Ījī was a Shāfiʿī judge, legal theorist, and one of the foremost authorities of the post-classical Ashʿarī school of theology. His monumental theological work, *Kitāb al-mawāqif fī ʿilm al-kalām*, serves as one of the main references for the Ashʿarī

definition. He states that knowledge is "an attribute which necessitates in its substratum an [intelligible form] by which [the soul] distinguishes the forms [of other things], whose contradictory is not possible" (*al-ʿilm ṣifa tūjibu lī maḥalliha tamyīzan bayna al-maʿānī lā yaḥtamilu al-naqīḍ*) (al-Ījī 2023: 90).[6] This definition would benefit from some commentary. To illustrate this definition, let us take the proposition *P* is *Q*, where *P* is a subject and *Q*, the predicate, is being affirmed of *P*. A person having knowledge of this proposition entails four mental components: (i) a cognition of *P*, (ii) a cognition of *Q*, (iii) the affirmative predication of *Q* for *P*, and (iv) along with *Q* in reality being predicated of *P*. This definition also stipulates that, the person holds the contradictory proposition, in this case *P* is not *Q*, is not possible to be true. Returning to the definition, knowledge defined as an attribute identifies its genus also entailing that, as an attribute, it subsides in an entity which possesses or is qualified by that attribute producing in the soul a certain type of perceptive state. This perceptive state is constituted by a distinguishing intelligible form of the object of knowledge such that the contradictory of that object of knowledge is not possible. Understood as such, the intelligible form is a complete disclosure of the object of knowledge. For example, when a person who conceives the essence of human and in the intellect a form corresponding to the essence of human occurs, this form is referred to as the distinguishing form. It is distinguishing because, by it, the essence of humanity is distinguished from other essences or forms in the intellect. What this distinguishing form relates to is the external essence of human. The distinguishing form is either a conception (*taṣawwur*), such as the essence of human, or an assent (*taṣdīq*), such as "John is standing" or "John is not standing" (Mulla Zādah 2023).

One key aspect of al-Ījī's definition of knowledge is the phrase "whose contradictory is not possible." This qualification introduces the idea of certainty (*yaqīn*) to the conception of knowledge. In his works on logic, and his late work on legal theory, *al-Mustaṣfā*, al-Ghazālī provides us with an expansive account of certainty and contrasts it with convictions and presumptions. In examining the ways the term certainty is used, he

school. It was widely taught in seminaries (*madāris*) and was the focus of numerous commentaries and marginal glosses.

[6] The term *maḥall* literally denotes a locus where something inheres or dwells. In this context, it refers to the human soul. The distinguishing form represents the mental image of an external object in the soul.

explains that when the soul accepts the truth of a proposition, one of three possible conditions arises, which various scholars have all referred to as certainty:

> *Certainty 1*: this is where "one is certain and conclusively asserts the truth [of a proposition] and conjoins it with a second certainty: namely, that one conclusively holds that the conclusiveness of one's certainty (i.e., the first certainty) is correct" (al-Ghazālī n.d.: 1:65). "one does not grant the possibility of error in his certainty on the first proposition nor in his certainty on the correctness of his certainty. He is sure, confident, and conclusively certain that it is inconceivable his opinion will change [in the future] or that he will come upon a proof, unknown to him, which would lead him to change his opinion" (al-Ghazālī 2016b: 112).[7]
>
> *Certainty 2*: here "a person[8] conclusively asserts the proposition to be true with no dispute regarding it, and is not cognizant of the proposition's contradictory whatsoever. Were the person to be cognizant of its contradictory, acceptance of it would prove to be difficult on the person's part. However, were the contradictory of the person's belief made aware [to the person], while attentive to it, by someone who is the most knowledgeable of people according to this person, such as a prophet or a truthful saint, it would bring about in the person a suspension of judgement" (al-Ghazālī n.d.: 1:66).
>
> *Certainty 3*: "the person finds composure and stability regarding a proposition and assents to it whether perceiving its contradictory or not. However, in the case where the person does perceive [its contradictory], it's natural disposition would not refuse accepting it" (al-Ghazālī n.d.: 1:66).

[7] This is almost identical to his account of certainty in his earlier works of logic such as *Miʿyār al-ʿilm*, *Miḥakk al-Naẓar*, and his autobiography *al-Munqidh min al-ḍalāl*. In the *Miʿyār al-ʿilm*, al-Ghazālī describes a person having certain knowledge as him:

> knowing a thing with such and such description, accompanied with the judgement that the thing could not be other than with such description. Were you to attempt to bring to your mind the possibility of error or being unaware of an error, it would not result in any impairment in you [of your judgment]. However, were this to accompany any possibility of error, then it is not certainty.
>
> (al-Ghazālī 2016a: 287)

[8] The term he uses here is *nafs*, meaning soul, which I have translated as person.

In Certainty 1 (C1), al-Ghazālī, in addition to maintaining that the subject *S* must hold the proposition *p* to be true with certainty (Csp), adds that *S* must also hold that he has certainty of his certainty of *p* (CCsp). Thus, Csp is a first-order certainty about *p* as it is a statement which corresponds to reality which *S* has certainty of. CCsp is a higher order of certainty than Csp, a second-order certainty, as it is a certainty, not about the truth of the proposition, but about the judgment made by *S* impossible of being an error; the Csp cannot be otherwise. Second-order certainty introduces the idea of indubitability and infallibility in this account of certainty, namely, that *S* is also certain that Csp is not possible to be mistaken.[9] Important for al-Ghazālī is that the impossibility of error applies to Csp *and* CCsp. He describes indubitability as a person being in such a state of certainty that:

> were it conveyed that one of the prophets performed a miracle while claiming what contradicts the proposition, the person would not hesitate to consider the transmitter to have lied. In fact, he would be certain that the transmitter is a liar and that the person who made the claim was in reality not a prophet, and what he presumed to be a miracle was in fact a fiction.
> (al-Ghazālī n.d.: 1:65)

We can state that in C1,

S is certain of *P* if *S* is indubitably certain of *S*'s certain assent that *P* is true.

Examples of propositions that are certain in this sense are: three is less than six; a thing cannot be in one and the same moment both eternal and temporal, existing and not existing, or in motion and at rest.

The laity of believers whether Muslim, Jewish, or Christian including, al-Ghazālī argues, the majority of theologians possess certainty in the second sense. All in this category share their adoption of their religions, creedal doctrines, and theological schools based on the "good opinion" they hold regarding those with religious authority and theological expertise (public trust in scientists would fall into this category). Their good opinion stems from the praise they have heard of such figures along with the condemnation of those holding contrary views since they were young,

[9] Descartes and Wittgenstein also stipulate this in their accounts of certainty (see Reed 2022).

thereby leading them to adopt these convictions uncritically, without an analysis of proofs. In the case of the theologians, they adopted their theological school's position, proofs, and defended them when challenged, without having independently arrived at these positions and examined the proofs themselves. Al-Ghazālī would extend this sense of certainty to scientific fields. For example, scientists who accept the findings of other scientists, whether in their disciplinary area or outside their area, could have certainty in this second sense as they neither know nor are able to assess the proofs for the scientific findings. Due to the reasons cited above, they uncritically adopt the findings of other scientists, not arriving those findings based on their own research and experimentation.

Al-Ghazālī's describes the nature of certainty in Certainty 2 as a firm conviction (*iʿtiqād jāzim*). Here, the subject holds the proposition to be true, with proof yielding firmness and certainty. Yet, what separates Certainty 2 from Certainty 1 is the condition of indubitability and the nature of proof of the proposition. Al-Ghazālī stipulates indubitability as a necessary condition in Certainty 1 alongside both, first-order and second-order certainties, being of the highest epistemic strength, namely, grounded in conclusively certain proofs. This condition, however, is not stipulated in Certainty 2. Additionally, the truth value of the proposition is a necessary condition in Certainty 1, whereas the proposition may or may not be true in Certainty 2, more on this will be discussed later.

Al-Ghazālī describes Certainty 3, which he calls presumption (*ẓann*), as extending over many degrees of epistemological strength and weakness. The strength of such presumed knowledge is due to many factors such as the number of people who transmit such a proposition, a person having the same experience as the transmitters, as well as contextual factors such as the transmitter's "face turning yellow" and "being in an unstable state" when transmitting a particular report of terrifying content. If enough factors are present and known, the strength of such a presumption can reach a point yielding certain knowledge.

When it comes to how the term certainty is used, *ḥadīth* scholars, al-Ghazālī tells us, use the term certainty for Certainty 1, 2, and 3. The majority of people, with the exception of expert scholars, have in mind Certainty 2 when using the term certainty. "They do not distinguish," al-Ghazālī notes, "between condition one (C1) and condition two (C2). The truth, however, is that certainty is the first condition while the second [condition] is where one encounters error (*maẓinna al-ghalaṭ*)" (al-Ghazālī n.d.: 1:66). Thus, for al-Ghazālī, Certainty 1 is true certainty (*yaqīn*), whereas as Certainty 2 is the area of firm convictions (*iʿtiqād*), and Certainty 3 is the area of presumptive knowledge (*ẓann*).

Applying Certainty 1 to the definition of knowledge, the impossibility of the distinguishing form's contrary is not possible for the subject the form inheres in, not only at the present time, but the contradictory is also not a possibility at some point in the future. This is due to the causes and basis of the distinguishing form coupled with the form's correspondence to extra-mental reality which, when both exist, yield this certainty of the impossibility of the contradictory form. Thus, to be counted as knowledge in al-Ījī's account, two conditions must be met: (1) the distinguishing form must be based on necessary (*ḍarūrī*) or demonstrative (*burhānī*) premises; and (2) the form must correspond to reality. Of significance is that this definition of knowledge incorporates what contemporary epistemologists call epistemic and psychological certainty. A subject is psychologically certain when he is convinced on the truth of a proposition and for various reasons "the subject is incapable of giving it up" (Reed 2022). This sense of certainty focuses on the resultant state of the mind in relation to a proposition. On the other hand, epistemic certainty occurs when a belief "has the highest possible epistemic status" (Reed 2022). Both types of certainty are incorporated in al-Ījī's definition of knowledge and al-Ghazālī's account of certainty. Certainty 1 would be aptly called epistemic certainty as the subject holds the proposition true based on the highest epistemic warrant coupled with indubitability and infallibility. On the other hand, Certainty 2 and Certainty 3 would not warrant being classified as epistemic certainty due to the first-order certainty being based on non-conclusively certain proofs, thereby rendering these two certainties as dubitable and fallible. Consequently, Certainty 2 and Certainty 3 would be aptly called psychological certainty excluding of epistemic certainty. This being the case, it is important to keep in mind that psychological certainty does not necessarily entail that it is devoid of any epistemological justification and strength. On the contrary, in the case of psychological certainty, there might be strong justification, yet it is not of the highest status and thus has some possibility of error.[10]

[10] Extending this to empirical knowledge, one would say that empirical evidence based on randomized, controlled studies are considered a "gold" standard whereas simple observation leads to a lower level of certainty. Experience in the case studies adds certainty and may be analogous to psychological certainty.

Based on the two conditions for knowledge, one can now account for other kinds of human cognition such as beliefs, doubts, and other types of cognitions. When one holds a proposition to be true, yet it is not based on premises which are necessary or demonstrative, rather it is based on non-certain or probabilistic premises, this cognition cannot be called knowledge as defined by al-Ījī due to the first condition not being met and is consequently termed a conviction or belief (*iʿtiqād*). If this conviction corresponds to reality, it is a correct conviction (*iʿtiqād ṣaḥīḥ*); and, in the case that it does not correspond, it would be an incorrect conviction (*iʿtiqād fāsid*) (see Fig. 2.1).

Looking again at the definition, the phrase "whose contradictory is not possible," as we said introduces the concept of certainty in the definition of knowledge. Convictions held regarding a proposition whose contrary is held to be possible are classified, based on its probability or likelihood to be true, as presumptive, or likely (*ẓannī*), or plausible (*rājiḥ*), implausible (*marjūḥ*), or unlikely (*wahmī*), or the contrary is equally likely or unlikely, where a person is in a state of doubt (*shakk*). Accordingly, this definition of knowledge excludes propositions which are presumptively or unlikely to be held true, and propositions which a person holds to its contrary

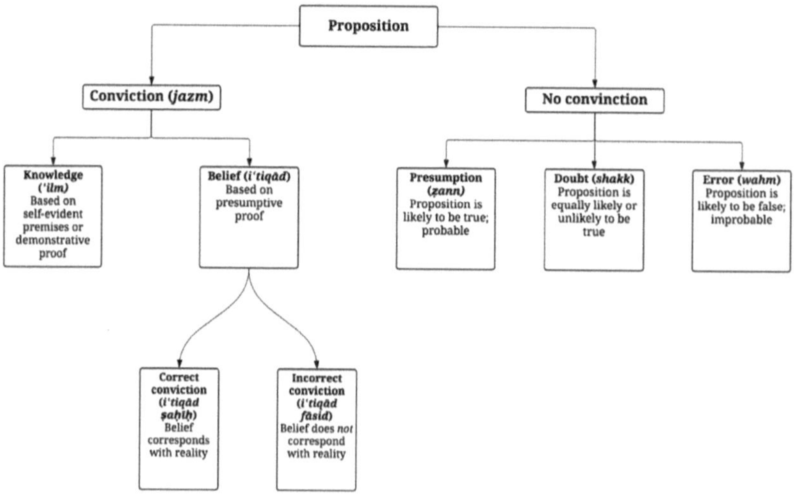

Fig. 2.1 Classification of propositions in terms of conviction. Imitation (*taqlīd*) is classified as a belief in this scheme

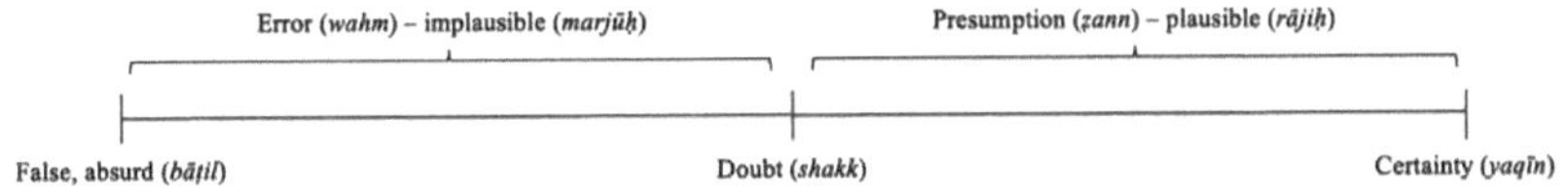

Fig. 2.2 Spectrum of degrees of conviction with regard to a proposition

to be equally likely to be true or false.[11] Also excluded are convictions which may or may not be true. Included in this are convictions where an expert's opinion is adopted uncritically, known as imitation (*taqlīd*) (Fig. 2.2).

Of importance is the identity of the type of possibility referred to in this definition of knowledge. What is not intended here is the mere logical possibility of the contrary being true. In the context of empirical observations, the logical possibility of the contradictory remains regardless of the evidence for the proposition under consideration and a person's certainty of it being true or false. This leads us to conclude that it is the empirical possibility which is intended in the phrase "whose contradictory is not possible," since all empirical matters are logically possible. For instance, the knowledge that P (e.g., "there is a mountain before me") entails that its contradictory ($\neg P$, "there is not a mountain before me") is, while logically conceivable, empirically impossible. The logical possibility does not negate or diminish my certainty of the proposition's truth.[12] The distinction between logical and empirical possibility provides important insights into the nature of scientific knowledge particularly in relation to the epistemic foundations of inductive and abductive reasoning within scientific inquiry. Within this framework, an empirical premise, insofar as it is directly known through observation, is known with certainty. However, when conjoined with a presumptive rational premise—as is the case in inductive or abductive inferences—the epistemic force of the conclusion is diminished, as it rests, at least in part, upon premises whose justificatory status remains presumptive rather than conclusive.

[11] Much of statistical analysis involves estimating the probability of observing the data, assuming the null hypothesis is true.

[12] This has relevance to the notion of falsifiability in scientific theories and their predictability. Although the conception of what is empirically impossible, discussed later, adds further considerations to the nature of scientific hypotheses.

Means of Human Knowledge

Having delineated a definition of knowledge pertinent to this chapter, it is now apropos to examine the sources of human knowledge in order to construct the foundational components of the epistemic framework. Throughout early to late works of Sunnī *kalām* theology, three sources of knowledge are identified: (1) the intellect (*al-ʿaql*), which provides humans with rational knowledge; (2) the five external senses, which provide humans sensory knowledge; and (3) testimonial knowledge in the form of true reports (*al-khabar al-ṣādiq*)[13] (see Ibn Fūrak 1987; al-Māturīdī 2023: 174–183; and al-Taftāzānī 2020). Limiting the sources of knowledge to these three raises the question of how to understand this list: Is it meant to be exhaustive of all sources of human knowledge? What is the status of numerous other obvious means by which we humans come to know things such as dreams, our inner sense (*wijdān*), Prophetic revelation (*waḥy*), and spiritual disclosures (*kashf*)? Do Sunnī theologians not consider these as sources of knowledge?

Understanding the nature of Sunnī theological discourse will help us shed light on this matter. One of the main purposes underlying the discipline of *kalām* theology is to establish and affirm true religious doctrine to interlocutors who do not hold these doctrines to be true. In enumerating the five goods of *kalām* theology, al-Ījī states, regarding the second good, "The second good is providing guidance to those who seek it through clarification of the correct path and compelling obstinate deniers by establishing apodictic arguments [against their claims]" (al-Ījī 2023). This captures one of the central aims of *kalām* theology, namely, to defend religious doctrines against arguments put forth by opponents with the hope of eventually winning over the interlocutors. Al-Jurjānī (d. 1413),[14] in his comments on this passage elaborating this aim, further states, "for compelling [the opponent], which includes disgracefully exposing the obstinate, might lead to him to acceptance [of true

[13] For example, al-Māturīdī states the following in *Kitāb al-Tawḥīd*, "the means by which knowledge of the reality of things (*ḥaqāʾiq al-ashyāʾ*) is reached are: the senses (*al-ʿiyān*), reports (*al-akhbār*), and rational inquiry(*al-naẓar*)" (al-Māturīdī 2023: 174).

[14] Sayyid Sharīf al-Jurjānī was a leading Sunnī scholar of the late post-classical Islamic period, who composed works on Ashʿarī *kalām*, logic, philosophy, grammar, rhetoric, and legal theory, which were well studied in the Islamic tradition. One of his most famous works is *al-Taʿrīfāt*, a concise lexicon of technical terms used in various Islamic sciences. His scholarship deeply influenced Ottoman, Safavid, and Moghul intellectual traditions.

doctrine] and [thus] being guided, and will, consequently, be of benefit for him" (al-Jurjānī 1893: 1:34). In order for *kalām* theology to carry out this role, there must be common principles held by all interlocutors for claims to be supported with evidence. Here, Sunnī theologians identify reason, something all humans share by virtue of being human, sense knowledge, and testimonial knowledge as shared principles based on which theological inquiries are conducted. Consequently, knowledge from dreams, personal inspiration, and spiritual disclosures are not, in themselves, sources of knowledge which can be invoked in the discipline of *kalām* theology, as the knowledge they produce is not shared by all humans. Knowledge claims from these sources must be verified and grounded in these three universally recognized sources. Thus, it is not the case that Sunnī theologians are restricting all human knowledge to these three sources. Rather, when they make their statement on the three sources, it is to identify valid sources of proof only within the discipline of *kalām* theology.

As further evidence of the claim made above, as well as shedding light on other sources of knowledge, Sunnī theologians have explored and identified such sources of knowledge. Illustratively, the Ottoman scholar, Mulla Zādah (d. 1655), in his *Mirqāt al-ʿulūm wa al-sirr al-maktūm*, has identified at least 13 sources of human knowledge:

1. The five external senses, when sound, provide knowledge of external particulars.
2. True reports in the form of mass-transmitted reports (*mutawātir*), which yield necessary knowledge, or solitary reports (*āḥād*) which yield presumptive knowledge.
3. The intellect perceives necessary knowledge and inferential knowledge based on self-evident premises.
4. The inner sense (*wijdān*) is that by which a person has self-knowledge and knowledge of a person's qualities such as joy, sadness, hunger, and satiation. These cognitions are not known with the aid of the intellect or by any external sense.
5. Immediate or primitive knowledge (*badāha*) is the "appearance of the reality of a thing to a human due to material constitution (*khilqa*) or natural intelligence (*fiṭra*) or due to intimate familiarity and constant practice such that [the thing's] essence and attributes are always present so that every time the thing is present its reality

is perceived necessarily" (Mulla Zādah 2023: 95). It is that knowledge which a person is not able to acquire due to it being innate, always present with a person, unable to not know.

6. The habitual course of events in the natural world (*al-ʿāda*) such as this mountain is made of stone.
7. Intuition (*ḥads*) is defined as the quick movement of the soul from principles to what is sought in one moment. This is usually due to the purity of the soul and the strength of a person's intellect.
8. Knowledge based on repeated observations (*tajrība*) and experiments. Most knowledge in the natural sciences results from repeated observations.
9. Rational inquiry (*naẓar ʿaqlī*) consists of ordering known concepts to arrive at something previously unknown.
10. Uncritical adoption of someone else's opinion (*taqlīd*). Usually this involves the adopting of expert opinion. It is uncritical in the sense of the person not knowing the evidence or reasoning used to arrive at the opinion adopted and the inability to discern if it is true or false.
11. Dream vision (*ruʾyā*) is a conviction which God creates into the heart of a person during sleep.
12. Inspiration (*ilhām*) is a conviction which God creates into the heart of a person during wakefulness.
13. Divine revelation (*waḥy*) always occurs by the intermediation of an angel to God's prophets (Mulla Zādah 2023: 81–105).

These 13 means of knowledge include the 3 mentioned in works of *kalām* theology yet enumerate the means of knowing reflective of the whole human experience. From this understanding of human knowledge, one can recognize the challenge—and perhaps the futility—of categorizing this epistemological approach strictly as rationalist (where reason is the most reliable source of knowledge), empiricist (where sense experience serves as the foundation of knowledge), or constructivist. For Sunnī scholars, all sources of human knowledge are identified and understood in relation to one another.

Necessary and Certain Propositions

Al-Ījī's definition of knowledge stipulates the two conditions: (i) of certainty, that the contradictory proposition is impossible to be true; and (ii) of truth, that the proposition corresponds to reality. This section will provide an account of what it means to meet the condition of certainty. Sunnī theologians hold that human knowledge is either necessary (*ḍarūrī*) or acquired. Al-Ījī reproduces al-Bāqillānī's (d. 1012) definition of necessary knowledge as referring to those cognitions "which persist in the soul of a created being such that a person cannot find a way to be disassociated from them." Necessary knowledge serves as the foundation for all other forms of acquired knowledge and is considered as their first principles (*al-mabādīʾ al-ʾūlā*) such that were there no necessary knowledge, knowledge in any form would not be possible.

Necessary knowledge is of three kinds: knowledge by virtue of our inner sense (*wijdān*), sensory knowledge (*ḥissīyāt*), and immediate or self-evident knowledge (*badīhīyyāt*). Knowledge produced by our inner sense is "of little use in the sciences for it is [knowledge] which is not shared between people and, consequently, cannot be used as evidence against others" (al-Jurjānī 1893). On the other hand, sensory and self-evident knowledge are "both the mainstay of the sciences" (al-Jurjānī 1893) due to being shared by all humans. Under sensory knowledge are propositions in which the senses play a role. Consequently, this category includes propositions based on experience (*tajrībīyyāt*), propositions based on mass-transmitted testimony (*mutawātirāt*), propositions based on sense observation (*mushāhadāt*), and propositions based on intuition (*ḥadsīyyāt*). Under self-evident knowledge are primary propositions (*awwalīyyāt*) and propositions with an embedded syllogism.[15] Importantly, self-evident propositions can be invoked as proof in all situations, whereas propositions based on experience, sense observation, intuition, and mass-transmitted testimony can be invoked as proof when all interlocutors share the same basis.

In their inquiries into what constitutes certainty, Sunnī theologians identify seven propositions that yield certainty.[16] Al-Ījī succinctly states:

[15] An example of such a proposition would be "the number four is even because it can be divided into two equal numbers."

[16] As al-Jurjānī clarifies, this entails: (1) a person believing *P* is *Q*; (2) the proposition corresponds to reality; and (3) that *P* is *Q* cannot be otherwise.

> Conclusive propositions (*qaṭʿīyya*) are seven: the first are primary propositions which the soul is never devoid of [and which come about] after [merely] conceiving the two parts [of a proposition].[17] The second are embedded syllogistic propositions such as "four is able to be divided into two equal parts, therefore it is an even [number]." Third are observational propositions which the intellect asserts based only on sense perception.[18] Fourth are experiential propositions which are judgements the intellect makes based on sense perception along with repetition [of that sense perception]. Fifth are intuited propositions such as knowledge of the Creator from the perfection of His acts. Sixth are propositions based on mass-transmitted testimony which [the intellect] judges on the basis of the mere testimony of a group of people whose agreement on a lie is impossible. Seventh are estimative judgements regarding sensibilia such as "every body must be in a direction."
> (al-Ījī 2023: 130)

In light of al-Ījī's definition of knowledge, we can now see that, for any judgment to be counted as knowledge, it must be from or based on one of these seven types of premises. Likewise, for a syllogism to yield a certain conclusion, and thus be demonstrative, its premises must also be of these seven types of propositions. Any proposition or syllogism composed of premises from other than these seven will yield knowledge of a presumptive nature.

It would be prudent to spend some time analyzing observational (*mushāhadāt*), experiential (*mujarrabāt*), which al-Ghazālī and Ashʿarī theologians refer to as the continuous natural course of events (*iṭṭirād al-ʿādāt*), and intuitional propositions as they relate directly to the scientific study of the natural world. Observational propositions extend to objects of knowledge known through the five external senses (*ḥiss*). Examples of such propositions are the sun exists, the sun is hot, and it gives off light. Observational propositions also extend to objects of knowledge known through the inner sense (*wijdān*) and includes knowledge of things such as the knowledge that we exist and of our own acts.[19] Such knowledge is always knowledge of particulars, that is, *this P* is *Q*. In order to

[17] Namely, the subject and predicate of a proposition.

[18] Such as our knowledge that the sun exists, and that it gives off light. This is also called sense knowledge (*maḥsūsāt*).

[19] Knowledge of the self is touched upon in Chapters 3 and 7, when discussing what it means to be human.

arrive at the universal proposition *every P* is *Q*, then this can only occur after observing many particulars and knowledge of the cause. Experiential propositions, while stipulating repeated observations, also require an unnoticed syllogism that the repeated phenomenon would only take place in the same manner, all or most of the time, if it had a cause (i.e., it is not due to coincidence), whether we know the essence of cause or not. Examples of such judgments provided by al-Jurjānī are "being hit with wood is painful" and that "scammony is a laxative." Lastly, intuitive propositions share with experiential propositions repeated observations and the unnoticed syllogism, however, it departs from experiential propositions in that, in intuitive propositions, the essence of the cause must be known, not just that there must be a cause which explains the phenomenon. An example of an intuitive proposition is our observation of the different states of the moon based on its various amounts of light which depend on the moon's position in relation to the sun. We then intuit that the moon's light is acquired from the sun's light (al-Jurjānī 1893).

As stated above, experiential propositions are universal propositions, while observational propositions are always particular, thus setting the two types of propositions apart. While observational propositions are certain and therefore yield knowledge, do Sunnī *kalām* scholars extend the possibility of certainty to experiential propositions, which are universal propositions about the natural world such as fire burns, stones fall to the ground, and wine is an intoxicant? Al-Ghazālī states that "experiential knowledge are certain for the person who experienced it. People have variegated knowledge of experiential knowledge due to their different experiences" (al-Ghazālī n.d.: 1:68). How does a person move from a particular proposition to a universal proposition? This move is made by the intellect with the aid of sensory knowledge coupled with repeated sense observations multiple times. One observation will not suffice to yield a universal proposition. Al-Ghazālī explains as follows:

> a person who experiences pain at a certain location of his body and pours over it a liquid after which the pain is removed will not have knowledge that it is this liquid which is the cause of pain being removed. This is because it's removal could have occurred by [mere] coincidence... However, if it occurs many times and in different circumstances, [then] certainty and knowledge that it is the agent of removal gets implanted in the soul as in the case of becoming warm by fire removes the cold and [eating] bread removes the pain of hunger.

(al-Ghazālī n.d.: 1:68)

The person arrives at this proposition with certainty by means of "a subtle syllogism" (al-Ghazālī 2016b: 120), which a person is not even aware of carrying out hypothetical conditional syllogism:

> Were *P* not the necessitating factor of *Q*, it would not be repeatedly observed together in every or a majority of cases.
> Yet we have observed *Q* in every or a majority of cases, with *P*, therefore *P* is the cause of the effect *Q*.
>
> (see al-Ghazālī 2016b: 120)

Another possible form would be:

> Were *P* and *Q* merely a case of coincidence, we would observe *Q* with other than *P*.
> Yet we have not observed other than *P* with *Q*, therefore *P* is the cause of *Q*.

What we observe in this conditional syllogism is a dilemma; here, it is between two possible explanations of an observed phenomenon: either what is observed is a coincidence, or there is a cause. What eliminates coincidence as a possible explanation is the repeated observation of the same phenomenon in every observed case. Of significance is that al-Ghazālī includes observing *P* and *Q* together in a majority of cases thereby lowering the empirical threshold somewhat for acquiring certainty. For ***kalām*** theologians, the repeated observations of *P* and *Q* together have enough epistemological strength to negate the coincidence as a possible explanation and consequently establish a causal relationship between *P* and *Q*—or that *P* is not the cause but concomitant with the cause as some kind as an explanation of the phenomenon. It is noteworthy that there are logicians who contest the thesis that experiential propositions and intuitive propositions yield certainty, instead claiming they only yield presumptive knowledge. These scholars hold that, "a phenomenon occurring according to one pattern time after time does not necessarily yield in a certainty which does not abate" (Mubīn 1909: 2:225). This is so because, it is possible, according to these scholars, that *P* is not the cause of *Q*, rather, it is merely associated with *Q* and that other factors other than *P*, such as time, place, and elevation were the actual cause of *Q*. Hence, this is no different than inductive reasoning, which can only yield presumptive knowledge. Yet, there is a significant difference between inductive reasoning and experiential propositions, and that is this "subtle syllogism," which is not present in mere inductive reasoning and consequently avoids the post hoc, *ergo propter hoc* fallacy[20] involved in inductive reasoning. This being said, the stronger position according to logicians is that intuitive and experiential propositions do result in certainty for the reason stated above. The unnoticed syllogism involved in an experiential proposition is composed of a premise which embeds a conception of the connection between a cause and effect. Al-Ghazālī does raise the question of the meaning of the implication between cause and effect (*maʿna talāzum al-asbāb wa-l musabbabāt*): Is it a necessary, logical implication or not? For this inquiry, he refers us to the seventeenth discussion in the *Incoherence of the Philosophers*, where he advocates an occasionalist doctrine on causality. This occasionalist metaphysics, where the cause and effect exist *with* each other does not imply that the effect exists *by* the cause, neither jeopardizes nor negates the possibility of certainty with regard to causes and effects.

This account presents a rather optimistic understanding of certainty as it relates to knowledge based on observations of the natural world. According to *kalām* theologians, the only way for a conclusion to be certain is if the premises on which it is based are certain and true. In this sense, it is in the premises where certainty lies essentially, not in the conclusion. The certainty found in the conclusion only results from the certainty found in the premises. This is contrary to the way most understand contemporary science, where certainty is to be found in the conclusion.

Further exploration into demonstrative reasoning is necessary to appreciate the place of the logical structure of scientific reasoning so as to analyze the truth claims related to the deliverables of science in terms of certainty. Sunnī *kalām* theologians, like the Shīʿī *kalām* theologians, the Peripatetics and Illuminationists of the *falāsifa* in their inquiries on demonstrative syllogisms, have identified two kinds of demonstrations which are relevant to scientific reasoning: why-demonstrations (*al-burhān*

[20] This is the fallacy of concluding that solely because event *B* occurred after event *A*, event *A* caused event *B*.

al-limmī) and that-demonstrations (*al-burhān al-innī*).[21] The difference between the two kinds of demonstration can be explained by looking at the following syllogism:

P1: This piece of iron's temperature is high.
P2: Every piece of iron whose temperature is high is malleable.
C1: Therefore, this piece of iron is malleable.

The key distinction between the two kinds of demonstrations rests on the nature of the middle term in the syllogism; in this case, the middle term is "temperature is high" or high temperature. In this example, high temperature is the cause and malleability the effect. Such syllogisms where the middle term is the cause are classified as *limmī* demonstrations. The reason they are classified as such is that in answer to the question "why[22] is this piece of iron malleable?" the middle term, the cause, is always given. This is an inference made from cause, high temperature, to effect, malleability of iron. *Innī* demonstrations, on the other hand, are inferences made from effect to cause. Such reasoning would take the following form:

P1: This piece of iron is malleable.
P2: Every piece of iron which is malleable has a high temperature.
C2: Therefore, this piece of iron's temperature is high.

In *innī* demonstrations, the syllogism starts with the fact of some observed phenomenon, in this case, the fact of the malleability of this piece of iron. However, the cause of this observation is not observed, but inferred. The middle term, in the case of *innī* demonstrations, is the effect.

While both kinds of demonstrations yield certainty, a *limmī* demonstration is of a higher epistemological strength than an *innī* demonstration. This is so because of the logical structure of each kind of demonstration. The *innī* demonstration resembles the logical form of inductive reasoning: [1] observed fact *Q*, [2] in order for *Q* to occur,

[21] For a detailed study on Avicenna's conception of demonstration (*al-burhān*), scientific knowledge, and the nature of scientific inquiry, see Strobino (2021). I am following him for the translations of these two demonstrations.

[22] From the Arabic interrogative particle *limā* (why).

P would have to be true, [3] therefore *P*. As constructed, the logical form is valid and can potentially yield certainty depending on the status of [2] "in order for *Q* to occur, *P* would have to be true." What inductive reasoning suffers from is the logical possibility that factors other than *P* could also possibly be the cause of the observed fact; these factors can work in combination with *P* or be other than *P* entirely in the case when one mistakenly identifies *P* as the cause of the effect *Q*. Thus, the most one can conclude is that *P* might be true for us to observe *Q*. The contrary is the case in *limmī* demonstrations as the cause is known but the effect is inferred, thus yielding certainty due to the knowledge of the cause. In their accounts of *limmī* and *innī* demonstration, *kalām* theologians and logicians have observed that the middle term plays two different roles in the two kinds of demonstration, either it serves as the cause or the effect. Taking account of the epistemological impact this has, they have observed that in *limmī* demonstrations, the middle term serves as the cause both in the mind as well as extra-mentally. In the case of *innī* demonstrations, the middle term serves as the cause only in the mind and not extra-mentally. Thus, what the above *innī* demonstration establishes is that the high temperature is the cause of iron's malleability in the mind of the observer only, not in extra-mental reality.[23]

[23] The impact of the epistemological variance of the two kinds of demonstration is seen in tracts of astronomy written by Muslim astronomers. In his *Sharḥ al-Tadhkira*, a commentary on an advanced theoretical astronomy manual by Nasīr al-Dīn al-Ṭūsī (d. 1274), al-Jurjānī considers astronomy as the study of the "external aspects of bodies" that utilizes *innī* demonstrations. It is in natural philosophy where the reality and true natures of bodies are studied, thus utilizing *limmī* demonstrations. This distinction results in two types of propositions, corresponding to *innī* and *limmī* demonstrations, "a proposition that is established 'according to the senses' (*bi ḥasab al-ḥiss*) and one that is established 'according to the reality' (*bi ḥasab al-ḥaqīqa*)" (Hasan 2017). Consequently, it is natural philosophy, and not astronomy, which investigates the causes of observations. For an excellent study on the two kinds of demonstrations and their impact on the demarcation of astronomy and natural philosophy, see Hasan (2017).

Furthermore, the epistemological variance is also relevant to critical realism and science education. In his *A Realist Theory of Science*, Ray Bhaskar claims the need for important distinctions when thinking about the activity of science which the scientist carries out. Bhaskar divides the physical world into three domains: "The real domain is the source of all mechanisms. The actual domain is the realm of events in which a particular mechanism is induced. The empirical domain is that in which we humans can observe actual events" (Blackie 2023: 1048). The distinction between what is established as a cause in the mind of the observer and both

In closing this discussion on certainty and demonstrative reasoning, it would be apropos to briefly relate the ***kalām*** theologians' account to the hypothetico-deductive (H–D) method, developed by Dutch scientist Christiaan Huygens (d. 1695) and often referred to as the scientific method in science and science education today (see Lawson 2000, 2010; Reichardt 2021: 8), as it relates to the question of certainty of scientific findings and theories. To be sure, the H–D method is used in disciplines other than those of the natural sciences. Reichardt has summarized the hypothetico-deductive method as follows:

1. Observe a phenomenon.
2. Formulate a hypothesis to explain the phenomenon.
3. Derive consequences or implications of the hypothesis.
4. Collect data to assess the consequences or implications of the hypothesis.
5. Conclude the hypothesis is either confirmed or rejected based on the results of the assessment.

(Reichardt 2021: 7).

The reasoning involved here can be put in the following logical form:

P1: If the hypothesis *H* was true, then we would observe *P*.
P2: We observe *P*.
C1: Therefore, hypothesis *H* is confirmed.

In the H–D method, we start with an observation and then develop a hypothesized explanation or cause. The epistemological status of this hypothesis is that it is confirmed, but we cannot claim that *H* is the actual cause. This is because the H–D method is a form of inductive reasoning, in which multiple competing hypotheses may still explain or be logically compatible with the observed phenomenon. The only case where we could infer hypothesis *H* with certainty is if *P1* were: "We would observe

in the mind and extra-mental world serves as a point of conversation to understanding the nature of scientific activity between the two intellectual traditions. Matthew Wilkinson (2013) has taken some steps toward engaging critical realism by grounding such an approach in the Qurʾan. It is hoped that the epistemological account provided in this chapter will allow this engagement to develop further.

P if hypothesis *H* were true." In this case, we could infer *H* with certainty because all other logically competing hypotheses would have been eliminated. What we can conclude with certainty, however, is that hypothesis *H* is false, which aligns with Karl Popper's falsificationist methodology. According to the confirmationist H–S method, the more confirmations a hypothesis receives, the more likely it is to be true. However, this method does not allow us to conclude with certainty that hypothesis *H* is true, even if we have many observations as it assumes that all other hypotheses, besides *H*, have been ruled out as possible explanations of *P*, which is not the case.

It is important to keep in mind that the goal of functional scientific discovery is not certainty, but rather establishing an adequate likelihood that the hypothesis is true or that the null hypothesis can be rejected. The level of certainty required for establishing theological doctrines is fundamentally different from the level required for scientific discoveries to provide practical utility.

Given the *kalām* scholars' understanding of experiential propositions and *innī* demonstrations, one might argue that it is justified to hold that *P* has a particular cause or causes with certainty. However, *kalām* scholars would likely argue that the confirmationist H–S method can, at best, yield presumptive knowledge of multiple hypotheses under certain conditions, but not certainty. This is because both experiential propositions and *innī* demonstrations involve explaining the relationship between two *observed* phenomena repeatedly. In contrast, the H–S method deals with only one observed phenomenon and infers a possible cause, rather than directly observing it. *Kalām* theologians argue that the "subtle syllogism" is where certainty emerges, as it explains the relationship between two observed phenomena. Based on this, *kalām* scholars would maintain that while the confirmationist H–S method may yield, in certain cases, presumptive knowledge of observed phenomena in the natural world, it does not offer certainty regarding the causes of these phenomena. The implications of this, particularly in relation to apparent conflicts with revealed truths, will be discussed in the final section of this chapter.

A number of observations, particularly relevant to the Islam and bioscience discourse, can be made from the preceding discussion on certainty and demonstrative reasoning. First, Ashʿarī theologians maintain that acquiring certainty over particular empirical observations is possible. Second, one can develop universal conclusions regarding empirical matters provided they are based on repetition and knowledge that is due to some cause and not mere coincidence. Third, developing a universal proposition on observed phenomenon is not solely an empirical matter but rather ultimately depends on syllogistic reasoning. According to these theologians, repeated observations must be coupled with a rational component in the form of a syllogism, to yield certainty. Fourth, identifying the exact cause is required for one to develop intuitive propositions. This implies that an error in identifying the exact cause will result in an error in the proposition. The above framework and observations are key components to developing a philosophy of science grounded in strong Islamic theological and epistemological commitments.[24]

Divinely Stipulated, Empirical, and Rational Judgments: Putting It All Together

In this section, I provide an account of propositional knowledge put forth by Ashʿarī theologians so as to address the relationships between rational, empirical, and Divinely stipulated judgments.[25] This account will draw heavily on the works of Muḥammad ibn Yūsūf al-Sanūsī (d. 1490)[26] as he

[24] For more on "reading" science through a theological lens, see Qureshi et al. (2022).

[25] These are renderings of the terms *al-ḥukm al-sharʿī*, *al-ḥukm al-ʿaqlī*, and *al-ḥukm al-ʿādī*. The translation of these terms requires some explanation. *Al-ḥukm* is the affirming (P is Q) or negating (P is not Q) of a predicate for a subject. Three kinds of judgment are identified depending on the basis of the judgment. If the basis of the judgment is the revealed law (*sharīʿa*), such that neither the unaided intellect nor repeated observations alone could lead a person to arrive at such judgments, then it is classified as a *al-ḥukm al-sharʿī*, which literally translates as a judgment based on the revealed law. I translate it as a divinely stipulated judgment since the judgment is a stipulation by an agent as the intellect alone cannot be the basis of such judgments, neither can any number of empirical observations serve as its basis. Hence, it must be stipulated by an agent and that agent can either be Divine or a human. I render *al-ḥukm al-ʿaqlī* as rational judgment as the basis of these kinds of judgments is the mere intellect, without any external aide. Lastly, *al-ḥukm al-ʿādī* is rendered as an empirical judgment. Its literal translation is "a judgment on the basis of the habitual course of events in the natural world." This is informed by the Ashʿarī conception of causality, where God creates both causes and effect directly, essentially negating any notion of secondary causes. The habit of God in the natural world is observed via the senses, which leads me to translate this term as an empirical judgment as the senses are the means of knowing the habit of God.

[26] Muḥammad ibn Yūsūf al-Sanūsī was a prominent North African theologian, logician, and ṣūfī. In addition to his significant contributions to these three disciplines, he also wrote on Qurʾanic exegesis, *ḥadīth*, medicine, and astronomy. He is best known for his "five creeds," which range from introductory texts for novices to advanced works for

has presented one of the more detailed analyses of propositional knowledge in Sunnī *kalām* works. Other theologians' works,[27] who comment on and extend al-Sanūsī's initial account will also be incorporated where appropriate.

The division of a judgment (*al-ḥukm*) into its three kinds is a division of a universal to its particulars. In order to understand the basis of such a division, let us look at the following propositions:

P1: Fasting on the Day of ʿArafa[28] is not an obligation.
P2: Twelve is an even number.
P3: Acetaminophen relieves pain.

A judgment involves the affirmation or negation of a thing of another thing. Explaining the basis of the division, al-Sanūsī states that:

> the affirmation or negation, which comprise a judgment, either (1) are both based on the Revealed law (*sharʿ*), such that the affirmation or judgement can only be known by means of it, or not. If not, then either (2) the intellect alone suffices to discern it without recourse to repeated observations or [empirical] examination or (3) it does not [depend on repeated observations].
>
> (al-Sanūsī 2019: 112)

The subject term in P1 (Fasting on the Day of ʿArafa), its predicate terms (an obligation), and its copula (that of a negation) together do not entail anything logically impossible, nor any logical necessity. Upon examining the basis on which these three judgments are made, we see that P1, while being rationally possible, is a negation which can only be made, neither on empirical evidence nor on the mere intellect, but only on the basis of an agent stipulating the nature of fasting on this particular day by negating the predication of obligation of fasting on such and such particular day; neither empirical evidence nor the intellect

more experienced students. Al-Sanūsī himself composed commentaries on these creeds. For more on his life and his impact on the Sunnī *kalām* tradition, see Olson (2020).

27 Theologians such as Aḥmad al-Dardīr (d. 1786), Muḥammad al-Amīr (d. 1816), and Ibrāhīm al-Bājūrī (d. 1860).

28 This refers to the ninth day of the twelfth month of the Islamic lunar calendar. It is a day when those who are not on pilgrimage at Makka are recommended to observe the fast.

on their own can confirm or reject this proposition. The agent is identified as God and knowledge of the stipulation is transmitted through the Prophet Muhammad (pbuh) and is thus classified as a Divinely stipulated judgment. P2 can be discerned and assessed in terms of its truthfulness without depending on any transmitted knowledge in the sense of Divine revelation, nor empirical evidence, in the sense of repeated observations or experimentation. Consequently, P2 is categorized as a rational judgment. P3 is an empirical judgment since knowledge of the subject, predicate, and copula are dependent solely on empirical knowledge.

Each of the three kinds of propositions is further divided into: (1) immediate (*ḍarūrī*) or inferred (*naẓarī*); and (2) necessary (*wājib*), possible (*jāʾiz*), or impossible (*mustaḥīl*) (see Figs. 2.3 and 2.4). A proposition being immediate entails that all its terms are apprehended by a person without any inquiry or investigation, whereas an inferred proposition entails apprehension after some degree of inquiry. The terms necessary, possible, or impossible are modal terms which describe the ontological status of the relationship of the subject and predicate in the proposition. From this understanding, "necessary" means "that whose existence cannot be negated," or, less preferably, as "that whose non-existence cannot be conceived"; "impossible" means "that whose existence cannot be affirmed," or "that whose existence the intellect cannot be conceived"; whereas, "possible" refers to that "whose existence or non-existence can be affirmed or negated" (al-Amīr 2023: 379–403; al-Dardīr n.d.: 41–43; al-Sanūsī 2019: 113–114).

To further clarify these modal terms, let us consider the following propositions in light of the definitions provided above:

P4: 32 divided by 8 divided by 4 is 1.
P5: Gold is a usurious item.
P6: Fire burns.

P4 is a rational judgment which is necessary yet inferred. It is rational due to it not depending on a revelation or stipulation from an agent, nor on observations. It is necessary since the negation of the predicate is impossible and inferred due to the inquiry required when considering the subject, predicate, and copula relationship. P5 is an inferred, Divinely stipulate judgment which is necessary. P6 is an immediate, empirical judgment which is necessary.

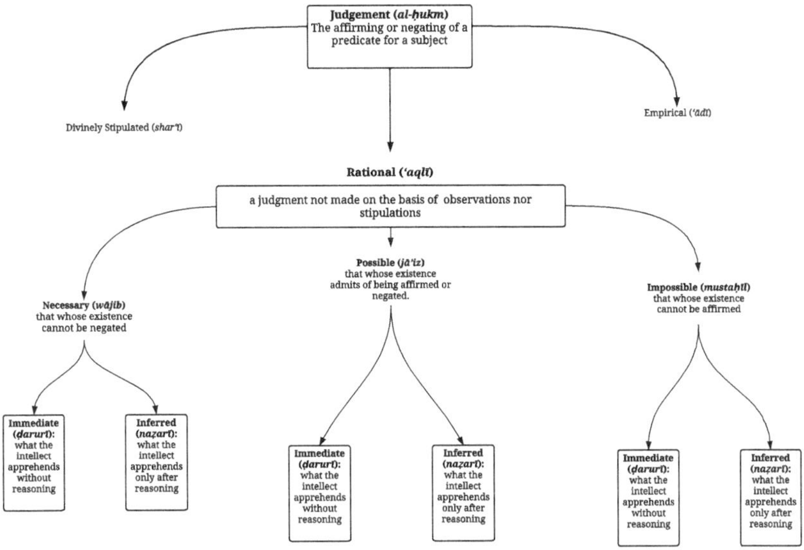

Fig. 2.3 Classification of rational judgments. The same classification applies to empirical and divinely stipulated judgments

Thus far, we have explored the basis for each kind of judgment. What has not been explored is the relationship between Divinely stipulated, empirical, and rational judgments. This is pertinent to the aim of this edited volume as it brings to light how empirical judgments, which involve judgments about the natural world, relate to rational and Divinely stipulated judgments, which include statements about the natural world. Understanding such relationships guides both inquiry in the natural sciences as well as theological inquiries, as both are informed by rational judgments. Furthermore, some rational and theological judgments inform empirical judgments by qualifying, informing, and guiding such inquiries. Similarly, empirical judgments will inform and guide theological inquiries the details of which will be explored later in the chapter.

P1 and P5 are Divinely stipulated, whereas P3 and P6 are empirical judgments. Each judgment does not entail a rational impossibility, that is, they are not themselves a contradiction nor entail a contradiction. Nor is each judgment a rational necessity. The proposition "Acetaminophen relieves pain" intrinsically does not entail a logical contradiction nor a

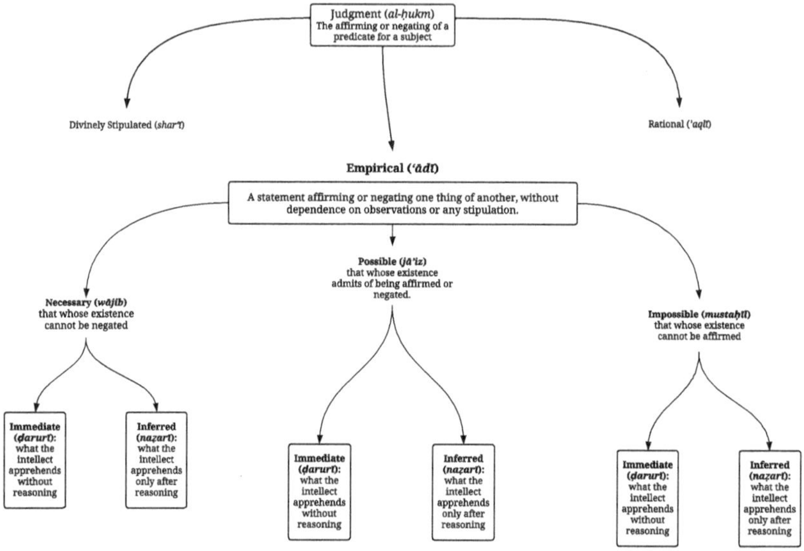

Fig. 2.4 Empirical judgments: all fall under rationally possible judgments

logical necessity; this makes it open to being true or false. Only empirical observations may furnish proof of these propositions' truth value. All Divinely stipulated or empirical judgments must fall under rationally possible judgments. The ontological implication of this is that no theological or scientific proposition will entail or itself be a rational impossibility.

Arguments: Its Kinds and Principles When in Conflict

How findings from the natural sciences can be understood in light of theological truths and commitments is a core concern for Muslim science educators, scientists, and theologians. This issue is not new; rather, it had drawn the attention of *kalām* theologians throughout the history of the Islamic intellectual tradition (see Griffel 2015 and 2018) and is thus not a problem unique to contemporary Muslims. The previous discussions on the definition of knowledge, certainty, and necessary and certain propositions serve as foundational elements to address this question. This section builds upon those concepts and terms to delineate a set of principles developed by *kalām* theologians for navigating apparent conflict between propositions. Before addressing this particular aim, understanding the nature of rational and transmitted proofs is of central importance for it clarifies the epistemological framework in which the principles are embedded.

Sunnī theologians examine the nature of proof (*al-dalīl*) from different considerations. One consideration is the distinction between arguments whose premises are rational (*ʿaqlī*) and those which are transmitted (*naqlī*). In this context, the term "rational" extends to premises beyond primary propositions to include experiential, observational, estimative, and intuited propositions, while "transmitted" extends to premises that are based on revelation, history, or any other type of testimonial knowledge. For *kalām* theologians, their main concern is with transmitted knowledge such as, but not limited to, the Qurʾan, statements of Prophet Muhammad (pbuh), statements of the early Muslim community, and claims of scholarly consensus (*ijmāʿ*).

Extending the rational–transmitted premise dichotomy, Ashʿarī theologians outline two conceptions of proof based on the nature of remote and proximate premises: a twofold and a threefold division. Al-Ījī tersely states the twofold division as follows:

> Proof (*dalīl*) is either rational with regard to all its premises, or transmitted with regard to all of them, or a composite of both. The first is rational, whereas the second is inconceivable due to truthfulness of the informer being a necessary condition, which is only established by the intellect. The third is what we name transmitted proof.
> (al-Daljī 2023: 180–181; al-Ījī 2023: 131; Al-Taftāzānī 1861: 1:39–40).

Clarifying this passage, al-Jurjānī notes that this classification is based on every premise, whether remote or proximate, as being rational, transmitted, or a composite. Keeping this in mind, there can be no proof that is transmitted whose proximate and remote premises are solely based on testimonial knowledge. More concretely, in the context of theology, it means truth claims based on the Qurʾan, statements of Prophet Muhammad, statements of the early Muslim community, and claims of scholarly consensus all have a rational premise, namely, knowledge of the speaker's truthfulness (i.e., a prophet of God) which is established rationally. The claim to prophethood, and hence unassailable truthfulness, relies on a person rationally investigating evidence of claims to prophethood, which signify the speaker's truthfulness. Thus, a remote premise would be something along the lines of:

P1: Muhammad evidenced miracles.
P2: Miracles demonstrate the truthfulness of Muhammad's claim of being a prophet of God.
C1: Prophets of God only speak the truth.

P1 is a premise based on sense observation, while P2 is a rational premise. Thus, any transmitted proof is based on, at least remotely, a rational proof due to the evidence of the truth claim resting on the signification of the evidence, which is rational in nature. Al-Jurjānī explains further that, "were its (i.e., the truthfulness of prophets) establishment sought via transmitted proofs, it would lead to a circular [argument] or continue on infinitely" (al-Jurjānī 1893: 1:208). Based on this exposition, there are only two kinds of proofs: purely rational and a composite of rational and transmitted premises; there is no category of a purely transmitted proof. This account presents the rational nature of theological proof as conceived in the Sunnī *kalām* tradition. This is a noteworthy point as it disrupts contemporary Western notions of science and religion vis-à-vis rationality, where the realm of science is considered rational, and religion is the domain of the irrational. This conception of the domains of each contributes critically to the conflict thesis.

The threefold division of proofs is based on consideration only of the proof's remote premises. In this division, a rational proof is one whose proximate premises are all rational premises such as:

PR1: The cosmos undergoes change.
PR2: Everything which undergoes change is temporally originated.

A transmitted proof is one whose proximate premises are all transmitted premises such as:

PT1: One who does not perform a command is disobedient based on God's statement "Did you disobey My command?" (Q.20:93)
PT2: Every disobedient person deserves punishment based on God's statement "And whoever keeps disobeying God and His messenger, verily his shall be the fire of hell abyss" (Q.72:23).

A composite proof is one whose premise is rational, while the other is transmitted such as:

> PC1: So and so has not performed what he is commanded to do. (rational)
> PC2: Every person who has not performed what he is commanded to do is disobedient. (transmitted) (al-Jurjānī 1893)

Understanding the rational–transmitted proof distinction raises the question of possible conflict between rational and transmitted proofs. How is one to proceed when facing such a situation? What are the metaphysical and theological implications of such a possibility? Successively answering these questions allows us to address the perceived Islam and science conflict thesis with the Sunnī *kalām* account of epistemology as our point of departure.

There are four possible ways rational and transmitted proofs can conflict: (1) both proofs are conclusively certain (*qaṭʿī*); (2) both are presumptive (*ẓannī*); (3) the transmitted proof is conclusively certain, while the rational proof is presumptively known; and (4) the rational proof is conclusively certain, while the transmitted proof is presumptively known (Mulla Zādah 2023: 68; al-Tahānwī 1996: 74). Two proofs are said to be in conflict when "the truth of one of two judgments entails the falseness of the other" (al-Tahānwī 1996: 74). In other words, it is necessary that propositions *P* and *Q* will be in conflict if and only if: (1) *P* is true then *Q* must be false; or (2) *P* is false then *Q* must be true. Importantly, what follows from this is that it is not possible for two certainly true propositions to ever conflict.

When considering transmitted proofs, other levels of consideration arise. There are two aspects of transmitted proofs which must be investigated: its mode of transmission (*thubūt*) as well as its mode of signification (*dalāla*). A transmitted proof's mode of transmission refers to how certain can we be that the proof can be attributed to the speaker. This depends on *how* the proof has reached us, which takes into consideration the number of transmitters at each stage of the report as well as the reliability of each transmitter. A transmitted proof's mode of signification takes us beyond the reliability of the testimonial knowledge in terms of its authenticity to the meaning of the testimony itself. This aspect involves an interpretive methodology applied to the interpretation of statements.[29]

Based on the above-mentioned aspects, transmitted proofs divide into four categories. A transmitted proof is categorized as conclusively certain (*qaṭʿī al-thubūt*) if its mode of conveyance is beyond question, whereas if it has been conveyed in a manner whose authenticity is only presumptive then it is categorized as a proof whose authenticity is presumptively held (*ẓannī al-thubūt*). A similar categorization applies to the meaning of a transmitted proof. A transmitted proof which can only be interpreted in its plain sense is categorized as conclusively certain (*qaṭʿī al-dalāla*), whereas if it can be interpreted in other than its plain sense it is categorized as having presumptive significance (*ẓannī al-dalāla*). Consequently, and ranked in order of highest epistemological strength, a transmitted proof maybe:

1. transmitted knowledge that has been conveyed with conclusive certainty and whose meaning is conclusively certain (*qaṭʿī al-thubūt wa qaṭʿī al-dalāla*). Examples of such transmitted knowledge which are conveyed with conclusive certainty are the Qurʾan in its entirety and mass-transmitted statements (*mutawātir*) of the Prophet Muhammad (pbuh). Verses in the Qurʾan dealing with tenets of faith such as God's oneness, the prayer, and fasting are examples of texts whose meanings are conclusively certain (*qaṭʿī al-dalāla*).
2. transmitted knowledge that has been conveyed with conclusive certainty yet whose meaning is presumptively authoritative (*qaṭʿī al-thubūt wa ẓannī al-dalāla*). Examples of this would be verses of the Qurʾan and mass-transmitted Prophetic traditions which are open to multiple lexical possibilities.
3. transmitted knowledge that has been conveyed with presumptive authenticity whose meaning is known with conclusive certainty (*ẓannī al-thubūt wa qaṭʿī al-dalāla*). Solitary Prophetic traditions with only one possible meaning are examples of this.

[29] These concepts are of particular importance to the science of *ḥadīth* interpretation. Both Chapters 6 and 7 build upon this brief exposition and delve into these aspects of Prophetic statements relevant to their topics.

4. transmitted knowledge that has been conveyed with presumptive authenticity and whose meaning is also presumptively authoritative (*ẓannī al-thubūt wa ẓannī al-dalāla*) (see Zysow 2013: 52–58).

Importantly, all four categories of transmitted knowledge possess epistemic strength and are of material significance when interpreting transmitted knowledge and resolving conflicts between propositions.

We are now in a position to explore the principles scholars use to resolve conflicts between proofs. Sunnī theologians such as al-Rāzī (1981), al-Ījī (2023: 132), al-Taftāzānī (1861: 1:40), and al-Jurjānī (1893: 1:201–211) have identified key principles for addressing such conflicts and have elaborated on the epistemological foundations of their approach. Their discussions, however, were dispersed throughout various works and remained fragmented and incomplete. Building on their discussions, Ashraf ʿAlī al-Tahānwī (d. 1943) provided a more systematized account by outlining four scenarios of possible conflict and their resolution. He presented these ideas in *al-Intibāhāt al-Mufīda fī Ḥall al-Ishtibāhāt al-Jadīda* and *Dirāya al-ʿIsma*, a commentary on a manual of peripatetic philosophy (al-Tahānwī 1996, 2014). It is essential to note that these principles are designed for religious scholars to use when assessing whether empirical evidence justifies the reinterpretation of transmitted revelatory propositions. I emphasize empirical evidence because multiple other factors such as logical coherence and scholarly consensus can also justify departing from the plain sense of these propositions, and empirical proof is just one among many considerations. These principles are not intended to guide scientists on how to conduct their research or present their findings. They will conduct their investigations according to the standards agreed upon by scientists themselves.

The four scenarios in al-Tahānwī's account are as follows:

Scenario One: A conflict between a conclusively certain rational proof[30] and a transmitted proof where the transmitted proof is conveyed with conclusive certainty and whose meaning is conclusively certain (*qaṭʿī al-thubūt wa qaṭʿī al-dalāla*).

Scenario Two: A conflict between a presumptive rational proof and a presumptive transmitted proof.

30 In this context, rational proof extends to all non-transmitted proofs, including empirical proofs.

Scenario Three: A conflict between a presumptive rational proof and a conclusively certain transmitted proof.
Scenario Four: A conflict between a conclusively certain rational proof and a presumptive transmitted proof.

Al-Tahānwī observes that Scenario One is:

absurd and will not occur, ever. For if two proofs are true with conclusively certainty, how is it then possible for them both to be in conflict? If there is a conflict, then it is not possible that both the proofs are true. Nor is it possible to provide an example [of this scenario] to consider, for it is impossible, whose extra-mental existence is inconceivable.

(al-Tahānwī 1996: 76)

In his reasoning here, al-Tahānwī is invoking the principle of non-contradiction, a first principle of rational thought, as to why there can be no conflict between two conclusively certain proofs. For them to be in conflict would violate the principle of non-contradiction and is thus rationally impossible. Consequently, a particular example of this scenario cannot be provided.[31] His reasoning is also reflective of the idea of the unity of truth from which the maxim "truth does not contradict truth" is derived (see Taylor 2000).

In Scenario Two, where both proofs are presumptive, the transmitted proof is given precedence over rational proofs.[32] Here, al-Tahānwī explains why this is at some length. The sciences of *kalām* theology and legal theory (*ūṣūl al-fiqh*) have soundly established the necessity to accept a transmitted proof which is presumptively true. At the same time, the necessity to accept a rational proof which is presumptively true is not soundly established, and a rational proof being presumptively true only means that is it possible for the proposition to be false. This being the case, judging it to be false would not entail violating any rational principle. Yet, al-Tahānwī considers another way of resolving the conflict only

[31] For educators and philosophers, I will provide the logical forms of each scenario. Scenario One can be formally represented as $D(r) \wedge D(t) \rightarrow \neg \bot$, where t: transmitted proof, r: rational proof, $D(r)$: a definitive rational proof, $D(t)$: a definitive transmitted proof, $P(r)$: a probabilistic rational proof, $P(t)$: a probabilistic transmitted proof, and $\bot$: a contradiction (Chowdhury 2023).

[32] Scenario Two can be formally represented as $P(r) \wedge P(t) \rightarrow P(t)$.

to later reject it. He considers the possibility of interpreting the transmitted text in a figurative manner, not according to its apparent, plain sense (*ẓāhir*). For al-Tahānwī, this is not an option due to the lack of warrant for such an interpretive gesture. He explains that:

> interpreting a text (*taʾwīl al-naṣṣ*) outside the limits of its plain sense is only undertaken when there is a countenanced need, and there is no such need in this case for the rational proof is only presumptive. Therefore, to interpret a text contrary to its plain sense in this scenario constitutes a heretical innovation, which the Sacred Law does not countenance, nor does the intellect deem to be good. [This is based on] what we previously mentioned that a rational proof being presumptive means it can possible be false.
>
> (al-Tahānwī 1996: 77)

Informing this resolution of conflict is the confluence of rational principles, *kalām* theology, and legal theory. The presumption of interpreting any transmitted proof is that it is to be understood in its plain, literal sense. This is based on the hermeneutical maxim "the presumption of the literal meaning" (*al-asl fī al-kalām al-ḥaqīqa*), which scholars such as Fakhr al-Dīn al-Rāzī (d. 1210) have reported scholarly consensus on (al-Rāzī 1981: 30:94). Accordingly, interpreting a transmitted proof in its figurative sense, which entails departing from this initial presumption, while allowed in principle, can only be undertaken when the literal signification of the text is impossible to maintain. Expressing this important, well-known maxim, al-Taftāzānī states, "The texts of the Qurʾan and the Sunnah are understood according to their literal signification (*tuḥmal ʿalā ẓawāhirihā*) as long as no conclusively certain proof [requires] them to be interpreted in other than their literal sense" (al-Taftāzānī 2020: 345). No such justification exists in this scenario, since the rational proof is of equal epistemological strength in conjunction with the absence of requiring one to accept a presumptive rational proof, while such a requirement does exist for a presumptive transmitted proof. Finally, al-Tahānwī notes that the transmitted proof is a statement of one for whom "error is not possible, while the other (rational proof) is the judgement of one whom error is possible. Therefore, the judgement of one who is infallible is given precedence over one who is fallible" (al-Tahānwī 2014: 116). In this last principle, we can see how a theological truth, that is, the infallibility of prophets, informs how one should resolve any conflict between proofs.

As a case study for Scenario Two, al-Tahānwī examines the Qurʾanic verse "And it is He who created the night and day, and the sun and moon: Each coursing in a mighty round" (Q.21:33) as being in conflict with the findings of some scientists, such as Avicenna (d. 1037), who held that the planets only have positional motion (*al-ḥaraka al-waḍʿīyya*), where an object rotates while staying in the same place, not a motion of place (*al-ḥaraka al-aynīyya*), where a body moves from one place to another place.[33] The plain sense of the verse affirms the movement of planets as a motion of place and is, consequently, in conflict with the empirical proofs which certain scientists bring in support of the positional motion of planets. Now, one can interpret the verse in a figurative sense to resolve the conflict, however, al-Tahānwī objects to this approach saying:

> they do not have any conclusively certain rational proof which negates the motion of place of planets. Therefore, the position [affirming] the motion of place [for planets] is obligatory [to uphold], as is rejecting the position of those scientists who negate the motion of place of planets.
>
> (al-Tahānwī 1996: 79)

What is significant to note here is that the verse can be interpreted in multiple ways and also be understood figuratively. Yet, this mere possibility is not enough to take recourse to a figurative interpretation because there is the absence of a proof of equal epistemological strength to warrant such a hermeneutical move. Again, as in this scenario, transmitted proofs may admit several possible interpretations. Rational proofs, insofar as they possess greater epistemic force, may serve to adjudicate between competing presumptive interpretations of transmitted proofs. However, any such conclusion about the interpretation would be presumptively held and open to further discussion and debate.

The method of resolving the conflict in Scenario Two is the same path for Scenario Three.[34] In fact, it is so, *a fortiori*, due to the transmitted proof being of a higher epistemic strength, namely, it being conclusively certain, rather than the rational proof, which is presumptive in

[33] The positional motion of the spheres, not to be found in Aristotle's works, is understood to be a unique contribution of Avicenna in the field of cosmology. For more on this, see McGinnis (2006) and Arif (2012).

[34] Scenario Three can be formally represented as $P(r) \wedge P(t) \rightarrow P(t)$ (Chowdhury 2023).

this case. In the final scenario, Scenario Four, where the rational proof is conclusively certain, and the transmitted proof is presumptively true, both proofs should be taken into consideration, and neither one should be discarded. This is because the rational proof enjoys certainty, and nothing with certainty should be ignored. At the same time, the transmitted proof, while presumptively true, must also be considered for the reasons explained above. In light of the epistemological variance in the strength of both proofs, the transmitted proof should be interpreted in a manner that does not conflict with the rational proof due to its strength.[35] As an application of this principle, al-Tahānwī looks at the Qurʾanic verses describing Dhul Qarnayn's journey. These verses describe one segment of his journey in the following manner: "When he reached the farthest shore upon which the sun set, he found it set in a spring of bemired water" (Q.18:86). Now understood literally, this verse means that the sun was physically in the spring of water. This conflicts with our empirical knowledge that the sun is a separate body from the earth, and it is impossible for the two bodies to come in physical contact with each other, let alone the larger body being physically in the smaller body. Al-Tahānwī parses his reasoning as follows:

P1: The sun is a separate body from the earth.
P2: Both bodies do not ever come into contact with each other.
P3: According to the literal meaning of the verse, the sun set in a spring of bemired water.
C1: The literal meaning of the verse conflicts with our conclusive certain rational proof.

"It is only in this one scenario," al-Tahānwī summarily highlights, "where the rational proof is given precedence over the transmitted proof. The position of granting precedence to rational proof in scenarios two and three and proceeding with it [alone] is incorrect" (al-Tahānwī 1996: 77–78). Applying the principles of resolving this conflict, it is required to interpret the verse in a figurative manner in accordance with principles of the Arabic language and Qurʾanic exegesis. Consequently, al-Tahānwī reinterprets the verse as conveying what the observer, Dhul Qarnayn, initially observed. It is "as if the sun set in a spring of bemired water,

35 Scenario Four can be formally represented as $D(r) \wedge P(t) \rightarrow D(r)$ (Chowdhury 2023).

which was not the case in reality. This is similar to when people on a boat in the sea observing as if the sun sets in the sea" (al-Tahānwī 1996: 79; see also al-Rāzī 1981: 21:77–78). Importantly, in resolving the conflict, discarding the verse or saying that the verse is mistaken are not interpretive possibilities for theological reasons related to the truthfulness of the prophets mentioned previously (Tables 2.1 and 2.2).

This structured process ensures that no scholarly community discards its proofs while maintaining rational coherence and epistemological integrity.

From the foregoing account on resolving any conflict between rational and transmitted proofs, the focus and guiding principle throughout the analysis is not on the source of the proof, whether it is rational (which includes the empirical) or transmitted (which includes propositions based on revealed sources). Rather, the guiding principles are on epistemological strength, starting with first principles of thought (e.g., the principle of non-contradiction) after which degrees of epistemological strength of each proof are considered. Of significance is that in no case is any conclusively certain proof, whether empirical or transmitted, to be abandoned, nor is any transmitted proof to be ignored as it is only the plain sense of a transmitted proof which may be interpreted, yet the transmitted proof itself is not to be rejected. Also of significance is that, in this framework, a conflict between conclusively certain rational and transmitted proofs is rationally impossible due to violating the principle of non-contradiction,

Table 2.1 Key assumptions informing the Sunnī *kalām* approach

#	*Assumption*
A1	Both scientific inquiry and theological inquiry have their foundations in rational principles, specifically the principle of non-contradiction
A2	In the Sunnī *kalām* framework, all empirical (scientific) propositions are rationally possible, as are transmitted revealed propositions
A3	As a consequence of A1 and A2, there can be no genuine conflict between a transmitted proof and a rational (empirical) proof
A4	Rational (empirical) propositions fall on a range of epistemological strengths. Some are known with certainty, while others are only presumptively known to varying degrees
A5	Transmitted propositions, in our case, revealed propositions, fall on a range of epistemological strength. Some are known with certainty, while others are presumptively known to varying degrees
A6	In no scenario should the proofs of either scholarly community be discarded

Table 2.2 Outline of steps of al-Tahānwī's framework

Step	*Process*
Step 1: The initial investigation	The scholar must investigate the epistemological strength of the transmitted proof as well as the rational (empirical) proof. The scholar engages the scientific community to assess the epistemological strength of the rational (empirical) proposition
Step 2: Assessing for claims of certainty—eliminating scenario one	If both communities claim certainty and conclude that there is a genuine conflict, then one or both communities must be mistaken regarding the level of certainty due to A3. In light of this, both communities should reevaluate their claims to determine if either proof has been incorrectly assessed
Step 3: Assessing for varying strengths	If both scholarly communities differ in the epistemological strength of their propositions, then • If the rational (empirical) proposition is presumptive and the transmitted proposition is presumptive, then one follows the path to resolution in **Scenario Two** • If the rational (empirical) proposition is presumptive and the transmitted proposition is conclusively certain, then one follows the path in **Scenario Three** • If the rational (empirical) propositions are conclusively certain and the transmitted proofs are presumptive, then one follows the path in **Scenario Four**

(continued)

Table 2.2 (continued)

Step	*Process*
Step 4: Continued engagement	In the case where one is unable to assess the epistemological strengths of one or more proofs, the scholar refrains from making any conclusions and presumes the literal interpretation of transmitted proof. He continues to engage both communities of scholars until the epistemological strengths of possible positions become apparent[36]

thereby anchoring this entire framework in rational knowledge. Thus, no conclusively certain scientific discovery will, in principle, conflict with conclusively certain transmitted knowledge. This certainly disrupts the religion and science "conflict thesis" as we know it, in its strong and weak forms in that it asks theologians *and* scientists to be aware of the epistemological strength of their truth claims. From the perspective of *kalām* theologians, the concern is with the epistemological strength of each proposition, not in the actual source of the proposition, as this framework also applies to instances where two rational proofs or two transmitted proofs are in apparent conflict. Finally, it should be kept in mind that this framework is a guideline for theologians when considering various scenarios for interpreting revealed, transmitted knowledge in light of empirical knowledge. It is not intended to prescribe how scientists should carry out their discipline-specific inquiries.

[36] This position is sometimes referred to as a deferral of judgment or *tawaqquf*. It is important to note that this deferral is not due to a lack of evidence, such as the silence of revealed sources on an issue. Rather, it describes the subjective epistemic state of the investigator, in that at a particular moment of inquiry he finds himself at equipoise regarding which proofs possess greater epistemic strength. Accordingly, one defers rendering judgment and continues examining the relevant evidence until the strength of one truth claim becomes preponderant. Crucially, *tawaqquf* is a temporary stage within an inquiry, not a permanent stance of non-commitment.

Conclusion

One of the major aims of this chapter is to unburden ourselves of and move beyond the religion–science conflict thesis. By doing so, we open ourselves to a variety of possible relationships that uphold the epistemic integrity of each discipline. One such possibility is known as science-engaged theology. In recent years, scholars studying the relationship between religion and science have advocated for science-engaged theology as a new approach to understanding the connection between the deliverables of these two disciplines of knowledge. One of the most notable recent works on this topic is *Science-Engaged Theology* by Perry and Leidenhag (2023), in which they rightly observe that revealed texts of all religions make numerous statements about the natural world, and due to this fact, scientists are not the sole inquirers into natural phenomena. In light of this, they propose "science-engaged theology [as] a way to place scientific research alongside—not in competition with—biblical exegesis, the study of history, and philosophical considerations as a ready resource for theological reflection" (Perry and Leidenhag 2023: 1). One consequence of this understanding is that when a theologian engages the scientific community and its claims, he must inquire about the best and most accurate understanding of where the scientific community stands on various issues. This entails understanding not just the terms which scientists use, but also the nature of scientific discovery and findings. In doing so, scientists and theologians will need to be philosophically and historically aware of the nature of their work to effectively engage with practitioners from various scholarly communities.

A remaining critical question that has not been addressed in the scholarship on science-engaged theology concerns the philosophical and epistemological basis for the interaction that science-engaged theology envisions. It is this question to which the Sunnī *kalām* framework for resolving any conflicts provides an answer. Sunnī *kalām* theologians see their discipline and the disciplines which study the natural world grounded in the first principles of rational thought and it is these principles from which fruitful engagement can take place. In light of these shared epistemological foundations, not only is science-engaged theology possible, but the field of theology-engaged science also becomes a real possibility. Here questions arise such as:

how does theological discussion of x shed new light on, correct a distortion of, or corroborate the discussion of y in the sciences? Examples of this might be Andrew Davison's discussion of analogy in describing machine learning and artificial intelligence, which is informed by discussion of analogy in Thomas Aquinas, Thomas Cajetan and Francisco Suárez.
(Perry and Leidenhag 2023: 63)

The remaining chapters of the book aim to contribute to these two possible modes of science and theology.

References

Abrahamov, 1993 Abrahamov, Binyamin. 1993. "Necessary Knowledge in Islamic Theology." *British Journal of Middle Eastern Studies* 20 (1): 20–32. https://doi.org/10.1080/13530199308705567.

Adamson, Peter, and Fedor Benevich. 2023. "Fakhr al-Din al-Razi." *The Stanford Encyclopedia of Philosophy*. Edited by Edward N. Zalta and Uri Nodelman. https://plato.stanford.edu/archives/spr2023/entries/al-din-al-razi/.

Al-Amīr, Muḥammad. 2023. *Ḥāshīyya al-Amīr ʿalā Itḥāf al-Murīd bī Jawhara al-Tawḥīd*, vol. 2. Edited by Anas Muḥammad ʿAdnān al-Sharfāwī. Damascus: Dār al-taqwā.

Arif, Syamsuddin. 2012. "The Universe as a System: Ibn Sina's Cosmology Revisited." In *New Perspectives on the History of Islamic Science: Volume 3*. Edited by Muzaffar Iqbal, 53–72. London: Routledge.

Benevich, Fedor. 2022. "Knowledge as a Mental State in Muʿtazilite Kalām." *Oriens* 50 (3–4): 244–279. https://doi.org/10.1163/18778372-12340016.

Blackie, Margaret A.L. 2023. "Implications of a Critical Realism Approach to Chemistry Research and Education." *Nature Chemistry* 15 (8), 1047–1050. https://doi.org/10.1038/s41557-023-01274-2.

Chowdhury, Safaruk Zaman. 2023. "Kalām Jadīd from the Indian Context: A Philosophical Examination of Mawlānā Ashraf ʿAlī Thānwī's (d. 1943) Text *al-Intibāhāt al-Mufīda*." Handout, American Society of Islamic Philosophy and Theology Conference, Harvard University, December 1–3.

Al-Daljī, Muḥammad ibn Muḥammad 2023. *Sharḥ Maqāṣid al-Maqāṣid*. Edited by Aḥmad ibn Suhail al-Mashhūr, vol. 2. Syria: Dār al-Shaykh al-Akbar.

Al-Dardīr, Aḥmad. n.d. *Sharḥ al-Kharīda al-Bahīyya*. Edited by ʿAbd al-Salām Shannār. Damascus: Dār al-Beiruti.

Al-Ghazālī, Muḥammad ibn Muḥammad. 2016a. *Miʿyār al-ʿilm*. Jeddah: Dār al-Minhāj.

Al-Ghazālī, Muḥammad ibn Muḥammad. 2016b. *Miḥakk al-Naẓar*. Jeddah: Dār al-Minhāj.

Al-Ghazālī, Muḥammad ibn Muḥammad. n.d. *al-Mustaṣfā min ʿIlm al-Uṣūl*, vol. 2. Edited by Aḥmad Zakī Ḥammād. Riyadh: Dār al-Maymān.

Griffel, Frank. 2015. "Al-Ghazâlî at His Most Rationalist: *The Universal Rule for Allegorically Interpreting Revelation (al-Qânûn al-Kullî fî t-Taʾwîl)*." In *Islam and Rationality: The Impact of al-Ghazâlî. Papers Collected on His 900th Anniversary*, vol. 1. Edited by G. Tamer, 89–120. Leiden: Brill.

Griffel, Frank. 2018. "Ibn Taymiyya and His Ashʿarite Opponents on Reason and Revelation: Similarities, Differences, and a Vicious Circle." *The Muslim World (Hartford)* 108 (1): 11–39.

Hasan, Moiz. 2017. "Foundations of Science in the Post-Classical Islamic Era: The Philosophical, Historical, and Historiographical Significance of Sayyid al-Sharīf al-Jurjānī's (d. 1413) Project." Ph.D. Dissertation, University of Notre Dame.

Ibn Fūrak, 1987. *Mujarrad Maqālāt al-Shaykh Abī-l-Ḥasan al-Ashʿarī.* Edited by Daniel Gimaret. Beirut: Dar al-Mashriq.

Al-Ījī, ʿAḍud al-Dīn. 2023. *al-Mawāqif fī ʿIlm al-Kalām.* Cairo: Maktab Iḥyāʾ al-Turāth al-Islāmī.

Al-Jurjānī, ʿAlī ibn Muḥammad. 1893. *Sharḥ al-Mawāqif*, vol. 3. Istanbul: Dār al-ṭabāʿa al-ʿāmira.

Lawson, Anton E. 2000. "The Generality of Hypothetico-Deductive Reasoning: Making Scientific Thinking Explicit." *The American Biology Teacher* 62 (7): 482–495. https://doi.org/10.2307/4450956.

Lawson, Anton E. 2010. "Basic Inferences of Scientific Reasoning, Argumentation, and Discovery." *Science Education* 94 (2): 336–364. https://doi.org/10.1002/sce.20357.

Al-Māturīdī, Abū al-Manṣūr. 2023. *Kitāb al-Tawḥīd.* Edited by Aḥmad Saʿd al-Damanhūrī. Amman: Maktabat al-Ghānim.

McGinnis, Jon. 2006. "Positioning Heaven: The Infidelity of a Faithful Aristotelian." *Phronesis* 51 (2): 140–161.

Mubīn, Muḥammad 1909. *Mirʾāt al-shurūḥ sharḥ Sullam al-ʿulūm*, vol. 2. Cairo: Maṭbaʿ al-Saʿāda.

Mustafa, Zainun, Azizan Baharuddin, and Shaikh Mohd Saifuddeen. 2021. "Islam, Science and Education: Delving into the Progress, Collaboration and Biases." *Journal of Islamic Thought and Civilization* 11 (2): 44–68. https://doi.org/10.32350/jitc.112.03.

Mulla Zādah, Muḥammad ibn Mulla Abu Bakr. 2023. *Mirqāt al-ʿUlūm wa al-Sirr al-Maktūm.* Amman: Maktaba al-Ghānim.

Olson, Caitlyn. 2020. "Beyond the Avicennian Turn: The Creeds of Muḥammad b. Yūsuf al-Sanūsī (d. 895/1490)." *Studia Islamica* 115 (1): 101–140. https://brill.com/abstract/journals/si/115/1/article-p101_4.xml.

Perry, John, and Joanna Leidenhag. 2023. *Science-Engaged Theology: Elements of Christianity and Science.* Cambridge: Cambridge University Press.

Qureshi, Omar, Afifi al-Akiti, and Aasim I. Padela. 2022. "Islam and Science: Reorienting the Discourse." In *Islam and Biomedicine*. Edited by Afifi al-Akiti and Aasim I. Padela, 245–254. Cham: Springer International Publishing.

Al-Rāzī, Fakhr al-Dīn. 1981. *Al-Tafsīr al-Kabīr wa-l Mafātīḥ al-Ghayb*, vol. 30. Beirut: Dār al-Fikr.

Reed, Baron. 2022. "Certainty." *The Stanford Encyclopedia of Philosophy*, Spring ed. Edited by Edward N. Zalta. https://plato.stanford.edu/archives/spr2022/entries/certainty.

Reichardt, Charles S. 2021. *The Method of Multiple Hypotheses: A Guide for Professional and Academic Researchers*. New York: Routledge.

Al-Sanūsī, Muḥammad ibn Yūsūf. 2019. *Sharḥ al-Muqaddimāt*. Edited by Anas Muḥammad ʿAdnān al-Sharfāwī. Damascus: Dār al-taqwā.

Strobino, Riccardo. 2021. *Avicenna's Theory of Science: Logic, Metaphysics, Epistemology*. Oakland, CA: University of California Press.

Al-Taftāzānī, Saʿd al-Dīn. 1861. *Sharḥ al-Maqāṣid al-Ṭālibīn fī ʿIlm Uṣūl al-Dīn*, vol. 2. Istanbul: Dār al-ṭabāʿa al-ʿāmira.

Al-Taftāzānī, Saʿd al-Dīn. 2020. *Sharḥ al-ʿAqāʾid al-Nasafiyya*. Edited by Anas Muḥammad ʿAdnān al-Sharfāwī. Damascus: Dār al-taqwā.

Al-Tahānwī, Ashraf ʿAlī. 1996. *al-Intibāhāt al-Mufīda fī Ḥall al-Ishtibāhāt al-Jadīda*. Translated by Nūr al-Bashar Muḥammad Nūr al-Ḥaqq. Karachi: Maktaba Dār al-ʿUlūm al-Karatshī.

Al-Tahānwī, Ashraf ʿAlī. 2014. "Dirāya al-ʿIsma." In *al-Talkhīṣāt al-ʿAshr*, 90–107. Karachi: Maktaba al-Bushrā.

Taylor, Richard C. 2000. "'Truth Does Not Contradict Truth': Averroes and the Unity of Truth." *Topoi* 19 (1): 3–16. https://doi.org/10.1023/A:1006343305914

Wilkinson, Matthew L.N. 2013. "Introducing Islamic Critical Realism: A Philosophy for Underlabouring Contemporary Islam." *Journal of Critical Realism* 12 (4): 419–442. https://doi.org/10.1179/1476743013Z.00000000014.

Zysow, Aron. 2013. *The Economy of Certainty: An Introduction to the Typology of Islamic Legal Theory*, vol. 2, 1st ed. Atlanta, GA: Lockwood Press.

PART II

Core Chapters: Bridging Islamic and Bioscientific Knowledge to Address "Big Questions" about the Human Being

CHAPTER 3

Scientific Advances in Understanding Human Nature: An Islamic Occasionalist Perspective

Nazif Muhtaroğlu and Macksood Aftab

Introduction

Neuroscience has provided profound insights into the nature of human beings. We now understand that certain neural activities in the brain are associated with certain mental phenomena.[1] Similarly, various mental states and events are increasingly being linked to specific neural activities. Damage to particular brain regions can lead to significant changes in personality, the loss of certain mental functions, or the emergence of new mental phenomena. These remarkable discoveries in neuroscience have

[1] By "mental phenomena," we mean all the occurrences and processes (states, events, etc.) in the mind such as perception, various kinds of feelings like pain and pleasure and consciousness and willing.

Present Address:
N. Muhtaroğlu (✉)
Brandeis University, Waltham, Massachusetts, USA

M. Aftab
Michigan State University, East Lansing, Michigan, US

Central Michigan University, Mount Pleasant, Michigan, US

© The Author(s), under exclusive license to Springer Nature Switzerland AG 2026

A. I. Padela (ed.), *Bridging Bioscience and Islam*, Islam & Science,
https://doi.org/10.1007/978-3-032-09996-9_3

led many to adopt a naturalist and physicalist perspective on the human mind.[2] However, this viewpoint may seem to conflict with Islamic theological views, which posits that humans possess a spiritual aspect, a soul, which is believed to hold individuals accountable for their actions after death and ensures their revival on Judgment Day.[3]

In this chapter, we will explore the problem of understanding human nature by considering contemporary neuroscientific research, its dominant philosophical interpretation of naturalism, and Islamic theological perspectives. In accordance with scriptural teachings, we regard humans as the hereditary descendants of Adam and Eve. The Prophet Adam was distinguished from the angels by his unique abilities, most notably his capacity to state the names of things—an indication of advanced cognitive faculties (Q.2:30–33). Humans are defined by the union of body and soul, though the nature of the soul remains open to various interpretations. In modern scientific discourse, this distinction aligns not merely with anatomical Homo sapiens but with behaviorally modern humans—those capable of advanced cognitive functions and complex symbolic thought. In other words, with both what evolutionary biologists consider anatomically modern Homo sapiens, who they claim emerged approximately 300,000 years ago, and Homo sapiens capable of advanced

[2] In this context, "naturalism" roughly refers to the philosophical view that all of reality is confined to natural entities, denying the existence of so-called "supernatural beings" such as souls, angels, and God. Physicalism is the view that reality fundamentally consists of the entities and processes recognized by the physical sciences. Physicalism is usually considered a form of naturalism. However, as we will see, many Muslim theologians considered "created reality," including souls and angels, to be made of physical atoms. Thus, they might be regarded as physicalists in a limited sense, as they also acknowledge the existence of God as a non-physical, supernatural being.

[3] In Islamic theology, there are various views on the nature of the human soul. In this chapter, we will present two main perspectives: 1) the atomist physicalist model of the human soul and 2) the view that the human soul is a distinct substance separate from the human body. Despite the differences between these models, all mainstream views agree that the human soul survives after death and will be accountable on Judgment Day. As the soul survives death, it serves as the locus of identity and personality, acting as a conscious and willful agent. Each model explains the process differently. The first, the atomist physicalist model, holds that the human soul is composed of subtle atoms that leave the body after death. According to this view, the mental properties relevant to identity and personality can be created by God even within a single atom. The second model posits that the human soul is a distinct substance separate from the body. This soul leaves the body after death and can survive independently. However, on Judgment Day, it will be reunited with the body, which will be recreated and revived.

cognitive functions and complex symbolic thought, the modern subtype, which they claim emerged around 50,000 years ago. In essence, we argue that there is nothing in neuroscientific research that inherently conflicts with Islamic perspectives. The conflict arises from the naturalist interpretation of these scientific findings when compared to Islamic viewpoints. To address this issue, we propose that Islamic theological perspectives on human nature offer more plausible models for understanding the link between the human brain and the mind, effectively avoiding the drawbacks that naturalist models encounter.

So in what follows, first, we present the current state of neuroscientific research. Next, we explore the philosophical mind–body problem, tracing its origins from Descartes's dualism of mental and material substances. This historical background provides a foundation for understanding contemporary naturalist and physicalist approaches. Then, we examine the prevalent naturalist interpretations of neuroscientific findings concerning the link between the human brain and mind. We will argue that these naturalist philosophical models are not the only way to interpret neuroscientific findings and that they either come to ignore obvious mental phenomena or encounter an updated version of the mind–body problem. Then, we will introduce two models from the Islamic intellectual tradition that can accommodate neuroscientific findings without falling into these problems. These models are occasionalist property dualism within physicalism and occasionalist substantial dualism. Finally, we apply these models to the Artificial Intelligence (AI) debate, assessing their compatibility with the thesis of Strong AI, which posits that AI-based machines can truly exemplify all mental properties humans possess, such as consciousness and sentience.

Neuroscience and Human Mind

The primary research enterprise in modern neuroscience is focused on understanding the relationship between the brain and consciousness. Consciousness is the ability to be aware of one's self and one's experiences. It is the ability of a subject ("I") to experience the world and to be aware of its own existence and feelings. As the American philosopher Thomas Nagel defines it for something to be conscious, there must be something "it is like" to be that thing (Nagel 1974).

Contemporary neuroscience is making tremendous advances in understanding consciousness as research reveals a very close relationship between occurrences in the brain and consciousness. For most neuroscientists, consciousness is a feature of the brain. Explaining the brain would lead to a complete explanation of consciousness. As contemporary American author and philosopher John Bickle states, "Historically, and especially in its present form, neuroscience is a *reductive* enterprise" (Bickle 2003: 322). It seeks to reduce the phenomenal experience of consciousness to physical processes of the brain.

The principal methods of neuroscientific investigation are empirical. Neuroscience uses scientific methods based in biochemistry to understand cellular mechanisms, functional Magnet Resonance Imaging (MRI) to assess areas of brain activation associated with experiences, electroencephalography (EEG) to evaluate electrical activity of the brain and so forth. Based on these investigative tools, neuroscience attempts to provide a complete account of human consciousness. In other words, it seeks to study the "subjective" and "first-person" experience of consciousness by "objective" and "third-person" methods of investigation.

If neuroscience is successful in this endeavor, it would suggest that the entirety of what it means to be human and all human experiences are fully founded upon neurobiological substrates. Hence, neurobiology, with its focus on neurons, axons, connectivity, and the complexity of the brain, holds the key to understanding human conscious experience. This positions neuroscience within a specific ontological framework, where all aspects of human reality are situated within the domain of empirical science. Francis Crick, the renowned Nobel laureate, who played a pivotal role in the discovery of DNA, writes:

> "You," your joys and your sorrows, your memories and your ambitions, your sense of personal identity and free will, are in fact no more than the behaviour of a vast assembly of nerve cells and their associated molecules. (Crick 1995: 3)

Such a perspective carries significant implications for ethics, philosophy, and theology. It challenges substance dualism—the view that the brain and mind (or body and soul) are composed of two distinct substances—a position still defended by many theologians (Swinburne 2009). Yet, it remains unclear whether a physicalist framework necessarily entails the denial of the soul or proves incompatible with the attribution of moral responsibility to humans. As we will see in later sections, certain Muslim theologians adopted a form of limited physicalism, regarding the created universe as physical while considering the human soul itself to belong on the spectrum of the physical order.

Physicalism also bears relevance for contemporary debates on Artificial Intelligence (AI). On this view, the difference between human beings and AI systems lies primarily in their mechanisms of operation: humans function through neurobiological processes, whereas AI systems operate on electronic computational processes. This raises the important question of whether advances in AI might eventually yield agents comparable to humans.

With this background in mind, we may now turn to contemporary neuroscientific research on the relationship between the brain and the human mind.

Brain as the Organ of Consciousness

The confidence of many neuroscientists in taking a physicalist approach to the human mind stems from the recent monumental successes of biomedical science in understanding the human body and its functions.

In many instances, previously mysterious elements of human function have been explained entirely by the physical characteristics of the human body. We have learned that the function of the human heart is to pump blood to the lungs and the body, in a remarkably reliable and consistent manner. We have learned that function of the kidneys is to filter waste from the bloodstream and produce urine. Similarly, the liver, pancreas, adrenal glands, and essentially all human organs have a specific definable function that, at times, can be reproduced by synthetic models. A mechanical heart can theoretically and practically replace the human heart. In other words, there is nothing more to these organs than their function; this function can be replaced by another device, and the human continues to operate as normal.

Similarly, this view of neuroscience assumes consciousness is a function of the human brain. Consciousness then must be reducible to brain science. The multi-billion-dollar research enterprise associated with this field is called "the study of the neural correlates of consciousness." As Italian philosopher Riccardo Manzotti points out:

> Like the stomach is the organ for digestion, like the lungs are the organ for respiration, like the heart is the organ for blood circulation, so the brain

> has to be the organ for the mind. At least, this is the prevailing expectation in science nowadays.
>
> (Manzotti and Moderato 2013: 85)

Contemporary neuroscience research operates under the assumption that the same sort of investigation which has been successful in explaining the functioning of the heart will be successful in explaining brain function. This particular view of neuroscience has philosophical and methodological limitations on what it can accomplish.

A key brain function is consciousness. Unlike the functions of other organs, however, consciousness is not the same type of thing as the function of other organs. Unlike other functions, consciousness is not quantifiable. Rather, it is a qualitative, subjective experience, which is ontologically distinct from the brain. Therefore, the epistemic method of neuroscience runs into the limitations of its own ontological framework. This will be explored in more detail later in this chapter.

For the purpose of giving the dominant neuroscientific view its due, the current state of neuroscientific research with a specific emphasis on its use in healthcare will be presented. This will help inform the degree to which some neuroscientists are justified in their physicalist worldview. It will also help us understand what exactly neuroscience tells us about consciousness and what its relationship is to the brain.

Limitations on the Method of Neuroscience

Neuroscience, like other fields of modern science, uses empirical study methods to evaluate the brain. These include reproducible observational studies which seek to correlate brain activity with mental processes. The empirical methods used include functional MRI, EEG, Neurostimulating probes, Positron Emission Tomography (PET), and so forth. Each of these methods has its own intrinsic limitations. For example, researchers have questioned many functional MRI studies on consciousness due to potentially flawed statistical methods and MRI analyses utilized (Eklund, Nichols, and Knutsson 2016). In addition, the study design, the selection and management research data, along with the methods employed for statistical analysis will affect the strength of the research findings.

As an empirical science, the propositions of neuroscience would fall in the category of sense-experience (*tajrībīyyāt*). We do not consider the epistemic status of these propositions as conclusively certain. Sense-experience (*tajrībīyyāt*) is repeated experience, and repeated experiences, in our view, cannot provide conclusive certainty because of the problem of induction. The problem of induction, as famously articulated by the Scottish philosopher David Hume (d. 1711), refers to the philosophical challenge of justifying inductive reasoning—the process of drawing general conclusions from specific observations (e.g., that the sun will rise tomorrow because it has always risen in the past). Hume pointed out that there is no logical necessity connecting past experiences to future outcomes; assuming that the future will resemble the past is itself an inductive assumption, which makes the reasoning circular. In sum, one cannot solve the problem of induction just by looking at the universe and the regularities within it assuming no agent behind such regularities for it leads to circularity. As will be extended later in this chapter, occasionalists have an advantage in offering a solution here: Since God creates everything and sustains the order, we accept a knowing and voluntary ultimate agent behind regularities. His intentions to sustain regularities so far give us evidence for His future sustenance, but He may will to change order at any moment. Thus, the expectation of order is not certain but likely. Thus, the certainty associated with this type of knowledge is presumptive or probabilistic (*ẓannī*), which may strengthen over time as multiple additional studies from various sites independently corroborate the findings but cannot be conclusive.[4]

This type of knowledge is actionable in the sense that it has practical implications and can be used in medicine to improve diagnosis and therapy for patients. In the clinical sciences, the level of evidence provided by research is graded by level (quality) of evidence from Class A (high quality) to Class C (limited) and the resulting strength of recommendation for therapy is graded Class 1 (strong)–Class 3 (no benefit) (Halperin, Levine, and Al-Khatib 2016).

In sum, these scientific observational studies present a type of knowledge achievable probabilistically (*ẓannī*) and not definitively (*qaṭʿī*). This

[4] Due to this problem, repeated experience (*tajrībīyyāt*) does not yield conclusive epistemic certainty. This stands in contrast to the epistemological framework outlined in Chapter 2, where certainty (*yaqīn*) is extended even to empirical experience which is categorized under conclusive propositions (*qaṭʿīyya*) (see Aftab 2005).

has important implications for mind–brain interaction and the occasionalist solution proposed later in this chapter.[5]

Consciousness in Medicine

When assessing a patient's consciousness, two essential elements must be considered: wakefulness and awareness. Wakefulness refers to the patient's ability to be alert and respond to sensory input from the environment, while awareness pertains to their capacity for orientation and higher-level cognitive functions. These two elements form the foundation of the medical understanding of consciousness and its disorders. The Glasgow Coma Scale (GCS) is the most commonly used clinical tool for evaluating these aspects. More recently, advanced Magnet Resonance Imaging (MRI) is being used to evaluate components of consciousness.

Wakefulness is a result of the physiological activation of the brain in which the patient is awake and arousable. This is clinically assessed by looking for spontaneous opening of the eyes by the patient. The neuroscientific mechanism of wakefulness is primarily mediated by the brainstem. These subcortical pathways are referred to as the reticular activating system or human ascending arousal network (AAN) (Edlow et al. 2023). Advanced MRI techniques such as Diffusion Tensor Imaging (DTI) are being used to map this network.

The AAN primarily involves the brainstem and deep brain white matter tracks,[6] not the cerebral cortex (the outer lining of the brain). This network mediates[7] wakefulness, whereas awareness is mediated by the cerebral cortex.

Awareness refers to the contents of consciousness and assesses if the patient is alert, oriented, and responsive. It is assessed by checking if

[5] A similar occasionalist solution had been successful earlier in Islamic history which allowed empirical science to flourish (see Aftab 2014).

[6] These tracks are composed of neuronal cells called axons, which transmit information from the cerebral cortex to the brainstem through the brain parenchyma which is composed of white matter. The cerebral cortex is composed of white matter and forms the outer lining of the brain.

[7] Activation of this network and its associated components in the brainstem are strongly associated with wakefulness. The network can now be identified with advanced MR imaging.

the patient can provide an appropriate verbal response, if they can articulate their name, time, and location, and if they can perform motor functions based on a given command such as raise your arm and follow other instructions (Qazi et al. 2018). Awareness is primarily mediated by midline cerebral cortical structures, and the primary network associated with awareness is known as the cortical default mode network (DMN) (Alves et al. 2019).

No single or bilateral cortical infarct, unless involving most of the cortex, leads to permanent unconsciousness (lack of wakefulness) (Friedman, Turk, and Budson 2023). Neurological disorders that disrupt the deep brain networks and brainstem affect the level of consciousness (wakefulness), whereas cortical disorders from seizures and migraines to strokes and dementia disrupt phenomenal consciousness (content of conscious experience) (Friedman, Turk, and Budson 2023).

Awareness of conscious experience requires the activation of several parts of the cerebral cortex. There are many different aspects to awareness including memory, sensations, perceptions, intentionality, agency, and the sense of "I" or self. For the purpose of this study, the last concept, that of the "self" or "I" is at the essence of the problem of consciousness. It is the ontologically distinct first-person experience which had previously been unexplainable by empirical science.

In regard to this particular aspect of phenomenal consciousness, the midline cortical structures of the default node network play a pivotal role. This is inferred from the disorders of these regions which affect the human sense of self-awareness.

> Dysfunction of these brain structures can lead to illusory states such as out-of-body experience, motion of the whole body or a body part (floating or falling), dis-ownership or estrangement of a body part (alien hand syndrome), distortion of viewpoint on the world, feeling of not being able to control or initiate actions, or misperceptions about one's own body image.
>
> (Lyu et al. 2023: 2502)

In particular, there is an area of the brain, known as "the precuneus," located in the posteromedial brain which is most intimately associated with our sense of the "self" or "I." Recent studies using electrical stimulation and neuroimaging suggest stimulating this subregion results in feelings of "self-dissociation—defined as feelings of detachment from one's self and the reality of the outside world" (Lyu et al. 2023: 2502).

A simplistic view of the neuroscience of consciousness may assume that various mental states can simply be mapped on the brain and thus correspond to specific physical states. However, this view is false and misleading. For example, in the case of visual perception, a visual stimulus activates the occipital cortex (a specific area in the very back of the brain). However, this activation has to then be processed by the default mode network for the individual to have a conscious experience of it. As Andrea Nani[8] and colleagues explain referring to the regions of the brain associated with visual activity:

> It is more likely that the activity of this region might be essential but not sufficient for the conscious perception of color. According to a number of fMRI studies, in order to have a full-fledged visual conscious experience something further needs to be added, namely, a complex and dynamical interaction among other brain areas, especially those, it has been claimed, of the fronto-parietal network. (Nani et al. 2019: 5).

This has been corroborated by multiple other studies which demonstrate that consciousness arises from connections among many areas of the brain (Vishe et al. 2023). This has given rise to several different models of consciousness and information processing, and computation which correspond to and purport to explain human consciousness.

As can be seen, tremendous progress has been made in neuroscience in recent years. The parts of the brain associated with consciousness have been reduced just to a few areas and to their interconnectivity. Large parts of the brain, including the cerebellum, play no role in conscious experience. This refined understanding of the neural correlates of consciousness indicates an intricate link between the brain and consciousness and has important implications for medicine. It can be used in the treatment planning and prognostication of patients with brain tumors, stroke, seizures, and so forth. The neural correlates of consciousness under discussion do not appear to be merely accidental. When specific neural states are altered, corresponding changes reliably occur in mental states. In this way, the variability of the former is systematically linked to the variability of the

[8] Andrea Nani is a researcher in the department of psychology at the University of Turin, Italy.

latter. We now turn to several significant empirical cases that illustrate and confirm this relationship.

Altering Brain Alters the Mind

Neuroscience indicates that if the brain is damaged, the mind is altered. In a famous case in 1848, Phineas Gage suffered traumatic brain injury when an iron rod was driven through his skull. This changed his personality:

> Gage was a polite, responsible and industrious young man until an accidental explosion drove a tamping iron through his prefrontal cortex, resulting in profound disturbances including poor judgment, lack of planning, disinhibition, socially inappropriate behavior, emotional dysregulation, and insensitivity.
>
> (Barrash et al. 2018: 2)

This change in personality gave solid evidence of the relationship between the brain and the mind for the first time. Since then a wide array of personality changes have been documented in patients undergoing frontal lobe resections due to brain tumors or abscesses. Many of these changes related to impulse-control and moral judgments (Koenigs et al. 2007).

Damage to particular brain regions is associated with specific cognitive function abnormalities. For example, hippocampal infarcts are associated with short-term memory loss; amygdala damage is associated with emotional regulation (Yang and Wang 2017).

A particularly interesting case occurs with the temporal lobe. Studies have shown that stimulating an area at the junction of the temporoparietal junction can lead to extraordinary experiences. This includes out-of-body experiences, hallucinations, and multisensory integration of various stimuli to simulate religious experiences. Swiss researchers Walter and Koening summarize the scientific basis of this phenomenon as:

> The temporal lobe appears to be involved in extraordinary sensations, including the weighing and integration of these potentially differing signals by connecting them to religious beliefs about God's intents and emotions.
>
> (Walter and Koenig 2023: 1201)

Patients with temporal lobe epilepsy (TLE) have been known to have intense religious experiences. In a series of experiments, Koren and

Persinger demonstrated[9] stimulating the temporal lobes elicited experiences of "sensing presence" and mystical experiences (Ruttan, Persinger, and Koren 1990).

Hyperreligiosity associated with temporal lobe epilepsy has also been reported in the literature in the case of a Muslim patient brought to the ED by his wife. A 40-year-old man presented to the ED with history of epilepsy and recent tonic-clonic seizure. After the seizure he:

> was constantly making religious remarks, saying "God is with me and I do not need doctors or medications." He would interpret every question asked to him as questioning his faith and at times attempted to convert doctors and staff to Islam. He believed everyone around him was preventing him from obtaining salvation.
>
> (Garcia-Santibanez and Sarva 2015: 1)

After treatment with anti-epileptic medication, he returned to baseline and the hyperreligiosity resolved. This and other studies associate the temporal lobe with religious experiences. This seems to suggest that some spiritual and mental states are a result of physical processes of the brain.

Traumatic Brain injuries (TBI) occur when there is multifocal injury to the cerebral cortex and/or the underlying neural connecting fibers. TBI is a spectrum of phenomenon that can affect consciousnesses. It can affect functions associated with consciousness such as those responsible for regulating emotional and cognitive processes. TBI can also cause difficulty with memory, concentration and attention, and decision-making. It can also disrupt mechanisms associated with the expression and regulation of emotions, behaviors, and self-awareness (Fleminger 2008).

In the cases of TLE and TBI, physical alterations in the brain result in changes in the content of consciousness. It seems the physical processes of the brain are intimately linked with the mental experiences. Let us now consider some of the practical findings concerning neural correlates of consciousness that have emerged from neuroscientific and medical research.

[9] Their particular method of using a helmet producing magnetic fields and results have been subsequently questioned. There is, however, additional corroborative evidence by multiple additional studies linking the temporal lobe to religious experiences.

Mind Reading Using Technology

Knowledge of neural correlates of consciousness has been used for practical insights into the human mind. Mind-reading technology is at the cutting edge of both neuroscience and artificial intelligence research. This example could be seen as reproducing the subjective experience of the human based on a study of the functional brain. If subjective experience can be reproduced, then does it not imply that it is reducible to the physical processes which produce it, since no separate substance or properties are required to account for it?[10]

In a recent study (Tang et al. 2023), patients are individually placed in a functional MRI machine, and a passage is read to them. The patients' pattern of brain activation is observed during the reading and is then run through an Artificial Intelligence algorithm. Based on these results, the original stimulus is reconstructed without the patient having to articulate it. Other studies have shown significant success and promise in reconstructing images seen by human participants from brain activity measured in fMRI machines (Koide-Majima, Nishimoto, and Majima 2024). The results are fairly impressive, with future advances in technology it seems perfectly reasonable to expect this type of semantic and visual reconstructions to be possible.

Traditionally one of the characteristics of the human mind has been that our thoughts and ideas are essentially private. As the influential South Asian Muslim philosopher Muhammad Iqbal (d. 1938) writes:

> My pleasures, pains and desires are exclusively mine, forming a part and parcel of my private ego alone. My feelings, hates and loves, judgements and resolutions are exclusively mine.
>
> (Iqbal 2013: 80)

If neuroimaging technology continues to advance it seems these private experiences may no longer be private and would be able to be represented to an outside observer.[11]

[10] This research is intimately connected to the aspirations of transhumanists to be able to "upload" one's brain state, that is, mind, and download it onto a mechanical device and thus maintain life and identity. For a greater discussion on this idea, see Chapter 7.

[11] Even in this case, however, the subjective nature of one's personal experience would not be reproducible, as is discussed later in this chapter.

There have been other experiments in which the use of AI and fMRI has been successful in reconstructing images presented to the patient. Images are shown to the patient then using data from brain scanning and a human visual decoding algorithm these images are fairly successfully reconstructed on a screen (Zarina et al. 2021). This remarkable fusion of neuroscience, physics, and artificial intelligence may give the impression that science is at the cusp of cracking the mystery of consciousness.

This type of technology is already in experimental clinical use in assessing for covert consciousness in patients unable to communicate via motor functions (such as use of speech, eye, or limb movements) (Owen et al. 2006). Patients are asked to imagine themselves engaging in spatial navigation or motor imagery tasks that correspond to known patterns of activity on fMRI scans. These include activities such as walking through their home or playing tennis (Beauregard, Aftab, and Sajid 2016). This can help establish the patient's level of consciousness, potentially upgrading a patient from a vegetative state (VS) to a minimally conscious state (MCS) (Owen et al. 2007). An MCS patient is likely to receive more aggressive rehabilitative care than a patient thought to be in a VS. In medicine, this type of interrogation of the brain with neuroimaging technology will likely play an important role in the future. So far, we have examined in detail the close relationship between neural states and mental states. We now turn to consider the broader implications of these findings: what they reveal overall, what conclusions may reasonably be drawn, and what claims they do not yet justify.

Is Neuroscience Successful in Explaining Consciousness?

Does the empirical study of neuroscience successfully reduce human consciousness to the aspects of the brain? Not exactly. Human consciousness is a first-person, subjective phenomenon. It is being studied by a third-person, objective[12] method. Although neuroscience can tell us a lot about the physical manifestations of consciousness, it tells us little about the phenomenon itself. As Manzotti[13] explains:

[12] Objective here implies a reference to the scientific method, which requires reproducible experiments whose data can be quantified. Whereas human consciousness is a personal experience, it evades being completely captured by the scientific method.

[13] Riccardo Manzotti is a Professor of Philosophy at IULM University in Milan, Italy.

> This misalignment between the means—science—and the end—explaining consciousness—gave rise to what has become a productive workaround: the search for "neural correlates of consciousness" (NCCs). Science can sidestep trying to explain consciousness and instead focus on characterizing the kind(s) of neural activity that are reliably correlated with consciousness. (Schurger and Graziano 2022: 1)

What consciousness is, is a first-person experience. Methods of modern science (of which neuroscience is a branch) were deliberately designed to exclude first-person experiences from science. One of the key "achievements" of the scientific revolution was the representation of the world in mathematical and quantifiable terms. Key figures who contributed to this revision in our understanding of science included René Descartes, Galileo Galilei, and Isaac Newton. For example, take the title of Newton's book, *The Mathematical Principles of Natural Philosophy.* Galileo writes:

> The Universe... is written in the language of mathematics, and its characters are triangles, circles and other geometrical figures without which it is humanly impossible to understand a single word of it. (Galileo, *The Assayer*, quoted in Goff 2019: 16)

This understanding of the universe had great benefits as it allowed material science to progress rapidly; however, it left out important elements of the natural world. These included all the qualitative elements of sensations and perceptions, such as color, taste, touch, feelings, and more. These were removed from the natural world and assigned to the soul (and later to the mind). So, for example, the sweetness of the mango exists not in the mango, but rather in the mind of a person tasting the mango. This important separation between the quantifiable world (elements of the world which can be represented mathematically) and the qualia (the subjective features of world requiring a first-person experience) has important implications. The former (the quantifiable) became the subject of modern science, whereas the latter (qualia) became the subject of psychology.

The quantifiable methods of modern science have been tremendously successful in explaining much of the natural world. Given this success, in the contemporary context, many philosophers and scientists are applying the methods of quantifiable methods of modern science to the qualia. In doing so, they are forgetting that modern science (as currently conceived) necessarily excludes qualia from its area of study. Therefore, this approach is bound to run into roadblocks.

In order to study consciousness via the brain, neuroscience uses methods of physical science and empirical research. To account for the qualia, some scientists are creating terminology and concepts which will help explain it. These includes terms such as information, computation, modeling, and representation, among others. The DMN (default mode network) is one such epistemic entity used to explain the neural correlates of consciousness. Manzotti calls these terms "epistemic entities." An epistemic entity is just a made-up concept used to explain and categorize. An example of such a concept is the "center of mass." Center of mass is just an artificially created concept to simplify gravitational computations. Such epistemic entities are just placeholders which do not have an independent existence. He writes, "we suggest that information (or computation, modeling, representing, and so forth) is like the notion of the center of mass. In fact, where the center of mass is supposed to be, there may be an empty space" (Manzotti and Moderato 2013: 89).

Consciousness, of course, is not empty space; it involves rich, real, phenomenal experience. The confusion here is that epistemic methods are being used to investigate something it cannot study. The ontology created by Galileo had two distinct elements which require different modalities for study. This necessarily means that it excludes qualia (or subjective experiences) from its methods. Manzotti explains:

> To make a long story short, neuroscience cannot tackle consciousness because the essential properties of the mind have been programmatically and selectively set aside from the physical world. The list of such properties is not exactly the same although there is a consensus on the main ones: quality, unity, duration, intentionality, and first-person perspectives. At present, neuroscience does not have a clue as to what is the relation between neural activity and consciousness.
>
> (Manzotti and Moderato 2013: 89)

There is an obvious ontological gap between conscious experience and the physical processes of the brain. The epistemic methods of modern science fail to bridge this gap simply because the subject matter of consciousness is a different type of thing—a type of thing not reducible to the mathematical language of modern science.

Similarly, in the case of mind-reading, the reproduction of the contents of thought does not reproduce the experience of consciousness. In fact, there is no possible way to do so, since it is a first-person experience. To interpret these studies as an account of conscious experience is to conflate the observation of brain mechanism with the phenomenal experience itself.

Similarly, the fMRI techniques for assessing consciousness via imagery tasks say nothing at all about the nature of the first-person experience. The information provided is quantitative, whereas the experience is qualitative. They simply provide an objective physical correlate of the subjective experience.

Within the framework of *kalām* epistemology, we can say that the correlations revealed by contemporary neuroscientific research gives us sensory knowledge based on repeated observations, namely, experience (*tajrībīyyāt*) which relate to the bodies.

However, it is crucial to distinguish between the correlations discovered by neuroscientific research and the philosophical interpretations of these correlations. There are various philosophical views on the nature of this relationship, with naturalist physicalism being the most dominant among scientists and philosophers today. However, it is not the only perspective, nor necessarily the most plausible one. Its epistemic status is based on presumptive arguments. As will be apparent in the later sections, when we compare the epistemic strength of naturalist physicalism with that of Islamic models, we find that Islamic models fare much better. To understand the inherent problems within the naturalist view, we will explore different versions of contemporary physicalism and highlight their drawbacks in the next section.

In summary, advances in neuroscience have significantly refined our understanding of the tight correlations between the human mind and neural activity in the brain. They have established an intricate relationship between various parts and pathways of the brain and consciousness, demonstrating that the brain is the essential substrate for conscious experience, not the heart or lungs.[14] Specific areas and networks in the brain are integral to particular conscious experiences.

[14] Other organs of the body such as the heart and the gut have a nervous system as well. This includes autonomic functions of the gut and heart. There have been reported cases of memories being transferred in patients with heart transplants. However, only the cerebral cortex of the brain is associated with phenomenal consciousness.

The Mind–Body Problem

Studying nature using the methods of modern science has allowed us to make remarkable progress. We have built vehicles that explore the depths of the sea and soar through the air. Not stopping there, we ventured beyond earth into space, reaching the moon. Now, our gaze extends to the distant regions of the universe and the depths of our own nature. Human conscious experience remains as mysterious as the uncharted macrocosm, even more so. As we have seen, neuroscience has made significant strides in understanding the human brain, nervous system, and their activation patterns. Yet, the human mind resists complete comprehension through this model. This resistance is not merely due to an immature scientific theory but suggests an ontological barrier between mind and brain. To grasp this issue's philosophical roots, we must begin with Descartes's substantial dualism and the ensuing mind–body problem.

Descartes's Substantial Dualism and the Mind–Body Problem

Descartes considers the body an extended substance (*res extensa*) that cannot think and the mind a thinking substance (*res cogitans*) that is not extended. As such, they have nothing in common (Descartes 1984: 54). This radical division, known as "substantial dualism," postulates two distinct substances: body and mind.[15] The body is subject to physical laws, whereas the mind is not. Yet, within the human being, they correlate in a tightly connected manner. When we open our eyes (a bodily state), we perceive a colorful world with red roses and green trees (a mental state). When someone hits us (a bodily state), we feel pain (a mental state). Similarly, when we want to drink water and plan how to reach the bottle (a mental state), our hands move and grab the bottle (a bodily state). How can these two different states be related to each other? If the mind affects the body and vice versa, how can two distinct substances interact if they belong to completely separate ontological categories? This issue is known as "the mind–body problem."[16]

[15] The radical formulation of substance dualism differs from softer versions of dualism found in the Islamic tradition (see Aftab 2015).

[16] Famously, Princess Elisabeth of Bohemia drew Descartes's attention to this problem in a letter she wrote to him in 1643 (see Anscombe and Geach 1969: 274–275).

The Hard Problem of Consciousness

The dominant view among neuroscientists and philosophers today is to deny the existence of a distinct substance called the "mind." This is in part due to the tremendous success of the physical sciences in explaining natural phenomena, the assumption is that it must also be able to explain the "mind." They assume that there is only one substance described by contemporary science, which is the physical body. Thus, it might seem that the mind–body problem disappears if the mind as a separate substance is denied. However, this problem is not so easily dismissed; it reappears within the materialist or physicalist ontological framework in a different form.

David Chalmers draws attention to human consciousness by distinguishing between the "easy problem" and the "hard problem" of consciousness. The easy problem involves identifying the neural mechanisms that form the basis of specific mental phenomena. Although the neural correlates of many mental phenomena, including consciousness, have not been fully identified, significant progress has been made in this area, making it a continually advancing field in neuroscience. The hard problem of consciousness, however, involves more than just identifying neural correlates. It seeks to understand the true nature of consciousness and conscious experience, aiming to explain the relationship between conscious experiences and their neural correlates, if any, to provide a comprehensive, holistic picture (Chalmers 1996: xi–xii). Conscious experience, or qualia, appears categorically different from neural activities in the brain. For instance, when we see a red apple, certain neurons fire in specific patterns, but the visual experience of seeing a red apple seems entirely different from the physical processes occurring in the brain. The challenge for materialists or physicalists is to explain how qualia, the subjective experience of consciousness, arises from the physical brain.

Explanatory Gap

There is an explanatory gap between the neural activity in the brain and conscious experience.[17] To see this gap more clearly consider the following thought experiments:

[17] As for the formulation of the problem of explanatory gap in this context, see McGinn (1989 and 2003).

Leibniz's Mill[18]: Imagine a machine capable of producing activities of the conscious mind, such as perception and thinking. Now, let us scale up this machine, maintaining the proportions of its parts, so that it becomes large enough for us to enter, like a mill. Once inside, we would only observe the machine parts and their interactions with each other; we would not observe any mental activity. If we liken Leibniz's hypothetical machine to the brain, upon examining the brain's tissue and nerve cells, we would only encounter physical structures, not the subjective conscious experiences of the individual. How can we explain this discrepancy? (Leibniz 1965: §17, 150).

Nagel's Bat: Thomas Nagel[19] invites us to imagine what it's like to be a bat, with their unique echolocation techniques. While we can understand how bats perceive and navigate, Nagel argues that we cannot truly know what it's like to be a bat unless we ourselves have the experience (Nagel 1974).

Mary's Room: Frank Jackson[20] asks us to imagine a girl named "Mary" who has lived in a black-and-white world since childhood and has never experienced color. Despite this, Mary is an expert in physics and knows all the physical phenomena related to seeing colors. She understands everything about light, its different wavelengths, the properties of objects, and the brain's mechanisms involved in color perception. However, will Mary learn something new if she is taken out of her monochromatic environment and shown a red apple? (Jackson 1982). Although she possesses all the physical information about the apple and its interaction with light, does her knowledge include the subjective experience of the apple's redness and its taste? Or is this information only obtainable through conscious experience?[21]

[18] Gottfried Wilhelm Leibniz (d. 1716) was a German polymath and philosopher.

[19] Thomas Nagel is an American philosopher, best known for his contribution to the philosophy of mind.

[20] Frank Jackson is an Australian analytic philosopher.

[21] Interestingly, a similar experiment was first proposed by the Spanish Muslim philosopher Muḥammad ibn ʿAbd al-Malik Ibn Ṭufayl (d. 1185) in which he points to the distinctness of personal experience from mere knowledge by giving the example of a blind man who knows all there is to know about his environment, but then actually experiences sight. He purports the "delight" and "clarity," which seeing brings is wholly missing without the personal experience (see Ibn Ṭufayl 1972: 107).

Chalmers's Zombies: David Chalmers[22] presents an intriguing thought experiment. Imagine two individuals with identical brain structures and neurochemical conditions. Despite having the same healthy brain structure, they differ in one important way. One of them has conscious experiences, while the other lacks such a state of consciousness. We can refer to the second individual, who lacks consciousness, as a "zombie." If we can conceive of these two people without falling into contradiction, then such entities are possible. Consequently, the neurochemical states and the conscious states in their brains cannot be considered identical, as two identical things cannot be separated. It implies that consciousness is something additional or extra which can be added to the individual. Therefore, does this not suggest that consciousness is distinct from the brain? (Chalmers 1995 and 1996).

These thought experiments demonstrate that the mind–body problem does not disappear easily; it persists in the form of the hard problem of consciousness. In response to this issue, contemporary physicalists have developed several positions: eliminative physicalism, reductive physicalism, and non-reductive physicalism. Let us examine what these positions propose and consider their potential drawbacks.

Contemporary Views of Naturalist Physicalism

Eliminative Physicalism

Eliminative physicalism seeks to eliminate the discourse of mental phenomena altogether, replacing it with neurophysical or neurobiological properties described by neuroscience. In other words, as neuroscience progresses, there will be no need to refer to visual experiences, beliefs, hopes, fears, pains, or even consciousness itself. Instead, these mental phenomena will be replaced by descriptions of the neural mechanisms truly associated with them (Churchland 1981). This approach is supported by examples from the history of science. For instance, the phlogiston theory of combustion posited that all combustible substances contained a fire-like element called "phlogiston," which was released

[22] David Chalmers is an Australian philosopher and cognitive scientist best known for coining the term "the hard problem of consciousness."

during combustion, leaving behind ash. This theory was completely abandoned by Joseph Priestley and Antoine Lavoisier, who rejected the existence of phlogiston. Instead, modern chemistry explained the phenomena of combustion through the role of oxygen (Churchland 1986: 281–282).

Although there are successful examples of eliminating certain categories in the history of science, not all phenomena can be successfully eliminated. Comparing the elimination of phlogiston to the proposed elimination of mental phenomena presents a weak analogy or disanalogy. Phlogiston was a non-observable theoretical element believed to exist, whereas the conscious experience of qualia is categorically different. Qualia are directly observable, albeit only from a first-person point of view. The subjective experience of qualia cannot be easily dismissed, unlike the concept of phlogiston. Consider someone who could not sleep last night due to a sharp toothache. Does any neurobiological explanation that completely ignores the felt pain make sense to that person? Qualia exist for those who perceive them and are fundamentally different from their corresponding neural activities. While one can observe the firing of certain neurons in a subject's brain, the conscious experience itself remains unobservable and distinct from this neural activity.

Reductive Physicalism

The second approach, reductive physicalism, does not categorically deny mental phenomena but seeks to reduce them to a neurobiophysical base. Patricia Churchland[23] describes the reduction of mental states to brain states in terms of whether a theory about mental states can be reduced to a theory about how neuronal ensembles function (Churchland 1986: 278–279). To clarify this point, she provides several examples of successful reductions from the history of science. For instance, light has been successfully reduced to electromagnetic radiation, meaning that the theory of optics has been subsumed under the theory of electromagnetic radiation, making it appropriate to identify light with electromagnetic radiation (Churchland 1986: 279). In other words, it has been theoretically demonstrated that light is identical to electromagnetic radiation. Light is not a phenomenon distinct from electromagnetic radiation. Notice that this approach does not deny or ignore light as a phenomenon;

[23] Patricia Smith Churchland is a Canadian-American analytic philosopher best known for her work on neurophilosophy.

rather, it identifies light with another phenomenon—electromagnetic radiation—that is subject to Maxwell's equations and scientific analysis.

Reductive physicalism, which seeks to identify mental phenomena with a neural base in the brain, comes in two forms. The first form, type-identity theory, identifies specific kinds of mental phenomena with specific kinds of neurophysical or neurobiological phenomena. The second form, token-identity theory, identifies particular mental phenomena with particular neurophysical or neurobiological phenomena (Westphal 2016: 65–67). Type-identity theory has been criticized due to the concept of multiple realizability. This concept suggests that a certain kind of mental state, such as conscious visual perception, can be realized in a physical base that does not have a neurophysical or neurobiological structure. For example, such a mental state might emerge in a silicon-based machine. If this is possible, then mental states cannot be identified solely with neurophysical or neurobiological states (Putnam 1967). To address this issue, token-identity theory is proposed. It allows for the multiple realizability of certain kinds of mental states in different physical media by positing the identity of particular mental states with specific neural phenomena in the brain, rather than with kinds of states. In other words, token-identity theory states that a particular visual experience is identical to a specific neural phenomenon in the brain, not that all visual experiences are identical to a single type of neural phenomenon.

Both forms of identity theories fall short of explaining the clear difference between mental phenomena and neurophysical or neurobiological phenomena. Defenders of this reductive approach might argue against this apparent difference by citing historical examples, such as the relationship between light and electromagnetic radiation. Before the theory of electromagnetism, these two phenomena were considered distinct. However, the theory revealed their true identity. Why, then, could not neuroscience similarly enable us to identify mental phenomena with neurophysical or neurobiological phenomena in the future?

The problem with this analogy is that it is fundamentally flawed. Scientifically speaking, the link between light and electromagnetic radiation is not strict identity. Light was discovered to be a specific type of electromagnetic radiation visible within the spectrum, while other types, such as X-rays or infrared radiation, are not visible to the naked eye. Visible electromagnetic radiation has particular wavelengths and frequencies distinct from those of other types. Thus, the statement "light is electromagnetic radiation" is analogous to saying "whales are mammals." Whales are categorized as mammals according to Linnaeus's taxonomy, a classification accepted in the eighteenth century. Previously, they were considered fish. This reclassification is a scientific advancement but does not eliminate the crucial differences between whales and other mammals, such as humans and apes. Similarly, identifying light as electromagnetic radiation does not erase the significant differences between light and X-rays or infrared radiation.

Therefore, even if we accept that consciousness or conscious experiences fall under the category of physical brain states, they possess distinctive qualities that set them apart from other neural states, whether unconscious or subconscious. In summary, reductive physicalism cannot easily circumvent the hard problem of consciousness.

Non-Reductive Physicalism

The third approach, non-reductive physicalism, accepts that mental states and neurophysical-biological states form two distinct properties, though both are properties of the same physical substance. These two kinds of properties are irreducible to each other ontologically but do not require us to formulate two distinct substances. This view is also known as "physicalist property dualism." A growing number of secular philosophers and scientists advocate for various forms of non-reductive physicalism, which are described below. While they maintain that humans are fundamentally physical beings, they acknowledge the existence of non-physical characteristics such as consciousness. Currently, science lacks the means to incorporate first-person experiences into its framework. These thinkers argue for an expansion of the scientific toolkit to accommodate these experiences. Theories that fall under the non-reductive physicalist umbrella include:

Panpsychism posits consciousness as a fundamental property of the universe that should be added to the basic building blocks of nature such as mass and charge. Philip Goff[24] has championed this cause arguing that the toolkit of physics cannot explain the intrinsic nature of its subject. Consciousness must then constitute the fundamental nature of matter (Goff 2019).

24 Philip Goff is a British author and philosopher, and a prolific champion of panpsychism.

The orchestrated objective model introduced by the mathematician Roger Penrose introduces phenomenal experience to science (Hameroff and Penrose 2014). The theory postulates that neuronal microtubules act as foci for quantum computing resulting in consciousness.

The Integrated information theory (IIT) is one of the major scientific models for consciousness which posits that it is the organization of the brain that makes it conscious, and there recognizes consciousness as an emergent property[25] of the physical system that gives rise to it. Koch[26] describes its conclusion as: "Consciousness depends on a physical substrate but is not reducible to it" (Koch 2014). Koch, who describes himself as a "romantic reductionist," states:

> Experience, the interior perspective of a functioning brain, is something fundamentally different from the material thing causing it and that it can never be fully reduced to physical properties of the brain … I believe that consciousness is a fundamental, an elementary, property of living matter.
> (Manzotti and Moderato 2013: 82)

Biological naturalism is a theory put forth by the philosopher John Searle,[27] in which he similarly argues for consciousness being causally related to the biological brain, yet ontologically distinct, as an emergent phenomenon (Searle 2007).

Expanding the scientific toolkit to include the qualia of consciousness alongside the quantifiable entities of physics holds promise. This

[25] In philosophy of mind, emergence is the idea that mental properties are qualitatively distinct from the physical properties which make up the brain. An analogy would be how the "wetness" of water is distinct from the chemical make-up of the water molecule. A related concept is supervenience. If the mind supervenes on the brain, each physical brain state would correspond to a specific mental state. The same physical state of the brain could not be associated with two distinct mental states. This would seem to contradict the growing evidence from neuroscience in favor of multiple realizability of the same mental states via different physical states. For example, pain can be processed and felt differently from person to person, or even the same person in different times. For details, see Lyu et al. (2023).

[26] Christof Koch is an American neurophysiologist and computation neuroscientist advocating for a modern variant of panpsychism.

[27] John Searle is an American philosopher, best known for inventing the "Chinese Room" thought experiment which teases out the difference between human and artificial intelligence.

approach could, at least theoretically, offer an ontologically comprehensive view of the observable universe by acknowledging mental phenomena, consciousness, and conscious experiences as integral components. Note that non-reductive physicalism, although making room for the existence of consciousness, does so without positing a human soul. This non-reductive form of physicalism faces the challenge of explaining the causal interaction between these distinct properties, an revised version of the Cartesian mind–body problem. It also faces the conundrum of explaining how a subjective experience can exist without positing a non-physical subject.

Problem of Causal Interaction

Let us explore further the causal interaction between two kinds of properties. In property dualism, mental properties are considered distinct from their neurological correlates, with the latter often viewed as the causal basis for the former. This means that neurophysical or neurobiological states cause mental states. If one believes that mental states are entirely causally inert, this view is known as "epiphenomenalism." Epiphenomena are effects of certain phenomena but do not causally influence other phenomena (Walter 2009: 91). For example, the smoke from a steam engine is an epiphenomenon; it is caused by burning coal but does not cause the locomotive to move. Conversely, if one also holds that mental states can causally influence other mental states as well as neurophysical or neurobiological states, this view is known as "downward causation" (Clayton 2004: 4–6).

Both reductive and non-reductive physicalism emphasize a strong link between conscious mental states and neurophysical or neurobiological states. This link is neither coincidental nor a simple correlation. This view is typically supported by detailed neuroscientific findings. For instance, drug or alcohol use alters neural activity in the brain, leading to changes in mental states. Similarly, damage to specific parts of the brain results in the loss or diminishment of certain mental activities or significant changes in personality traits. It can be argued that these findings cannot be mere correlations, as discussed earlier. The physicalists claim that this demonstrates that mental phenomena are identical to their neural correlates or that neural states causally influence mental states. The evidence for this claim is limited.

Even if there is a strict link between mental states and physical states, observation alone falls short of justifying their identity or causal interaction. No matter how constant the link between two types of events or objects, it remains a correlation. We do not observe efficient causality in nature, only correlations. These correlations can be loose and accidental or tight and regular. While we may discover deeper and more general correlations behind the apparent ones, they remain correlations. Inferring efficient causality from correlations involves at best an abductive inference (a presumptive kind of argument) to explain the constancy of those correlations in the best way. But, this inference only makes sense if one initially accepts the truth of naturalism or physicalism and completely rules out the causal influence of a supernatural entity or God. In addition, even the cogency of this abductive inference is questionable. The naturalist either comes to ignore obvious mental phenomena or encounters an updated version of the mind–body problem—the problem of explaining how these two distinct properties causally interact. The constancy of correlations can also be explained within different metaphysical frameworks.

Does Consciousness Require a Soul?

An additional potential problem with considering consciousness a fundamental property of nature is the question of, to which entity does the experience belong? Conscious experience cannot exist in isolation; it must belong to someone or somebody. This experience requires a subject; a conscious experience cannot exist without someone experiencing it, as Descartes eloquently articulated in his *cogito ergo sum* thought experiment. Namely, the "I" of the "I think" also must exist. What does the "I" refer to? To the physical brain, the soul, or the conscious mind?

This differentiation between the subject and subjective experience is obscured in the neurophilosophical discourse, which conflates the "I" with subjective experience. Non-reductive physicalism seems to sidestep this problem. Whereas, within an Islamic worldview, the "I" is of the utmost importance and points toward the soul. As we shall see, some Muslim scholars regard the soul as a substance distinct from the body, while others describe it as a *subtle body* (*jism laṭīf*). In either case, they draw a clear distinction between the soul, understood as the subject or the "I," and subjective experiences such as qualia and perception (*idrāk*). These experiences are classified as accidents that inhere in either spiritual or bodily substances, resulting in the close correlations between physical and mental phenomena.

In the next section, we will discuss two occasionalist models from Islamic intellectual history that offer explanations for these correlations without falling into the pitfalls of naturalist perspectives.

Two Models of Islamic Occasionalism

Let us begin with the following verse from the Qurʾan:

> *And they ask you of the soul (rūḥ): Say: the soul proceeded from my Lord's Amr Command: but of knowledge, only a little to you is given* (Q.17:85).

In this verse, it is implied that the soul[28] is not well known or clearly revealed to humans. It is challenging to derive a definitive view on the nature of the soul from an Islamic perspective based solely on this verse. Muslim theologians proposed different models of the human soul—some formulating views consistent with this verse without explicitly presenting them as theological claims and others grounding their positions in alternative scriptural evidence. In what follows, we will examine two significant models: the physicalist conception of the soul and the (substantial) dualist conception of the soul.

The Physicalist View of the Soul and Occasionalist Property Dualism

Early Ashʿarī theologians, drawing on some other verses, developed a materialist conception of the soul from the Qurʾan. For instance, al-Ashʿarī refers to the Qurʾanic verse "And We have counted all things in a clear registry" (Q.36:12) and argues that everything in the finite universe is composed of indivisible particles. He reasons that it is impossible to count something infinite. If a single entity could be divided infinitely, its parts would be uncountable. Thus, division must stop at some point,

[28] Although "the soul" in this verse is often understood to be referring to the human soul, some scholars also have interpreted it as referring to the archangel Gabriel. In any case, whether it refers to the human soul or not, it refers to a soul, which is important for our purposes in this chapter. The verse says that little is revealed to us about the nature of the soul.

which is the indivisible particle, or atom.[29] Applying this to all finite beings, al-Ashᶜarī concludes that souls are also composed of atoms. We will further explore how al-Ashᶜarī elaborates on his materialist conception of the soul, but the key point here is to understand how he connects this specific model of the soul to Islamic revealed sources. This approach would be considered a transmitted argument. Whether it is conclusive or presumptive needs to be assessed in light of subsequent developments within the Ashᶜarī school of theology.

Later Ashᶜarīs began to recognize that al-Ashᶜarī's argument for atomism was flawed. One significant reason was that it overlooked the distinction between potential and actual infinity. While dividing something into smaller pieces can theoretically continue indefinitely, this does not mean that physical objects contain an infinite number of parts. The division process is only potentially infinite and does not imply the existence of actually infinite quantities. Therefore, al-Ashᶜarī's argument is presumptive and lacks strong cogency.

Although later Ashᶜarīs continued to support atomism, their reasons were more pragmatic than theologically binding. For instance, al-Taftāzānī found a theoretical advantage in adopting atomism: it made it easier to reject the doctrines of Hellenic philosophers, such as the eternal universe, the fixed and necessary order of the universe, and the unchangeable motion of heavenly spheres. Additionally, atomism aligned better with other Ashᶜarī doctrines like creation *ex nihilo*, the afterlife, and bodily resurrection (al-Taftāzānī 1950: 32).[30] These arguments, though still presumptive, were more cogent than al-Ashᶜarī's original argument.

It became increasingly evident that a definitive view on the nature of the soul could not be derived from revealed sources. Since God did not reveal specific knowledge about this issue, it remains undecided in terms of transmitted sources and arguments. Therefore, one might suspend judgment (*tawaqquf*) on the nature of the soul within the context of revealed sources. However, the exploration of the soul's nature through rational arguments, as exemplified by later Ashᶜarī theologians, continued. A variety of positions were defended: some upheld a physicalist

[29] The indivisible particle here is not like atoms of contemporary physics which are divisible. Al-Ashᶜarī's atoms are even conceptually indivisible (see al-Ashᶜarī 1953: 127–192).

[30] For more information about the evolution of atomism within Ashᶜarism, see Malik and Muhtaroğlu (2022).

atomist model, while others embraced substantial dualism and regarded the human soul as non-physical. From these perspectives, we can derive models that engage meaningfully with the mind–body problem and the so-called "hard problem" of consciousness. In what follows, we will first examine the occasionalist physicalist model of property dualism, rooted in atomism, and then turn to the occasionalist model of substance dualism, aligned with substantial dualism.

The occasionalist physicalist model of property dualism can be derived from Ashʿarī atomism. The Ashʿarī school of Islamic *kalām* predominantly holds that finite reality consists of indivisible physical particles, or atoms. Each atom is homogeneous but exists with certain accidents, which account for the diversity among them. All finite beings are composed of these atoms.[31] According to al-Ashʿarī, considered to be the founding father of the school, indivisible particles can densely cluster to form compact bodies such as humans, animals, trees, plants, and rocks. They can also come together more sparsely to form subtle bodies such as air, spirits, *jinn*, and angels (Ibn Fūrak 1987: 207). Thus, the soul (*rūḥ*) is not a substance distinct from matter; rather, it is a subtle matter consisting of fewer atoms than more massive bodies.

Al-Ashʿarī compares the human soul to air, noting that the Arabic term *rūḥ* (soul) is related to *riḥ* (wind). He likens the spirit to the wind. As the wind may travel within dense bodies, human soul enters the denser part of the human body and departs at death (Ibn Fūrak 1987: 271). This etymological similarity interestingly has an ontological correspondence in his metaphysics. He further explains that life is not the same as the soul; rather, life is an accident created in the atoms of the soul when it enters the human body, the person becomes alive (Ibn Fūrak 1987: 257). Mental phenomena, such as perception and knowledge, are also considered accidents. Whereas it is a matter of discussion whether the soul is a type of physical entity; conscious experience is clearly a non-physical mental phenomenon. As noted earlier, theologians distinguish between the soul, understood as a subtle body (*jism laṭīf*), and its accidents, which include conscious experience.

Al-Ashʿarī considers life a necessary precondition for the existence of mental states. For instance, perception (*idrāk*) can only occur in bodies

[31] For more detailed information about Ashʿarī atomism, see Fakhry (1958); Wolfson (1976); and Bulgen (2024).

that are alive. He also underlines a crucial distinction between the accidents of life and perception. While life can exist in the atoms of the entire body, perception exists only in specific parts of the body relevant to the type of perception. Additionally, there are contingent conditions for the creation of perception accidents. For example, visual perception is created by God when people have healthy eyes, open them, and the object of their perception exists in a luminous environment. However, these conditions are not strictly necessary for visual perception. God creates visual perception under these conditions due to His habitual practice of creation (*ʿāda*). In the actual world, God does not create states of affairs randomly but in an orderly manner, known as His habit of creation (Ibn Fūrak 1987: 131–132).[32] For instance, God typically creates the burning of cotton when fire is nearby, not water. Yet, fire has no causal influence over burning; God is the sole causal agent. This view, known as occasionalism, holds that fire is only an occasional cause, not a true efficient cause. Al-Ashʿarī also believes that God can change His habit of creation. It is possible for God to create visual perception in blind people or even create mental states in a single atom after bestowing it with the accident of life, if He wishes (Ibn Fūrak 1987: 131–132, 263). Al-Juwaynī, a later Ashʿarī scholar, posits that the accident of knowledge specifically inheres in a particular part of the human body, namely, the atoms of the heart, rather than elsewhere (al-Juwaynī 1969: 129–130, 410, 414, and 474).

In Ashʿarī atomism, all accidents are considered transitory, meaning they exist only momentarily. For continuity, they must be constantly recreated in atoms by God (Ibn Fūrak 1987: 237). As noted earlier, God creates accidents not arbitrarily but in a regular, orderly manner. This Divine habit is the basis of the order in the universe. Under certain conditions, specific events or phenomena occur consistently (Ibn Fūrak 1987: 131–132). In other words, finite events or phenomena are correlated and exhibit a constant regularity, but there is no causal interaction among them. To reiterate, God is regarded as the sole cause, and through His habitual practice, He causes all occurrences in a regular manner.

Within the Ashʿarī metaphysical framework, a scientist can work to identify the stable and consistent correlations that arise from the habitual act of Divine creation. In other words, scientists do not need to assume

[32] To al-Ashʿarī, God freely chooses to follow a habit (*ajra al-ʿāda*) in creating events.

efficient causation to explain natural phenomena. In the realm of scientific explanation, they rely on correlations and laws maintained by God. In the realm of metaphysical explanation, they do not attribute efficient causality to natural beings as the source of these correlations or laws. Instead, they refer to God's power as the sole efficient cause responsible for the origination and continuity of these correlations or laws. God could suspend or change these correlations or laws if His wisdom requires it, as in the case of miracles. However, this mere possibility does not give us reason to believe that God will suspend or change them in the future arbitrarily. Therefore, while the knowledge of correlations and laws of nature is metaphysically contingent, it is empirically solid. This experiential knowledge (*mujarrabāt*) is of a presumptive kind. Yet, since God is also considered the guarantor of order in the universe, the epistemic strength of this presumptive knowledge is very high.

From this metaphysical framework, we can plausibly formulate an occasionalist and physicalist model of property dualism. All finite beings are physical or material, as they are constituted from indivisible particles. Humans also possess mental properties (accidents). While some Ashʿarīs believed that mental properties inhered in the heart, in light of contemporary neuroscience, it is reasonable to hold that mental properties inhere in the brain or whichever organ science identifies. Additionally, we can posit that the neurophysical and neurobiological properties of the brain are correlated with these mental properties due to God's habitual way of creation. However, these two different kinds of properties do not causally interact; instead, God constantly creates them in perfect correlation. We now present the dualist model.

The Dualist View of the Soul and Occasionalist Substance Dualism

The second model is derived from substantial dualism within an occasionalist framework. Not all Ashʿarīs are strictly atomists. For instance, al-Ghazālī adopts atomism in some of his works but leans toward substantial dualism in others. In *Moderation in Belief* (*Al-Iqtiṣād fī al-Iʿtiqād*), he follows mainstream Ashʿarism and adopts atomism, asserting that the created universe consists of indivisible particles and their accidents (al-Ghazālī 2013: 27). Consequently, the human spirit and mental properties must be understood within this materialist framework. However, in the *Revival of Religious Sciences* (*Iḥyāʾ ʿulūm al-dīn*), when discussing human nature, al-Ghazālī uses four different terms: *Heart, Spirit, Soul, and Intelligence*. After noting that each of these terms has a literal and a secondary meaning, he writes:

> It is this second meaning, common to all four terms… may be defined as that subtle tenuous substance, spiritual in nature, which is the perceiving and knowing essence of man, and in reality is man. It is immaterial and immortal. It is created directly by God, capable of knowing Him, and is morally responsible to Him.
> (al-Ghazālī 2010: xv and xvii)

It is with the "immaterial and immortal" aspect of the human soul that a form of dualism between body and soul enters the discourse on human nature. Based on this perspective, we can plausibly attribute substantial dualism to al-Ghazālī. Nonetheless, we can still interpret this ontological view within the framework of occasionalism. In humans, God creates the spirit and soul and the material body all together. Although mental states inhere in the spirit or the soul, not in the body, these states might be perfectly correlated with the physical states in the human body. God constantly maintains this correlation without any causal interaction between the two substances in question.

To sum up, both the property dualism within a materialist atomist framework and the substantial dualism we have presented do not accept causal interaction between different substances or properties. Instead, God is considered the sole cause, maintaining perfect correlations between mental states and their neural correlates. These models recognize the reality of mental phenomena without reducing them to physical activity in the brain. They align with neuroscientific discoveries about the tight correlations between mental states and their neural correlates while avoiding the updated mind–body problem and the hard problem of consciousness. Thus, these occasionalist models avoid the issues inherent in naturalist perspectives. Additionally, occasionalist models recognize only one efficient cause (i.e., God) to explain physical and mental phenomena and the correlations among them. In contrast, naturalist perspectives assume numerous natural causes. Given the principle of parsimony, occasionalist models fare better in this respect.

In conclusion, while both occasionalist and naturalist perspectives rely on presumptive arguments, the epistemic strength of the former seems much higher than the latter in terms of explanatory power and parsimony. Within the framework of occasionalism, we may even compare property dualism and substance dualism. Although both occasionalist models appear equally acceptable from a theological standpoint, given the absence of decisive evidence in the revealed sources to privilege one over the other, one may rationally prefer the occasionalist model of property dualism on the grounds of simplicity. Whereas the substance dualist model posits two distinct types of created substances, the property dualist model requires only one, thereby avoiding an unnecessary multiplication of kinds of created beings. In any case, one of the key intellectual virtues of occasionalism is its ability to resolve the central challenge posed by the problem of interaction between the brain and the mind. Notably, the occasionalist framework also offers a distinctive approach to understanding scientific causality. By grounding causal relations in Divine will rather than in the inherent powers of created entities, occasionalism not only accommodates empirical science but actively supports it (Aftab 2014). From this perspective, the link between cause and effect is not merely an observed regularity but is sustained by the continuous will of God. In the context of neuroscience and the study of consciousness, occasionalism secures the correlation between mental states and physical brain states without positing problematic causal mechanisms between distinct substances. Instead, the observed interaction is upheld by Divine agency, offering a coherent theological account that aligns with empirical findings. We now turn to consider the implications of these two models for Artificial Intelligence (AI).

Artificial Intelligence and Human Mind

Now let's apply these two models to Artificial Intelligence (AI) and examine their implications for Strong AI. There are various perspectives regarding the progress and potential of AI. Narrow AI focuses on specializing in and executing specific tasks, such as language processing, facial recognition, or playing particular games, with high proficiency. In contrast, General AI aims to achieve or even surpass the level of human intelligence across all tasks. General AI is an ambiguous project. Some proponents argue that AI's simulation of human intelligence in all tasks would be sufficient. Here, "simulation" is the key term, implying that

AI does not need to truly replicate human mental experiences. If an AI-based machine behaves as if it is conscious, that would be enough.[33] However, others believe that AI can replicate human mental states and attain consciousness. This thesis is known as "Strong AI" (Searle 1980). The intriguing question is whether Strong AI is genuinely possible.

The two occasionalist models we have presented do not reject the possibility that God could create human-like mental states or consciousness in machines. Within the framework of these models, God is omnipotent and capable of creating any logically or metaphysically possible scenario. From the physicalist perspective, God can create mental properties inhering in the atoms of machines just as He creates mental properties inhering in the atoms of organic bodies. Ultimately, everything is composed of the same atoms. From the dualist perspective, God could create souls in machines as He does in organic bodies. There is no inherent contradiction in either scenario. However, whether the creation of intelligence in machines aligns with God's habitual way of creation and the nomological regularities He maintains in the universe is a separate issue.

We do not have any explicit statement in the revealed sources that may either reject or support this compatibility. God did not inform us about His plans in creating the world as to whether creating intelligent machines fall within His habitual way of creation or not. We do not have any transmitted argument about the empirical possibility of the emergence of machine intelligence in this world. Thus, within the context of revealed sources, the proper attitude would be to suspend judgment on this matter (*tawaqquf*).

In addition, neither do we have a rational argument to either reject or support this possibility. We do not fully understand the limits of Divine habitual creation. Our knowledge of God's habit of creation comes from observation, not from mere contemplation. The regularities we

[33] The view that underlies AI's simulation of human mental capacities traces back to Alan Turing. Turing proposed a conversational test, now known as the Turing Test, to determine whether machines can successfully simulate human intelligence. In the Turing Test, a human evaluator interacts with both a machine and a human through a text-based interface (such as a chat). The evaluator's task is to discern which of the two participants is the machine. If the evaluator cannot reliably distinguish the machine from the human, the machine is considered to have passed the test and demonstrated human-like intelligence. Turing argued that if a machine could convincingly simulate human conversation, it should be regarded as intelligent. For more details, see Turing (1950).

observe in the universe may not be the most fundamental regularities or foundational laws; there could be deeper and more profound regularities. Science progresses by discovering these deeper regularities and laws, but we cannot be certain that we have discovered the deepest or most fundamental ones.

Consider a primitive society that observes the regular movements of heavenly bodies and notes their consistent patterns. Imagine these people have never witnessed a solar eclipse. The first solar eclipse would appear to them as an exception to the regular behavior they associate with the heavenly bodies. However, this event still fits within certain patterns of God's habit of creation, even though the primitive people have not discovered this fact yet.

Similarly, today we might be like those primitive people in relation to the emergence of intelligence and human-like mental properties in machines. We cannot say for sure whether such an occurrence falls within God's habit of creation because we do not know the most fundamental laws or patterns of creation. Therefore, even within a context of rational argumentation, we have no compelling reason to either reject or support the thesis of Strong AI. Accordingly, we prefer to suspend judgment on this matter, remaining skeptical regarding the nomological possibility or impossibility of AI in machines.

However, it is important to note a significant caveat. Contemporary developments in AI predominantly rely on a computational approach, where AI is defined by its computational structure.[34] In this view, all intelligence is algorithmic, meaning data are processed according to a set of clearly defined rules. This applies regardless of whether the algorithm follows a symbolic or connectionist architecture. Consequently, contemporary discussions about AI are primarily focused on computers executing algorithmic software (Haugeland 1985; Rumelhart, McClelland, and PDP Research Group 1986; Russell and Norvig 2021).

The question of whether the human mind has a computational structure similar to a computer is a matter of debate. Some philosophers, like John Searle, reject this idea. Searle designed a thought experiment known as the "Chinese Room" to demonstrate that the human mind is not akin to a computer, arguing that certain aspects of the

[34] While contemporary advancements in AI largely follow a computational approach, it is by no means the only path being explored. For alternative approaches, see Navon (2024).

human mind transcend computation and cannot be captured by algorithmic data processing. In this thought experiment, a person locked in a room communicates with the outside world via messages written in Chinese. The individual does not understand Chinese but uses a reference book to respond to the messages. By following rules that match certain input characters to specific output characters, the individual processes the inputs and produces Chinese characters as outputs. An outside interrogator, observing the coherent responses, assumes that the person understands Chinese. However, according to Searle, the individual inside the room does not truly understand Chinese; that person is merely manipulating symbols according to an algorithm. The conscious experience of understanding the meaning of the symbols is absent (Searle 1980).

Searle's Chinese Room argument aims to illustrate that an electronic system functionally equivalent to a human cannot be said to possess understanding in the same way humans do. It highlights the distinction between mere symbol manipulation, which computers excel at, and genuine comprehension, which involves conscious experience. This thought experiment challenges the notion that human cognition can be fully replicated by computational systems and underscores the complexity of human understanding that goes beyond mere algorithmic processes.

Thus, if Searle is correct, the human mind possesses certain aspects that cannot be captured by algorithms. This implies that within the framework of the computational approach to AI, the emergence of a human-like mind in machines (i.e., Strong AI) is metaphysically or broadly logically impossible. Recall that for Ashʿarī theologians, such impossibilities do not fall within the scope of Divine power. Therefore, if Searle's view holds true, Strong AI within the computational approach to AI would be incompatible with an Ashʿarī theological perspective.

The central question is whether Searle is right and whether the difference between AI and humans lies in the former taking a computational approach and the latter not. This remains an ongoing debate. As this issue is inherently philosophical, theological perspectives, including Islamic ones, do not necessarily have to take a position in this debate. It is not primarily a theological issue. However, the stance one takes in this debate does have theological implications.

If one supports the computational approach to human mind, Strong AI within this framework would be possible from an Islamic theological perspective. Conversely, if one rejects the computational approach to human mind in line with Searle's arguments, Strong AI within this framework would be strictly impossible from an Islamic theological perspective.

In essence, the question of whether God can create consciousness or mental states assuming the framework of computational AI is distinct from the question of whether He can create human-like mental states in such machines. The former assumes that computation is essential to the structure of both the human mind and artificial intelligence. If the human mind is wholly computational, then there is no logical barrier to the idea that God could create a similar mind within algorithmically structured machines. If the human mind is not entirely computational, then attempting to recreate such a mind in AI systems built solely on computational principles would be a logical impossibility—one that lies outside even the scope of Divine power. In other words, while God might be capable of instilling a mind into any kind of machine, the machine capable of hosting such a mind could not be purely computational in nature. However, whether human consciousness, subjective experience, or volition are fully captured by algorithms—or whether they involve dimensions that transcend any computational framework—is not a question that can be definitively resolved through the revealed sources of Islam. There is no transmitted evidence for either position. Therefore, from a theological lens, we cannot make any claims regarding the possibility of human-like minds emerging from AI systems built on computational architectures. Nonetheless, rational arguments may lead individuals to favor one position over another, and any such position would inevitably carry theological implications. These implications, however, are binding only for those who accept the corresponding rational premises, and not for others who reasonably reject them.

Conclusion

The human being has a complex nature having a spiritual/mental dimension in addition to the material elements. While contemporary neuroscience advances our understanding by linking mental states to specific neural activities, it predominantly adopts a naturalist and physicalist perspective, which conflicts with Islamic discourse.

Starting with an analysis of Descartes's substantial dualism and the ensuing mind–body problem, we explored various naturalist interpretations, such as eliminative and reductive physicalism, which face inherent challenges in addressing the hard problem of consciousness. Non-reductive physicalism, while acknowledging the distinct properties of mental and neural states, struggles with the issue of causal interaction between these properties.

We introduced two models from Islamic intellectual tradition that offer alternative interpretations: the occasionalist physicalist model of property dualism and the occasionalist model of substance dualism. Although one model identifies the spiritual/mental dimension of human beings with a separate substantial soul, the other explains it within the limits of physicalist atomism. Nonetheless, these models propose that God, as the sole cause, maintains perfect correlations between mental phenomena and their neural correlates without assuming causal interaction between them. Thus, these models align with neuroscientific findings while avoiding the pitfalls of naturalist perspectives. While both models are equally viable from a theological standpoint, the physicalist model may be rationally preferred over substance dualism on the basis of ontological simplicity.

Applying these models to AI, we discussed the implications for Strong AI. Since these occasionalist models do not reject the logical possibility that human-like mental states or consciousness may emerge in machines, it is also theologically possible that God could create it. Yet, the compatibility of such creation with God's habitual way of creation and the nomological regularities of the universe remains an open question as we do not know what the most fundamental nomological regularities of the universe are. Within the framework of the computational approach to AI—the prevailing paradigm in contemporary discourse—the question of whether all aspects of the human mind and consciousness can be fully captured by algorithmic structures falls beyond the scope of Islam's revealed sources and can instead be explored through rational inquiry alone.

Accordingly, we propose suspending making any theological claims on the possibility of AI attaining human-like consciousness—on one condition: if an individual adopts a position in this debate based on rational argumentation, that person should also accept the theological implications that follow from that position. However, these implications apply only to those who accept the underlying rational premises, and not to those who reasonably reject them.

Ultimately, these discussions underscore the need for a nuanced approach that integrates scientific advancements with philosophical and theological insights, offering a more comprehensive understanding of human nature and consciousness. Islamic occasionalism offers a neat solution to the mind–body problem, which is compatible with both neuroscientific research and Islamic theological doctrines.

References

Aftab, Macksood A. 2005. Primer on Islam and the Problem of Causation, Induction, and Skepticism. *Journal of Islamic Philosophy* 1 (1): 95–100. https://doi.org/10.5840/islamicphil2005114.

Aftab, Macksood. 2014. "How Muslim Theologians Saved Islamic Science." *Virtual Mosque*, December 18. https://www.virtualmosque.com/society/education-and-science/how-muslim-theologians-saved-islamic-science/.

Aftab, Macksood. 2015. Is Islam Committed to Dualism in the Context of the Problem of Free Will? *Journal of Cognition and Neuroethics* 3 (1): 1–12.

Al-Ash'arī. 1953. Vindication of the Science of Kalām (Risāla fī Istiḥsān al-Khawḍ fī ʿilm al-Kalām). In *The Theology of al-Ashʿarī*, ed. Richard J. McCarthy, 127–192. Beirut: Imprimerie Catholique.

Al-Ghazālī, Abu Hamid. 2010. *The Marvels of the Heart, Kitāb sharḥ ʿajāʾib al-qalb.* Translated by Walter James Skellie. Louisville, KY: Fons Vitae.

Al-Ghazālī, Abu Hamid. 2013. *Moderation in Belief*. Translated by Aladdin Yaqub. Chicago, IL: University of Chicago Press.

Al-Juwaynī. 1969. Al-Shāmil fī usūli'd-dīn. In *al-Nashshar, Suhayr Muhammad Mukhtār, and Faisal Badr Awn*, ed. Ali Sāmi. Munshaāt al-Maarif: Alexandria.

Al-Taftāzānī, Saʿd al-Dīn. 1950. *A Commentary on the Creed of Islam*. Translated by E.E. Elder. New York: Columbia University Press.

Alves, Pedro N., Chris Foulon, Vyacheslav Karolis, Danilo Bzdok, Daniel S. Margulies, Emmanuelle Volle, Michel Thiebaut, and de Schotten. 2019. An Improved Neuroanatomical Model of the Default-Mode Network Reconciles Previous Neuroimaging and Neuropathological Findings. *Communications Biology* 2:370. https://doi.org/10.1038/s42003-019-0611-3.

Anscombe, Elizabeth, and Peter Thomas Geach, eds. 1969. *Descartes: Philosophical Writings*. London: Nelson.

Barrash, Joseph, Donald T. Stuss, Nazan Aksan, Steven W Anderson, Robert D. Jones, Kenneth Manzel, and Daniel Tranel. 2018. "'Frontal Lobe Syndrome'? Subtypes of Acquired Personality Disturbances in Patients with Focal Brain Damage." *Cortex* 106: 65–80. https://doi.org/10.1016/j.cortex.2018.05.007.

Beauregard, James, Macksood Aftab, and Amna Sajid. 2016. Consciousness, Neuroimaging and Personhood. *Journal of Cognition and Neuroethics* 4 (1): 1–11. http://jcn.cognethic.org/jcnv4i1.pdf.

Bickle, John. 2003. Philosophy of Mind and the Neurosciences. In *The Blackwell Guide to Philosophy of Mind*, ed. Stephen P. Stich and Ted A. Warfield. Oxford: Blackwell Publishing.

Bulgen, Mehmet. 2024. *Kelamın Nesne Kuramı: Cüveyni'de Cisim, Hareket ve Nedensellik*. Istanbul: Marmara Üniversitesi İlahiyat Fakültesi Yayınları.

Chalmers, David J. 1996. The Conscious Mind. In *Search of a Theory of Conscious Experience*. New York: Oxford University Press.

Chalmers, David. 1995. Facing Up The Problem of Consciousness. *Journal of Consciousness Studies* 2 (3): 200–219.

Churchland, Patricia Smith. 1986. *Neurophilosophy: Toward A Unified Science of Mind Brain*. Cambridge, MA: Bradford Books, MIT Press.

Churchland, Paul. 1981. Eliminative Materialism and the Propositional Attitudes. *The Journal of Philosophy* 78 (2): 67–90.

Clayton, Philip. 2004. *Mind and Emergence: From Quantum to Consciousness*. New York: Oxford University Press.

Crick, Francis. 1995. *Astonishing Hypothesis: The Scientific Search for the Soul*. New York: Touchstone.

Descartes, René. 1984. Meditations. In *The Philosophical Writings of Descartes*, ed. John Cottingham, Robert Stoothoff, and Dugald Murdoch, vol. 2. Cambridge: Cambridge University Press.

Edlow, Brian L., Mark Olchanyi, Holly J. Freeman, Jian Li, Chiara Maffei, Samuel B. Snider, Lilla Zöllei, J. Eugenio Iglesias, Jean Augustinack, Yelena G. Bodie, Robin L. Haynes, Douglas N. Greve, Bram R. Diamond, Allison Stevens, Joseph T. Giacino, Christophe Destrieux, Andre van der Kouwe, Emery N. Brown, Rebecca D. Folkerth, Bruce Fischl, and Hannah C. Kinney. 2023. "Sustaining Wakefulness: Brainstem Connectivity in Human Consciousness." *bioRxiv* [Preprint]. https://doi.org/10.1101/2023.07.13.548265.

Eklund, Anders, Thomas E. Nichols, and Hans Knutsson. 2016. Cluster Failure: Why fMRI Inferences for Spatial Extent have Inflated False-Positive Rates. *Proceedings of the National Academy of Sciences (PNAS)* 113 (28): 7900–7905. https://doi.org/10.1073/pnas.1602413113.

Fakhry, Majid. 1958. *Islamic Occasionalism and Its Critique by Averroes and Aquinas*. London: George Allen and Unwin.

Fleminger, S. 2008. Long-Term Psychiatric Disorders after Traumatic Brain Injury. *European Journal of Anaesthesiology* 25:123–130. https://doi.org/10.1017/S0265021507003250.

Friedman, Garrett, Katherine W. Turk, and Andrew E. Budson. 2023. The Current of Consciousness: Neural Correlates and Clinical Aspects. *Current Neurology and Neuroscience Reports* 23 (7): 345–352. https://doi.org/10.1007/s11910-023-01276-0.

Garcia-Santibanez, R., and H. Sarva. 2015. Isolated Hyperreligiosity in a Patient with Temporal Lobe Epilepsy. *Case Reports in Neurological Medicine*. https://doi.org/10.1155/2015/235856.
Goff, Philip. 2019. *Galileo's Error: Foundations for a New Science of Consciousness*. London: Rider Books.
Halperin, Jonathan L., Glenn N. Levine, Sana M. Al-Khatib, et al. 2016. Further Evolution of the ACC/AHA Clinical Practice Guideline Recommendation Classification System: A Report of the American College of Cardiology/American Heart Association Task Force on Clinical Practice Guidelines. *Journal of the American College of Cardiology* 67 (13): 1572–1574. https://doi.org/10.1016/j.jacc.2015.09.001.
Hameroff, Stuart, and Roger Penrose. 2014. Consciousness in the Universe: A Review of the 'Orch OR' Theory. *Physics of Life Reviews* 11 (1): 39–78. https://doi.org/10.1016/j.plrev.2013.08.002.
Haugeland, John. 1985. *Artificial Intelligence: The Very Idea*. Cambridge, MA: MIT Press.
Fūrak, Ibn. 1987. In *Mujarrad Maqālāt al-Ashʿarī*, ed. Daniel Gimaret. Beirut: Dar el-Machreq Editeurs Sarl.
Ibn Ṭufayl, and Muḥammad ibn ʿAbd al-Malik. 1972. *Ibn Tufayl's Hayy Ibn Yaqzān: A Philosophical Tale*. Translated with Introduction and Notes by Lenn Evan Goodman. New York: Twayne.
Iqbal, Muhammad. 2013. *The Reconstruction of Religious Thought in Islam*. Stanford, CA: Stanford University Press.
Jackson, Frank. 1982. Epiphenomenal Qualia. *Philosophical Quarterly* 32:127–136.
Koch, Christof. 2014. "Is Consciousness Universal?" *Scientific American*, January 1. https://www.scientificamerican.com/article/is-consciousness-universal/.
Koenigs, Michael, Liane Young, Ralph Adolphs, Daniel Tranel, Fiery Cushman, Marc Hauser, and Antonio Damasio. 2007. Damage to the Prefrontal Cortex Increases Utilitarian Moral Judgements. *Nature* 46 (7138): 908–911. https://doi.org/10.1038/nature05631.
Koide-Majima, Naoko, Shinji Nishimoto, and Kei Majima. 2024. Mental Image Reconstruction from Human Brain Activity: Neural Decoding of Mental Imagery via Deep Neural Network-Based Bayesian Estimation. *Neural Networks* 170:349–363. https://doi.org/10.1016/j.neunet.2023.11.024.
Leibniz, Gottfried W. 1965. *Monadology and Other Philosophical Essays*. Translated by Paul Schrecker and Anne Martin Schrecker. Indianapolis, IN: Bobbs-Merrill Educational Publishing.
Lyu, Dian, James Robert Stieger, Cindy Xin, Eileen Ma, Zoe Lusk, Mariel Kalkach Aparicio, Katherine Werbaneth, Claire Megan Perry, Karl Deisseroth, Vivek Buch, and Josef Parvizi. 2023. Causal Evidence for the Processing

of Bodily Self in the Anterior Precuneus. *Neuron* 111 (16): 2502–2512.e4. https://doi.org/10.1016/j.neuron.2023.05.013.

Malik, Shoaib A., and Nazif Muhtaroglu. 2022. An Ashʿarī Perspective on Theology of Nature. *European Journal of Analytic Philosophy* 18 (2): 5–35. https://doi.org/10.31820/ejap.18.2.9.

Manzotti, Ricardo, and Paolo Moderato. 2013. Neuroscience: Dualism in Disguise. In *Contemporary Dualism: A Defense*, ed. Andrea Lavazza and Howard Robinson, 81–97. New York: Routledge.

McGinn, Colin. 1989. "Can We Solve the Mind–Body Problem?" *Mind* 98 (391): 349–366. https://doi.org/10.1093/mind/XCVIII.391.349.

McGinn, Colin. 2003. "What Constitutes the Mind–Body Problem?" *Philosophical Issues* 13, *Philosophy of Mind*: 148–162.

Nagel, Thomas. 1974. What Is It to Be Like a Bat? *The Philosophical Review* 83 (4): 435–450.

Nani, Andrea, Jordi Manuello, Lorenzo Mancuso, Donato Liloia, Tommaso Costa, and Franco Cauda. 2019. The Neural Correlates of Consciousness and Attention: Two Sister Processes of the Brain. *Frontiers in Neuroscience* 13:1169. https://doi.org/10.3389/fnins.2019.01169.

Navon, Mois. 2024. To Make a Mind—A Primer on Conscious Robots. *Theology and Science* 22 (1): 221–241. https://doi.org/10.1080/14746700.2023.2294530.

Owen, Adrian M., Martin R. Coleman, Melanie Boly, Matthew H. Davis, Steven Laureys, and John D. Pickard. 2006. Detecting Awareness in the Vegetative State. *Science* 313 (5792): 1402. https://doi.org/10.1126/science.1130197.

Owen, Adrian M., Martin R. Coleman, Melanie Boly, Matthew H. Davis, Steven Laureys, and John D. Pickard. 2007. Using Functional Magnetic Resonance Imaging to Detect Covert Awareness in the Vegetative State. *Archives of Neurology* 64 (8): 1098–1102. https://doi.org/10.1001/archneur.64.8.1098.

Putnam, Hilary. 1967. Psychological Predicates. In *Art, Mind, and Religion*, ed. William H. Capitan and David D. Merrill, 37–48. Pittsburgh, PA: University of Pittsburgh Press.

Qazi, Faisal, Don Fette, Syed S. Jafri, and Aasim I. Padela. 2018. Framing the Mind–Body Problem in Contemporary Neuroscientific and Sunni Islamic Theological Discourse. *The New Bioethics* 24 (2): 158–175. https://doi.org/10.1080/20502877.2018.1438835.

Rumelhart, David E., James L. McClelland, and PDP Research Group. 1986. *Parallel Distributed Processing: Explorations in the Microstructure of Cognition, Volumes 1 and 2*. Cambridge, MA: MIT Press.

Russell, Stuart J., and Peter Norvig. 2021. *Artificial Intelligence: A Modern Approach*. 4th ed. Upper Saddle River, NJ: Pearson.

Ruttan, L. A., M. A. Persinger, and S. Koren. 1990. Enhancement of Temporal Lobe-Related Experiences During Brief Exposures to MilliGauss Intensity Extremely Low Frequency Magnetic Fields. *Journal of Bioelectricity* 9 (1): 33–54. https://doi.org/10.3109/15368379009027758.

Schurger, Aaron, and Michael Graziano. 2022. Consciousness Explained or Described? *Neuroscience of Consciousness*. https://doi.org/10.1093/nc/niac001.

Searle, John R. 1980. Minds, Brains, Programs. *The Behavioral and Brain Sciences* 3:417–424.

Searle, John. 2007. Biological naturalism. In *The Blackwell Companion to Consciousness. Max Velmans and Susan Schneider*, 327–336. Chichester: Blackwell.

Swinburne, Richard. 2009. Substance Dualism. *Faith and Philosophy* 26 (5): 501–513. https://doi.org/10.5840/faithphil200926551.

Tang, Jerry, Amanda LeBel, Shailee Jain, and Alexander G. Huth. 2023. Semantic Reconstruction of Continuous Language from Non-Invasive Brain Recordings. *Nature Neuroscience* 26:858–866. https://doi.org/10.1038/s41593-023-01304-9.

Turing, Alan M. 1950. Computing Machinery and Intelligence. *Mind* 59 (236): 433–460. https://doi.org/10.1093/mind/LIX.236.433.

Vishe, Gal, Edden M. Gerber, Robert T. Knight, and Leon Y. Deouell. 2023. Distinct Ventral Stream and Prefrontal Cortex Representational Dynamics during Sustained Conscious Visual Perception. *Cell Reports* 42 (7): https://doi.org/10.1016/j.celrep.2023.112752.

Walter, Henrik. 2009. *Neurophilosophy of Free Will: From Libertarian Illusions to a Concept of Natural Autonomy*. Translated by Cynthia Klohr. Cambridge, MA: MIT Press.

Walter, Yoshija, and Thomas Koenig. 2023. The Induction of Religious Experiences and Temporal Lobe Activation: Neuronal Source Localization Using EEG Inverse Solutions. *Psych* 5 (4): 1191–1206. https://doi.org/10.3390/psych5040079.

Westphal, Jonathan. 2016. *The Mind–Body Problem*. Cambridge, MA: MIT Press.

Wolfson, Harry A. 1976. *The Philosophy of the Kalam*. Cambridge, MA: Harvard University Press.

Yang, Ying, and Jian-Zhi Wang. 2017. From Structure to Behavior in Basolateral Amygdala-Hippocampus Circuits. *Front Neural Circuits* 11:86. https://doi.org/10.3389/fncir.2017.00086.

Zarina, Rakhimberdina, Quentin Jodelet, Xin Liu, and Tsuyoshi Murata. 2021. Natural Image Reconstruction From fMRI Using Deep Learning: A Survey. *Frontiers in Neuroscience* 15. https://doi.org/10.3389/fnins.2021.795488.

CHAPTER 4

Human Origins and Evolution in Light of Revealed Truths: Theological Anthropology, Scholarly Consensus, and Similarity

Omar Qureshi

Introduction

The question of Islam's compatibility or incompatibility with evolution is an ongoing and lively area of inquiry (Guessoum 2011: 271–324; Hameed 2015).[1] No doubt this impacts Muslim scientists, biology teachers, theologians, and students. The diversity of approaches to this topic and the wide spectrum of answers to the many questions this topic brings do not make the task of the teacher, student, and scientists any easier. Although important aspects of the evidence will be addressed,

[1] Another indicator of this trend is Volume 21, Issue 4 of *Theology and Science* of 2023, which was entirely dedicated to the topic of Islam and evolution.

O. Qureshi (✉)
Yaqeen Institute for Islamic Research, Carrollton, TX, USA
e-mail: omar.qureshi@yaqeeninstitute.org

Logic and Theology, Qalam Seminary, Carrollton, TX, USA

© The Author(s), under exclusive license to Springer Nature Switzerland AG 2026

A. I. Padela (ed.), *Bridging Bioscience and Islam*, Islam & Science,
https://doi.org/10.1007/978-3-032-09996-9_4

the goal of this chapter is not to assess the scientific evidence of the various points of evolutionary theory. Instead, this chapter should be seen as an intervention on the topic of humans origins in light of revealed truths, with curricular and pedagogical concerns. Specifically, questions grounded in the Islamic tradition will be raised on the nature of evidence invoked by evolutionary biologists on common ancestry, as well as identifying tensions and alternative interpretations of human–ape similarities in light of hermeneutical methodologies and epistemological commitments of Muslim scholars. This inquiry is also informed by my own experience as a biology teacher in both Islamic and independent schools where Muslim and non-Muslim students shared their questions on this topic.

To address these educational concerns and emerging from the theological approach to topics in Islamic and science outlined by Qureshi et al. (2022), this chapter will conduct this inquiry taking the Sunnī theological and legal traditions as its point of departure. Building on Chapter 2 of this volume, I will provide the necessary theological framework and core creedal doctrines Muslim scientists, science educators, and students will need in order to fruitfully engage topics in the Islam and evolution conversation. These will serve as the theological and metaphysical commitments which inform any science and Islamic theology curriculums and pedagogical approaches at the various sites of Islamic education. I will outline the doctrine of scholarly consensus (*ijmāʿ*) and the role it plays in Sunnī epistemology by exploring the issue of Adamic exceptionalism.

In 2014, Paul Bruggink developed a scorecard of Christian theologians' and scientists' views on the compatibility of a historical Adam and current evolutionary accounts of human origins. The question of the scorecard was: "Were Adam and Eve historical?" He compiled the perspectives of 86 individuals, out of which 44 views were affirmative, while 5 believed in human creation but not that Adam and Eve were the first humans. On the other hand, 30 individuals' views were categorized as "No" (Bruggink 2015). These results indicate lively scholarly debate on the impact of evolutionary understandings on Christian theology. This debate has generated a spectrum of views ranging from reaffirming traditional Christian doctrine to modified views that Adam and Eve were not the first humans, to denial of the historicity of the two individuals.

Since then, the conversation around this question has continued and has taken new, interesting directions. In 2019, Joshua Swamidass published his monograph *The Genealogical Adam and Eve: The Surprising Science of Universal Ancestry*. Regarding Adam and Eve, Swamidass holds that they lived in the Middle East recently (possibly around 6,000–10,000 years ago), are the genealogical ancestors of all humans across the globe, and are both de novo created by God. Importantly, he also asserts that there were beings prior to and contemporaneous with Adam, and that the progeny of Adam and Eve interbred with these beings (Swamidass 2019). This rests on a difference between genealogical versus genetic ancestry. He explains that "genealogical ancestry concerns the connections in family trees, pedigrees, and genealogies. It is an 'ordinary' definition of ancestry," whereas "[g]enetic ancestry, in contrast, traces the history of small stretches of DNA. Genetics is not an ordinary definition of ancestry, an anachronism in theology and Scripture" (Swamidass 2019: 31). In other words, genealogical ancestry, which does not address or take into account DNA, is the understanding of ancestry in Scripture. This contrasts with genetic ancestry, which is based on DNA samples and which is the concern of evolutionary biologists. From this key distinction, Swamidass puts forth the hypothesis that Adam and Eve are our *genealogical ancestors* and not necessarily our genetic ancestors. In this way, he is able to maintain explicit teachings of the Old and New Testament while also maintaining the evolutionary account of human origins.

Also noteworthy is the contribution of William Lane Craig in 2021, titled *In Quest of the Historical Adam: A Biblical and Scientific Exploration*. Craig's engagement with evolutionary accounts of human origins is based on an understanding that a human is a living being anatomically similar to us, with certain cognitive capacities, such as the ability to grasp symbolic thought and plan for the future. He also adds that the term "human" includes being created in the image of God. He does not identify humans with Homo sapiens, considering it too narrow a category. Instead, Craig argues that, if Adam and Eve are to be the progenitors of mankind, then they had to be ancestral to both Homo sapiens and Neanderthals. Consequently, Craig situates Adam and Eve with the earliest ancestor, *Homo heidelbergensis* (Craig 2021). This work represents Craig's attempt to examine theological teachings on human origins in light of scientific evidence on the same question.

While there is this lively scholarly discourse in the Christian community of scientists and theologians on the Prophet Adam and human origins in light of the evolutionary account, this issue has only recently received critical attention in Muslim circles, and has been particularly treated in Anglophone Muslim discourse. This chapter focuses on this body of Muslim scholarship. Furthermore, the chapter presents an alternative understanding of morphological similarities, a key point of evidence for human evolution, based in Sunnī metaphysics. This discussion will allow for the raising questions about the logical and metaphysical commitments informing evolutionary biologists' usage of morphological similarities to assert common ancestry between humans and other living beings.

Theology-Engaged Science: The Classification of Creedal Doctrines

Perhaps one of the most important aspects of the Islamic theology for science educators and scientists to keep in mind is the nature of creedal doctrines in terms of their obligation for a Muslim to know and believe in. Understanding the status of each doctrine helps define the nature of our inquiries when engaging with issues at the intersection of Islamic theology and bioscience. It allows those engaged in such questions to focus their inquiries on areas which the Islamic theological tradition delimits, thereby avoiding jeopardizing their faith by the adoption of positions which contravene agreed-upon doctrine or fall outside the area of valid differences as determined by the Islamic theological tradition.

A systematic account of the various types of creedal doctrines and their status is provided by the Ashʿarī theologian Muḥammad Ṣāliḥ al-Ghursī (b. 1950).[2] Gathering scattered statements throughout various theological works, al-Ghursī formulated a systematic classification schema based on the creedal doctrine and whether knowledge, lack of knowledge, or rejection of it constitutes an obligation, a sin, disbelief, a heretical innovation, or is simply a valid difference of opinion (al-Ghursī 2018). His fivefold classification of creedal doctrines is as follows:

> **Category 1**. Doctrines for which a *lack of knowledge on the part of a person* entails unbelief (*kufr*). Doctrines that fall in this category are the existence of God, His oneness, and His essential attributes such as knowledge, power, and will.

[2] Muḥammad Ṣāliḥ al-Ghursī is a contemporary linguist, legal theorist, and theologian located in the city of Konya, Turkey. He is a jurist of the Shāfiʿī legal school and a theologian of the Ashʿarī theological school. In Konya, he established a college of Islamic sciences called *Madarsa al-Falāḥ*. To date, al-Ghursī has authored over 28 works in the fields of theology, law, legal theory, logic, Islamic thought, and Islamic education.

Category 2. Doctrines whose *denial* entails unbelief, and whose lack of knowledge is considered a sin. Examples of doctrines in this category are God being eternal and everything other than God is temporally originated. Al-Ghursī explains this category stating, "From this category are majority of creedal doctrines whose proof is conclusively established (*thubūt*) and signification (*dalāla*) [is also] conclusive from the Qurʾan and mass-transmitted Sunnah. Some of [these doctrines] may not be known to the laity" (al-Ghursī 2018: 95).

Category 3. Doctrines whose rejection constitutes a grave heretical innovation and a sin. Doctrines of this category are established by the plain sense of the Qurʾan and Sunnah, which the early Muslim community (*al-salaf*) have accepted. "To understand the texts in other than their plain sense, and to interpret them in a manner not aligned with the early Muslim community constitutes a grave heresy. It could be unbelief if the doctrine which the Muslim community has transmitted has scholarly consensus and is necessarily known to be of the religion" (al-Ghursī 2018: 95).

Category 4. Doctrines whose rejection constitutes a minor heretical innovation and a sin. The ignorance of such doctrines is excused. This category of doctrine is established by solitary Prophetic reports (*al-akhbār al-āḥād*).

Category 5. Doctrines based on conflicting evidences derived from the apparent sense of the Qurʾan and Sunnah. This category is where the independent interpretations of scholars (*ijtihād*) belong, and hence the area of scholarly difference of opinion. Most of the differences between the three Sunnī theological schools belong to this category.

This classification incorporates important hermeneutical and epistemological principles regarding specific types of religious knowledge such as the epistemological strength of the doctrine, the plain sense of transmitted texts, mass-transmitted (*mutawātir*) and solitary (*āḥāḍ*) reports, and the authority of the doctrines and interpretations held by the early Muslim community (*salaf*). Knowing which category a particular doctrine belongs to provides clarity for students, teachers, researchers, and theologians regarding the nature of its proof and the corresponding moral obligation in Islam, including whether the doctrine must be accepted or rejected and whether it must be known or may remain unknown. In addition to knowing authenticated creedal doctrines, determining which category it belongs to becomes one of the initial, critical steps of any inquiry.

Doctrinal Teachings About Humankind

Summarizing current understandings of human evolution among biologists, Kostas Kampourakis and Kevin McCain state:

> Human evolution is a fact. We have evolved from an ancestral species who was our common ancestor with the apes—who are in turn our closest relatives. The skeletal and behavioral similarities between humans and apes are easy to see. DNA analyses have also confirmed our relatedness with all apes, the chimpanzees being our closest relatives—in fact, they are genetically closer to us than they are to gorillas. Different estimates are available about when the evolutionary divergence between chimpanzees and humans took place, putting it at around 5 million years ago, whereas the common ancestor of humans and chimpanzees separated from the gorillas about 7 million years ago.
>
> (Kampourakis and McCain 2020: 107)

Two important points from this quote are: (1) the factual nature of human evolution from a common ancestor with apes; and (2) similarities—skeletal, behavioral, and DNA—are cited as evidence for this fact. For Muslims—whether theologians, scientists, or educators—the implications of the claims as well as the evidence brought forth are significant as they impact creedal doctrines—some that are known with the highest epistemological strength while others known presumptively. This section provides scientists and educators with the creedal doctrines related to Adam and human beings which Sunnī theologians have identified as essential for a Muslim to uphold as true and will apply al-Ghursī's classification framework of creedal doctrines laid out above. For Muslim scientists and science educators, these doctrines are theological and metaphysical commitments that form the point of departure when engaging with statements such as the fact of human evolution from ancestral species and the supporting evidence of similarity by Kampourakis and McCain. Hence, they should play a crucial role when developing science curricula as well as in curricula for other disciplines. Herein, I rely primarily on the works of the Ashʿarī theologian Saʿīd Ramaḍān al-Būṭī (d. 2013)[3] as he had a number of works and lectures engaging with Darwinian evolution throughout his career.

Let me first clarify in broad terms what the term "human" refers to in this chapter. In accordance with Islamic teachings anchored in the Qurʾan, Prophetic teachings, and scholarly consensus, the term human refers to Adam and Eve and their genealogical descendants. The Prophet Adam is a special creation of God and is the first human to be created. Adam and Eve were real, historical persons and the progenitors of all humans. The Prophet Adam was distinguished from the angels by his unique abilities, most notably his capacity to state the name of things—an indication of advanced cognitive faculties (see Q.2:31–33). A human person is a union of body and soul. In modern biological discourse, this distinction aligns with anatomical Homo sapiens.[4] Based on this understanding, I will now identify and explain key doctrinal points regarding the Prophet Adam and human beings which form the core theological commitments of a Muslim.

1. **The place and rank of humans in God's creation**. The doctrine is that humans are the most superior and noble of God's creation. al-Būṭī substantiates this point with two transmitted proofs, that is, Qurʾanic verses: (i) "*Verily We have honored the offspring of Adam; and borne them on land and sea; and provided them of good, wholesome things; and greatly favored them over a multitude of those We have created*" (Q.17:70); and (ii) "*And remind them when We said to the angels, 'Fall down in homage to Adam,' and they all bowed down to him as one; but not Satan, he utterly refused, waxed proud, and was wholly of the unbelievers*" (Q.2:34). While, based on testimonial knowledge, it is conclusively known that humans are superior to all of God's creation, that humans are superior to angels is presumptively known (*ẓannī*). This is due to the last part of (Q.17:70),

[3] Saʿīd Ramaḍān al-Būṭī, of Kurdish origin, was "a traditionally trained *ʿālim* securing a large international audience through mass produced printed publications and audiovisual media" (Pierret 2015). His training from his father and other Damascene scholars, "infused al-Būṭī with an unshakable attachment to what he considered to be the tenets of traditional Sunnī Islam, that is, Ashʿarī theology, and observance of the teachings of a particular *madhhab* (a school of law, in this case that of al-Shāfiʿī) in matters of *fiqh*" (Pierret 2015).

[4] This echoes the definition offered by Nazif Muhtaroğlu and Macksood Aftab in Chapter 3 on human nature in this volume.

"*and greatly favored them over a multitude of those We have created.*" While there is discussion among theologians on the superiority of humans and angels, the majority of Sunnī theologians maintain that elect humans, namely, prophets and saints are superior to the elect angels, namely, those mentioned in the Qurʾan, while the righteous believers are superior to the generality of angels (al-Būṭī 2020: 245–47). The doctrine of human superiority over all creation, with the exception of angels, is conclusively known and belongs to Category 3, while the doctrine of human superiority over angels is presumptively known and hence belongs to Category 5.

2. **Human composition and origin.** The doctrine is that humans are created from dust (*turāb*) and are "multiple from their origin, the first human, Adam" (al-Būṭī 2020: 249). According to al-Būṭī's analysis, this doctrine falls under Category 2 as "the demonstration of this doctrine is based solely on mass-transmitted true reports. This doctrine is not of the type of issues which relate to sensory knowledge which would make it rest on empirical and observational evidence" (al-Būṭī 2020: 249). As this is a historical question, this doctrine can only be known through testimonial knowledge, specifically true reports about the past. The question of how Adam descended to the earth, the location of his descent, how he and Eve produced their offspring do not fall under any of the five categories of belief. Consequently, such questions can be explored and answered, while there is no obligation to affirm or deny any specific answer. The last section of this chapter will explore the issues of human and Adamic exceptionalism (Malik 2021) in more detail as they relate to the doctrine of the first human.
3. **The perfection of the physical human form**. The doctrine is that humans, since their creation, have been created by God in the best and most beautiful of physical and spiritual forms. This applies to each and every member of the human species based on the verses such as "*Verily, We have created man in the best, most beautiful, conformation*" (Q.95:4). Discussions and knowledge on the origins of human life cannot be based on "empirical demonstrations and sensory observations as they are, in general, historical discussions" (al-Būṭī 2020: 250–51). Of significance is that the human form did not result from the process of evolution from prior living beings with imperfect forms eventually leading to the current form that humans have. Rather, humans possessed this form from the very

first human. This doctrine falls under Category 2 based on the epistemological strength of the proof; namely, it is doctrine whose denial entails unbelief, and whose lack of knowledge is considered a sin.

4. **Humans did not evolve from another species**. Based on these three doctrines, it is necessary for Muslims to believe that "humans have not gradually transformed, throughout their entire history, from any species-evolution" (al-Būṭī 2020: 252). In other words, grounded in the essentialist metaphysical commitment of Sunnī *kalām* theology, which holds that the whatness of things is determined by virtue of their essences, a human is a human by virtue of it possessing the essence "human." A logical and relevant corollary of this essentialist metaphysics is that differences between humans and other living things are differences due to different essences or kinds, not a difference of degree (al-Būṭī 2020: 252; Ṣabrī 1981: 2:280). This is, for al-Būṭī, a certain and conclusive doctrine based on the doctrines of human rank, human origins, and perfection of the human form.

To be sure, there are many other doctrines that relate to the Prophet Adam and human beings not investigated here. Most would belong to Categories 3–5 of al-Ghursī's classification framework. Yet, the doctrines mentioned herein directly relate to Islam and evolution and the question of human origin, and should form the foundations of curriculum and research inquiries.

Scholarly Consensus About the First Human

This section provides an account of scholarly consensus (*ijmāᶜ*), its epistemological role, definition, types, and the consequences of contravening it. This will be followed by a study of some key doctrines relevant to human origins which enjoy scholarly consensus. The doctrine of scholarly consensus plays a central epistemological role in the Sunnī tradition and constitutes a key aspect of transmitted proofs.[5] Only those aspects of scholarly consensus which are pertinent to our inquiry will be explored.[6]

Despite not addressing the specific proof texts and arguments for scholarly consensus, it is apropos to shed light on the role it plays in the Muslim community and understanding God's guidance for the community. There have been many ways Muslim scholars have sought to prove the authoritative nature of scholarly consensus. Some scholars have invoked Qurʾanic verses and statements of the Prophet Muhammad, while other scholars have used rational arguments in support of scholarly consensus. Two oft-quoted statements of the Prophet in support of scholarly consensus are: "What Muslims consider to be good is good in the view of God," (Abu Amina Elias 2012) and "My community shall never agree on a falsehood" (al-Tirmidhī, n.d.). What these statements indicate is that the Muslim community's scholarly consensus is seen as one of the ways God continues to guide the community on various matters after the passing of the Prophet. Additionally, the knowledge yielded by scholarly consensus is vouchsafed from error by virtue of the Divine grace it has received. Muslim scholars recognize that the mere consensus of scholars, logically speaking, does not determine the truth of a proposition. This extends to the consensus of any scholarly community, such as the consensus of physicians, lawyers, engineers, or psychologists. The consensus of any community of scholars can possibly be on something that is false, as is often observed on many issues such as the geocentric Ptolemaic model and the Miasma theory of disease. Both were widely held among the

[5] The probative status of scholarly consensus and its definition differs widely between Sunnīs and Shīʿa. Gleave summarizes the major difference stating:

> For Sunnis, *ijmāʿ* is a source of law, generally, justified through revelatory sources (such as the famous Prophetic report, "My community shall never agree upon an error"). Put simply, for the Imamis, *ijmāʿ* is a source because when all agree, the Imam must have been present amongst those forming the consensus. *ijmāʿ* has no probative force (*hujjiya*) in itself; it is only a useful mechanism for uncovering the opinion of the Imam.
>
> (Gleave 2018)

For a Zaydi account of consensus, see Rajani and Gleave (2023). For more on a Twelver Shīʿī account, see al-Ṭūsī's (d. 1067) *Al-ʿUdda fī Usūl al-Fiqh* (1997: 2:601–43) and ʿAllāmah al-Ḥillī's (d. 1325) *Mabādiʾ al-Wuṣūl Ilā ʿIlm al-Uṣūl* (2016).

[6] For studies on the historical developments and comparative nature of consensus, see Aḥmad Ḥasan (1978) and Mairaj Syed (2015).

scientific community across civilizations only to be abandoned later due to new evidence. However, while the scholarly consensus of the Muslim community logically has no bearing on the truth of a proposition, it is protected from error through Divine grace mentioned in the proof-texts discussed above. Hence, scholarly consensus holds great weight in the Sunnī theological tradition, which views it as one of the many ways God continually guides the Muslim community. The role of scholarly consensus in Islam has analogs in Orthodox Judaism, where rabbinical consensus is emphasized, as well as in Catholicism, which upholds patristic consensus and the magisterium. Though there are different conceptions and conditions for consensus, they all play a role in the cohesion of the faith community and God's continued guidance.

In light of the epistemic status of scholarly consensus in the Sunnī theological tradition, many Sunnī scholars have cautioned against careless and unverified claims of consensus on a particular issue. Aḥmad ibn Ḥanbal (d. 855), the eponymous founder of the Ḥanbalī legal and theological school, is reported to have said, "Whoever claims there is scholarly consensus [on an issue] is a liar." He said this, despite using scholarly consensus as a source in his legal and theological methodology, in the context of warning against unverified claims of scholarly consensus. In response to these loose claims of scholarly consensus, numerous reference works were composed documenting verified claims of scholarly consensus on various legal and theological doctrines. Some of the more recognized reference works in this area are Ibn al-Mundhir's (d. 877) *Kitāb al-Ijmāʿ*, Muḥammad ibn ʿAbd Allah al-Ṣairafī's (d. 942) *al-Ijmāʿ*, and Ibn Ḥazm's (d. 1064) *Marātib al-Ijmāʿ*. Thus, claims of scholarly consensus, like other transmitted proofs, need to be referenced and verified when invoked in scholarly discussions.

Like many sources of Islamic knowledge, Muslim scholars have provided and debated various definitions of scholarly consensus, yet a prevalent core definition is apparent. For example, the Ḥanafī legal theorist Muḥibballāh ibn ʿAbd al-Shakūr al-Bihārī (d. 1707) defined scholarly consensus as: "the agreement of every *mujtahid* of this community during any time period on a legal matter" (al-Lucknawī 2002: 2:260). This is nearly identical to al-Ghazālī's (d. 1111) definition in his *al-Mustaṣfā min ʿIlm al-Uṣūl* (al-Ghazālī, n.d.). The Shāfiʿī jurist Tāj al-Dīn al-Subkī (d. 1370) states that scholarly consensus is "the agreement of every *mujtahid* of the Muslim community, [occurring] after the death of [the Prophet] Muhammad (God bless him and grant him peace), during any time period, on any issue" (al-Maḥallī 2005: 2:131). Later jurists of the Mālikī and Ḥanbalī schools have provided the same definition as that of al-Subkī (see al-Mardāwī 2000: 4:1521–22; Zaidān 1993: 295–96).

Let us now look closely at the definition to develop a clearer understanding of scholarly consensus. This will serve as the foundation for the subsequent inquiries as they relate to the question of evolution and specifically human origins. The term *mujtahid* in the definition refers to a person who is qualified to directly and independently interpret (*ijtihād*) the sources of the revealed law (*sharīʿa*).[7] It is the judgment of *mujtahid*s which constitutes the content of scholarly consensus. Consequently, the judgments and opinions of non-*mujtahid*s, namely, the laity, are not considered when determining scholarly consensus.[8] Also excluded is an expert in any discipline of knowledge if they are not a *mujtahid* as the judgment of an individual is considered only by virtue of him being a *mujtahid* (see al-Maḥallī 2005; al-Mardāwī 2000: 4:1522). Non-Muslims are also excluded from consideration as "being Muslim is a condition for independent reasoning (*ijtihād*)" (al-Maḥallī 2005: 2:132; Zaidān 1993: 297). It is important to note that the content of scholarly consensus is the opinions of *mujtahid*s. And because they, by virtue of their qualifications, interpret the sources of revelation directly and independently, this ensures that each opinion is reached freely and, almost always, through debate with other *mujtahid*s. In other words, a *mujtahid* cannot simply

[7] The scholarly community of *mujtahids* in Sunnī Islam is structured in a similar way to the rabbinical community in Judaism, and unlike the centralized authority of Catholicism. In the Sunnī tradition, scholarly authority is decentralized, plural, and community-based. Formal qualifications, such as possessing a license to teach or to issue legal responsa from living scholars, determine who is recognized as part of this scholarly community, and these qualifications are internalized within Muslim societies so that scholars are readily identified by laypersons. This model ensures that authority is both rigorous and socially embedded, balancing technical expertise with communal recognition.

Moreover, the decentralization of authority has historically fostered a rich diversity of opinions across the different Sunnī legal and theological schools (*madhāhib*), while still maintaining a shared framework of legitimacy. The authority of the *mujtahid* rests not in a centralized institution but in the mastery of the sciences of the *sharīʿa*, the recognition of peers, and the acceptance of the wider community. This differs sharply from Catholicism, where the magisterium and papal authority provide a single, final arbiter of doctrine. Instead, the Sunnī system operates on a principle of distributed legitimacy, where scholarly consensus and independent juristic reasoning (*ijtihād*) play complementary roles in guiding the community.

[8] For more on the qualifications of a *mujtahid* and the role of *ijtihād*, see Hallaq (1984).

follow the opinion of another *mujtahid*, as they have the ability to interpret the sources of revealed law independently. This aspect of scholarly consensus brings to light that it is not formed merely by conforming to the majority opinion of *mujtahids* or what may have traditionally been held by the Muslim community. However, once consensus is reached, it becomes binding for the rest of the community to adhere to it. Further discussion on this point will follow.

The next aspect of scholarly consensus to explore is the areas to which it extends. While most Ḥanafī definitions limit the scope of scholarly consensus to legal or religious matters (*al-amr al-sharʿī*), the other schools extend the probative nature of scholarly consensus to any matter. This is understood as extending from religious matters, such as the permissibility of marriage, prayer, or the alms-tax, to worldly matters such as the military, warfare, and the public good, to linguistic matters, such as the meaning of certain particles or rules of grammar, theological issues, such as the cosmos being temporally originated, and any area beyond these (al-Maḥallī 2005: 2:154; al-Mardāwī 2000; Zaidān 1993). For each discipline, only the opinions of those at the level of independent reasoning and recognized experts are considered. It seems that the Ḥanafī have restricted the scope of scholarly consensus when compared to the other legal schools. Yet, many jurists consider this to be a semantical difference. The Ḥanafīs, according to these jurists, intend an unrestricted, comprehensive meaning of religion to include matters that are explicitly and primarily legal as well as those which the law requires and necessitates; thus a broad scope is intended. Obviously, the meaning of religion in the other definition is constrained to where it refers only to matters of the law (Farghalī 1971: 54).

While all Sunnī jurists consider scholarly consensus to be a source of religion, the question of the epistemological valence of it being a source of religion is disputed. Specifically, the question is, while scholarly consensus is a source of religion and has probative status, do we know this conclusively (*qaṭʿī*) or presumptively (*ẓannī*)? This question impacts the status of one who denies scholarly consensus as a source of law as well as denying or contravening a judgment on which there is scholarly consensus. All four Sunnī legal schools—Ḥanafī, Mālikī, Shāfiʿī, and Ḥanbalī—agree that, in addition to scholarly consensus being a source of religion, that it is conclusively so. Fakhr al-Dīn al-Rāzī (d. 1210) and Sayf al-Dīn al-Āmidī (d. 1233), both Shāfiʿī legal theorists, are noted to have held that it is only presumptively known to be a source of religion. However, the Shāfiʿī jurist al-Maḥallī (d. 1459) states that, "the correct position is that it is conclusively probative" (al-Maḥallī 2005: 2:156). This is also the authoritative position of the Ḥanafī, Mālikī, and Ḥanbalī schools of law (see, e.g., al-Lucknawī 2002: 2:262; al-Mardāwī 2000: 4:1530).

In light of this, Sunnī jurists have explored many subsidiary questions on which the conclusiveness of the probative nature of scholarly consensus impacts directly. The specific questions of denying scholarly consensus as a source of religion and contriving scholarly consensus are pertinent to our inquiry. On the question of denying scholarly consensus as a source of religion, Sunnī jurists hold that such a person remains a Muslim but is deemed a heretical innovator (*mubtadiʿ*) (see, e.g., Zaidān 1993: 306). The reason why such an individual is still considered a Muslim is due to the transmitted proofs that the obligation to follow scholarly consensus has not been established to him directly. Consequently, his denial does not entail denying that which the Prophet Muhammad has communicated, which is unbelief (*kufr*). In terms of a person denying a judgment on which there is scholarly consensus, Sunnī jurists consider the contravention of scholarly consensus to be unlawful, a major sin, and an act punishable by God in the afterlife—yet, the person remains a Muslim (al-Maḥallī 2005: 2:157; al-Mardāwī 2000: 4:1681–82; Zaidān 1993: 306).

The last issue on scholarly consensus and perhaps one of the more salient ones related to our inquiry is the epistemological role it plays in transmitted proofs and how they are interpreted. We will look at the hermeneutical role scholarly consensus plays when interpreting the Qurʾan and the traditions of the Prophet Muhammad. As an example, we will look at the Qurʾanic verse, "*Forbidden to you are your mothers, and your daughters*" (Q.4:23). This verse is the proof text forbidding a man to marry his mother or daughter. However, the legal status of marrying one's grandmother or granddaughter and how this verse relates to it is an issue much discussed by Qurʾanic exegetes and jurists. Specifically, the verse mentions mothers, daughters, and other categories of people with whom marriage is forbidden. The ruling on these categories is conclusive (*qaṭʿī*) and certain (*yaqīnī*). Now the terms mothers (*ummahāt*) and daughters (*banāt*) signify biological mothers and daughters in their literal senses and extend to grandmothers, great-grandmothers, granddaughters, and great-granddaughters in a figurative sense. On the basis of this usage, a *mujtahid* can argue that the ruling extends to grandmothers and granddaughters, yet this can only be held presumptively (*ẓannī*). This is because it is *possible* that what is intended by the terms *ummahāt* and *banāt* includes grandmothers and granddaughters. We do not know if this is intended with certainty as the lexical possibility that it does not extend to them exists. It is equally possible that grandmothers and granddaughters are not intended, and only the literal signification of the terms is intended. "Based on this possibility, a *mujtahid* can argue that a person marrying their grandmother or granddaughter is not forbidden" (Usmānī 1983: 52). Thus, both positions become valid legal interpretations as they are both known presumptively. Yet, there is scholarly consensus that marriage to grandmothers, great-grandmothers, granddaughters, and great-granddaughters is forbidden. In light of the scholarly consensus, the reading of the verse forbidding such marriages is epistemologically strengthened from being presumptive to conclusive and certain. Importantly, the other possible reading of the verse allowing for such marriages is rejected and given no consideration. A *mujtahid* can no longer claim this as a possible reading and a valid judgment (Usmānī 1983: 53). In light of all this, the verse may be rendered as "*Forbidden to you are your female ancestors, and your female descendants*," which takes into account the various lexical significations and scholarly consensus of the ruling.[9] What we see in this example is the epistemological impact scholarly consensus has on transmitted proofs. It raises the epistemic strength of textual signification from being presumptive to being conclusive. Critically, in doing so, it also excludes possible presumptive readings of texts based on logical or lexical possibilities.

The Origin of Mankind and Scholarly Consensus

Having provided an account of scholarly consensus and explored some of its salient and relevant aspects above, I now provide a summary of the treatment of scholarly consensus by contemporary writers on evolution, Islam, and the issue of human origin. I will restrict this summary to writers who situate themselves within the Sunnī tradition. It is important to keep in mind that my treatment here is descriptive; I do not assess the strength of the claims regarding scholarly consensus nor the conclusions reached by the authors. David Jalajel, a consultant with the Prince

[9] Some other Qurʾanic verses this applies to are Q.5:6 on the requirement of ablution for the ritual prayer, and Q.5:90 regarding the prohibition of alcohol. The same principles extend to how Prophetic traditions are interpreted and legal analogies.

Sultan Research Institute at King Saud University and holder of a PhD in Arabic and Islamic Studies from the University of the Western Cape, South Africa, has written two relevant works on Islam and evolution: *Islam and Biological Evolution: Exploring Classical Sources and Methodologies* (2009), and an influential paper on titled *Tawaqquf and Acceptance of Human Evolution* (2018). Jalajel situates his works in the Sunnī theological tradition which he considers to be "the 'mainstream' of Islamic thinking" (Jalajel 2018: 4). Although Jalajel does not discuss the methodological aspects of the Sunnī theological tradition as it relates to scholarly consensus, one finds him invoking scholarly consensus throughout these two writings. In *Islam and Biological Evolution*, scholarly consensus is invoked on such issues as *mutawātir ḥadīth*, solitary *ḥadīth* on matters of the unseen, interpretation of Q.2:65 and Q.7:166 of the Qurʾan, and the issue of causation (Jalajel 2009). Looking at *Tawaqquf and Acceptance of Human Evolution*, Jalajel explicitly cites scholarly consensus on two occasions: once on the topic of *tawaqquf*; and the other on the scientific consensus on evolution (Jalajel 2018: 8 and 26).[10] Jalajel's analysis mostly focuses on Qurʾanic verses and *ḥadīth* with quotes from exegetes. Other than the two occasions mentioned, scholarly consensus is not invoked in any textual analysis. However, Jalajel does make a statement that *could* be interpreted as scholarly consensus on certain doctrines related to Adam and his origins. He states:

> Classical Muslim scholars consider the following to be apparent from this body of textual evidence: Adam was created by God directly from earth. Both Adam and his wife were created by God without parents and all human beings living on the Earth today are direct descendants of these two people. These are the conclusions that have been reached by all classical commentators on these texts.
>
> (Jalajel 2018: 11)

The claim that all classical commentators could imply that there is scholarly consensus on these doctrines, yet it could also imply that these are the positions of the majority of *mujtahid*s.

[10] The scientific consensus which Jalajel invokes is not a consensus as defined in the Islamic tradition. Hence, it does not count as an instance of scholarly consensus in the technical sense.

I will now look at how scholarly consensus figures in the works of Shoaib Ahmed Malik, a lecturer in Science and Religion at the University of Edinburgh and author of a number of articles and a monograph on Islam, evolution, and human origins: *Islam and Evolution: Al-Ghazālī and the Modern Evolutionary Paradigm* (2021). In a co-authored article, "Does Belief in Human Evolution Entail Kufr (Disbelief)? Evaluating the Concerns of a Muslim Theologian" (2020), Malik and Elvira Kulieva mention scholarly consensus often, focusing on whether there is consensus on the compatibility or incompatibility of evolution with Islamic doctrine. After surveying various scholars' views, they conclude that "it would be false to state that every theologian is against the (potential or actual) compatibility of evolution and Islam" (Malik and Kulieva 2020: 655). Rather, the literature notes "several positions ranging from complete acceptance to complete rejection with partial acceptance in between" (Malik and Kulieva 2020: 654). Of note is whether the entirety of evolution with all its metaphysical commitments and conclusions are intended or if only some core aspects of evolution are intended. This remains unclear when Malik speaks about the compatibility question.[11] In Malik's monograph, *Islam and Evolution* (2021), he does not dwell much on the role of scholarly consensus in al-Ghazālī's thought. This limitation is acknowledged within the text (Malik 2021: 292). The absence, however, is rather peculiar considering the significant and authoritative place scholarly consensus has in al-Ghazālī's hermeneutics in the areas of doctrine and law. In fact, al-Ghazālī's understanding of the probative nature of scholarly consensus as a source of religion (al-Ghazālī, n.d.: 1:260), the ruling of someone who denies it as a source of religion, and of someone who contravenes it,[12] is no different than the above

[11] Saʿīd Ramaḍān al-Būṭī, for example, understands that there are various theories of evolution. In his lectures, he also speaks about Darwinism versus Neo-Darwinism. In his analysis of evolution, he engages specific issues agreeing or disagreeing with them. For example, one of the main points of incompatibility for al-Būṭī is the claim that human beings evolved from prior living beings. Al-Būṭī engages other points such as natural selection and survival of the fittest with different assessments of each point. The point here is that in the works of many scholars, they do not treat evolution as a referring to a fixed theory that is one dimensional. Al-Būṭī and other scholars do not adopt a zero-sum approach to this issue. Their engagement clearly indicates it is a term which refers to a theory composed of many different claims (al-Būṭī 2013, 2017).

[12] Al-Ghazālī states in his *al-Mustaṣfā*, "The ruling [on scholarly consensus] is that it is obligatory to follow and unlawful to contravene" (al-Ghazālī, n.d.: 1:289).

account of Sunnī jurists. Moreover, in his explanation of the order of the sources of law, al-Ghazālī declares that in every inquiry, ***prior*** to looking into the Qur᾽an and Sunnah, the ***mujtahid*** must ***first*** investigate if there is any scholarly consensus on the issue. If scholarly consensus exists on the issue, then the ***mujtahid*** need not look any further into the other sources. The reason for this hermeneutical approach is due to the epistemological strength that scholarly consensus holds in the category of transmitted proof. Al-Ghazālī asserts that scholarly consensus is epistemologically higher in strength than all types of transmitted texts of the Qur᾽an and Sunnah, for they both "can be abrogated while scholarly consensus cannot [possibly] be abrogated" (al-Ghazālī, n.d.: 2:711).

While, Malik acknowledges that scholarly consensus is a "valid form of theological proof and can hold strong weight for grounding one's creedal and jurisprudential positions in Sunnī Islam" (Malik 2021: 292), the complete absence of scholarly consensus in his list of the seven principles that constitute al-Ghazālī's hermeneutic framework is quite conspicuous (see Malik 2021: 296–97). Furthermore, with respect to specific doctrines as they relate to human evolution, Malik does not cite any doctrine on which there is scholarly consensus in this particular work.[13] From the above, we can conclude that both writers are aligned with the Sunnī tradition in that they acknowledge scholarly consensus as a source of religion and its probative nature in principle. Additionally, scholarly consensus appears in their analysis of some issues related to Islam and evolution at a general level in the sense of compatibility or incompatibility of Islam and evolution. Yet, we have seen that no specific human origin related theological doctrine that they explicitly attribute scholarly consensus to in their analysis.

Having analyzed current scholarship on the treatment of scholarly consensus on the question of human origins, our attention will now focus on Prophet Adam and whether or not he is the first of the human species. The question has been treated extensively by both Jalajel (2018) and

[13] The phrase "***min nafsin wahidatin***" translates to "[from] one soul" and is in the context of the origin of the human species. Malik does cite, without further comments or analysis, scholarly consensus on "***nafsin wahidatin***" in the Qur᾽an as referring to Adam in the context of Daud Batchelor's interpretation of the Qur᾽anic passage and Malik does not comment on the consensus (Malik 2021: 317).

Malik (2021, 2023),[14] and has subsequently generated much discussion among both Sunnī and Shīʿī academics, scholars (*ʿulamāʾ*), educators, and laity (see Andani 2022; Inloes 2023; Iqbal 2023; Kocsenda 2022; Mansur 2023). I will examine what role, if any, scholarly consensus has played in Jalajel and Malik's treatment of this question; subsequently, I will examine if any doctrine on which there is scholarly consensus related to this question actually exists, and finally the impact of the doctrines on the positions which Jalajel and Malik advocate.

First, it would be apropos to clarify the position of Adamic exceptionalism (AE). It helps to contrast this with what Malik terms Human exceptionalism (HE). HE is a position which maintains that "Adam is considered the first member of humankind, and, since he was created miraculously, he and his entire progeny are exempted from the process of evolution" (Malik 2021: 262). According to HE, all life undergoes the process of evolution with exception of Adam and his entire progeny. Adam being the first member of the human species excludes the possibility of any non-Adamic human beings. On the other hand, AE "agrees with HE insofar that Adam's miraculous creation is retained. However, rather than assuming that Adam is the first member of humankind, AE states that Adam and Eve are isolated miracles, and there is nothing in Islamic scripture that explicitly and clearly suggests that they were the first members of humankind" (Malik 2021: 262). AE extends the process of evolution to all life forms, including all human beings. The only exception is Adam and Eve, whose creation Jalajel and Malik consider to be miraculous. An implication of this position is that all prophets, including the Prophet Muhammad, came about via the process of evolution and share common ancestry with non-human life forms (Mansur 2023: 8). The theological implications of this have not been explored by Jalajel and Malik, as the Prophet Muhammad's physical form is considered by Muslim theologians, based on various scriptural indicants, to have been the most perfect of all human forms. Another important assumption with

[14] The focus on Jalajel and Malik's scholarship is due to their strong association with Adamic exceptionalism. While the term "Adamic Exceptionalism" is Malik's contribution, the idea has its genesis with Jalajel (2009) and is further articulated by him in a later article (2018). Malik builds on Jalajel's contribution and introduces the term in his monograph (2009). I consider both to be the best articulators and strongest advocates of this position in the Islam and evolution conversation. To be sure, Malik considers HE and AE as both viable options for Muslims to adopt, although AE makes more sense to him (personal communication, March 13, 2023).

AE is that the Prophet Adam may not be the first human, and it is possible that there were non-Adamic humans before or contemporaneous with the Prophet Adam. AE "uniquely divorces the connection between the *creation of Adam* with the *start of humanity*" and consequently opens up "the possibility of all humans today having a lineage that predates Adam" all the way back to the origin of life (Malik 2021: 135).

Having identified Adamic exceptionalism, we can now analyze the reasoning behind this position. Since Malik relies on Jalajel's account, I will look at his account in *Tawaqquf and Acceptance of Human Evolution* and add any elaborations or additions from Malik as necessary. Jalajel's analysis consists of Qur'anic verses, Prophetic traditions, and commentary from Qur'anic exegetes. He summarizes his methodology stating: "The first step in answering this question is to look at the texts of the Qur'an and Sunnah that address the creation of Adam and Eve, particularly in the context of how classical scholars understood them" (Jalajel 2018: 9). In the two sections titled "The Textual and Interpretive Background" (Jalajel 2018: 9) and "Theological Implications for Human Evolution" (Jalajel 2018: 13), he looks at ten verses from the Qur'an along with five Prophetic traditions, with commentary on the verses from exegetes such as al-Ṭabarī, Ibn Kathīr, al-Rāzī, Ibn Taymiyya, and others. In looking at the sections of Jalajel's analysis which address the Prophet Adam's creation, based on Q.4:1 and Q.49:13, he states that human beings are descended from a single couple. Based on a Prophetic tradition, the male member of the couple is identified as the Prophet Adam. Next, he establishes the idea that all humans living on the Earth today are descendants of the Prophet Adam, stating that this is repeated throughout the Qur'an and cites, as an example, Q.7:27 and two Prophetic traditions. This analysis leads Jalajel to conclude the following doctrinal points: "Adam was created by God directly from earth. Both Adam and his wife were created by God without parents and all human beings living on the Earth today are direct descendants of these two people" (Jalajel 2018: 11). Although he further states that these points have "never been called into question by pre-modern Sunni scholarship" (Jalajel 2018: 12), it is not clear if he intends that this constitutes a scholarly consensus or not. Also, his reading of all the texts cited in a restricted sense as referring to human beings *living today* rather than all humans requires hermeneutical grounding.

Building on these conclusions, Jalajel carries on his analyses to answer other questions. In answer to the topic of other living beings at the time of the Prophet Adam's creation, he states that the "same texts are silent on what living organisms existed on Earth at the time. Therefore, nothing can be assumed about that on a theological level, and Muslims are not required to have any specific beliefs about it" (Jalajel 2018: 14). Malik further elaborates this position stating, "the Qurʾān doesn't affirm nor negate the idea of there being already-existing humans on earth. In other words, the Qurʾān is silent on there being humans on earth before Adam's descent" (Malik 2021: 135). If there were humans before the Prophet Adam, it would mean, according to AE, that: (1) these humans came about through the process of evolution; and (2) the Prophet Adam is not the father of all humans; he is only the father of a subset of human beings. That the Prophet Adam is the first human is a doctrine "(m)ost Muslims assume," according to Malik, while not explaining why they would assume such a doctrine (Malik 2021: 134, emphasis added). Jalajel's and Malik's account of AE is composed mostly on texts from the Qurʾan and Prophetic traditions. They leverage the works of Qurʾanic exegetes and commentaries on Prophetic traditions where they deem it necessary. Scholarly consensus is not invoked either in affirming or negating their interpretations of the Qurʾan and Prophetic traditions on doctrines related to the Prophet Adam and human beings.

At this point, we should ask if there are any such doctrines on which there is scholarly consensus that impact AE and the assumptions it rests on. There are some doctrines which have received scholarly consensus that relate to the Prophet Adam and human origins. One such doctrine is that God created the Prophet Adam directly with His hands. "All who are associated with Islam, Sunnis and heretics, hold this doctrine. Even the People of the Book hold this doctrine" (Ibn Taymiyya 1995: 4:365). Another doctrine is on the prophethood of Adam. While the Qurʾan does not use the word prophet (*nabī*) in regard to the Prophet Adam, there is scholarly consensus on his prophethood, though scholars differed on the question of the Prophet Adam being a messenger (*rasūl*) (Ibn al-Qaṭṭān 2011: 40; al-Taftāzānī 2020).[15]

Of specific interest here is scholarly consensus on the Prophet Adam being the first human, the father of each and every human being. On this doctrine, ʿAlī Ibn al-Qaṭṭān's (d. 1230)[16] reports relevant material

[15] Among the category of prophets are those who were sent with a specialized message, for example, a revelation, leading to the difference in terms.

[16] ʿAlī Ibn al-Qaṭṭān was a judge and *ḥadīth* scholar specializing in narrator criticism. Born in Fez, he moved to Marrakech, where he gained prominence as a scholar during the

in his catalog of creedal and legal doctrines on which there is scholarly consensus, titled *The Sufficiency: On Issues Regarding Scholarly Consensus* (*al-Iqnāʿ fī Masāʾil al-Ijmāʿ*) (Ibn al-Qaṭṭān 2011). In the section on creedal doctrines which scholarly consensus exists, Ibn al-Qaṭṭān documents the following doctrine: "Muslims are in agreement that God, the Exalted, distinguished Adam (God bless him) [from other prophets] in that He created him with His hand, commanded the Angels to prostrate to him, and made him the father of mankind, each and every one of them" (*bi an khalaqahu bi yadihi wa asjada lahu malāʾikatahu wa jaʿalahu aba al-bashar kullihim*) (Ibn al-Qaṭṭān 2011: 41). Three points need to be made in order understand this doctrine as it relates to HE and AE. First, the term *al-bashar* contains the determinative particle *al* or the determinative *alif* and *lām*. It is used to denote the species (*jins*), the nature or essence of the thing, generality (*ʿumūm*), and the totality of individuals (*istighrāq*) belonging to that species. An indication that the determinative *alif* and *lām* denotes the essence of the thing is that it cannot be replaced with the word "every" (*kull*), while if it denotes the totality of individuals in the genus, which is its literal usage, it can be replaced with the word "every" (*kull*) (al-Daqar 1986: 72). Here, the determinative particle in *al-bashar* can be replaced with the word "every" and can thus be read as "every human" (*kullI bashar*). Thus, the phrase would have to be interpreted as "and making him the father of every human." Here, father is understood to mean progenitor. Second, Ibn al-Qaṭṭān ended this sentence with "each and every one of them, all of them" (*kullhim*). In this sentence, *kullihim* serves as an emphatic, thus signifying "each and every one of them." In doing so, what is intended is that not a single, particular human thing to which the term *al-bashar* is predicated is excluded from Adam being the father. Third, linguists, lexicographers, and logicians in the Islamic tradition hold that the term *bashar* and its synonym *insān* are concepts which, when predicated of their particulars, are predicated univocally (*muwāṭaʾtan*) as opposed to being predicated analogically

rule of the Almohads (1130–1269). He composed numerous works in the areas of law, legal theory, and *ḥadīth* sciences. Jurists and theologians consider his work on scholarly consensus, *al-Iqnāʿ fī Masāʾil al-Ijmāʿ*, a mainstay reference in the field. Muslim scholars have recognized Ibn al-Qaṭṭān for his exhaustive use of reliable sources documenting scholarly consensus and the scope of topics on which there is scholarly consensus (Ibn al-Qaṭṭān 2011: 22).

(*mushakkikan*) or homonymously (*mutarādifan*).[17] Univocal predication of concepts to their particulars entails that every particular equally instantiates the concept such as the term "animal" used of human beings and cows. Applied to the term "human" (*bashar*, *insān*), this would mean that all particular human beings to which the term is predicated are all equally human.

The implication of these three points on AE above, especially the univocal predication of the term human, will now be addressed. This scholarly consensus entails that all particulars on which the term human is predicated are all equal in being human and the Prophet Adam is the father (i.e. progenitor) of that particular on which human is predicated (excepting himself). Now, if there were humans who were prior to the Prophet Adam or contemporaries of him, according to this scholarly consensus, they would have to be related to the Prophet Adam. Yet, a human being that was prior to the Prophet Adam or his contemporary would mean that the Prophet Adam could not be the father of that human, which would entail a contravention of this scholarly consensus. It may be possible to assume that Qurʾanic verses *could* be interpreted as allowing for the possibility of non-Adamic humans, that is, that there is no definitive signification for the contrary view. This would entail that both readings are presumptive. Yet, taking into account the transmitted proof of scholarly consensus, the reading which allows for the possibility of non-Adamic humans is excluded as a valid reading. The logical implication of this also excludes the possibility of a reading which restricts the verses as referring solely to all human beings living on the Earth today as being the direct descendants of Adam. Importantly, it is argued that the Prophet Adam being the first human, the father/progenitor of all (subsequent) humans, and the impossibility of non-Adamic humans takes on the epistemic value of being conclusive due to this scholarly consensus. Based on the foregoing, the claim that "there is nothing in Islamic scripture that explicitly and clearly suggests that they were the first members of humankind" (Malik 2021: 262) must be re-examined and revised. It seems that the term Islamic scripture refers to the Qurʾan and Sunnah in this case. While that may be apropos, scholarly consensus regarding the interpretation of the Qurʾan and Sunnah is overlooked. Scholarly consensus not only constitutes an additional source

[17] For more on types of predication, see al-Kātibī (2024), and for a more detailed discussion on Ashʿarī conceptions of the term human (*insān*), see Shihadeh (2012).

that Muslims are required to take into consideration when propounding doctrines, it also informs scriptural interpretation and thus cannot be ignored when conducting a full analysis of scripture.

What Does Similarity Signify? The Logical Leap to Common Ancestry

Having delved into the Islamic doctrines and scholarly sources related to human origin and the Prophet Adam, I now turn to bioscience. Evidence for evolution comes from different fields in biology, such as the fossil record, biogeography, and embryology. As it happens, one of the most frequently cited evidence is the phenomenon of similarity; it is the similarities between living things at the molecular biological level, such as amino acids and DNA, the similarities at the embryological level, or the similarities between morphological structures which are drawn upon to support the premise of common ancestry. Indeed, evolutionary biologists consider the existence of similarities between living things as the strongest proof for evolution and common descent of all living things.

Yet, similarities between living things has been something long observed in human history. Regarding the similarities between humans and apes specifically, the poet Ennius (*c.*239–169 BC), the father of Roman literature and author of the *Annales*, exclaimed, "how like us is that ugly brute, the ape! (*simia quam similis, turpissima bestia, nobis!*) but the two differ in habits" (Ennius 2018: 284). Here, it is likely the physical similarities between humans and apes is intended, while the distinction is made between their behavior and habits. This phenomenon has been observed and documented in the Islamic tradition as well. Numerous Muslim scholars have documented their understanding of the similarities between humans and apes. For our purpose at hand, I will look closely at entry in *Ḥayāt al-ḥayawān al-kubrā*, a zoological lexicon by the Cairene Shāfiʿī jurist and traditionist, Muḥammad ibn Mūsa al-Damīrī (d. 1405).[18] His entry on apes states:

> The ape[19] (*al-qird*). An animal that is well-known… It is an animal that is ugly, witty, intelligent, quick to understand things, and can learn the crafts.

[18] For more al-Damīrī, his work and Arabic zoology before him, see Somogyi (1950).

[19] Or monkey; the Arabic term *al-qird* extends to both.

> It is mentioned that a Nubian king gifted al-Mutawakkil a monkey which could sew and another which could mold gold. The Yemenites would teach apes to take care of their needs such that a butcher or a grocer will teach an ape to watch over their shop [while the owner left] until the owner would return. They would also teach them how to steal, whereupon the [monkey] would steal [for them]...
> This animal strongly resembles (*shabīhun*) humans in most of its qualities. Apes laugh, become joyous, sit on their posteriors, imitate, grasp objects with their hands. They have separated fingers up to the fingertips and nails. They can be trained and taught to do things, find intimacy with humans, walk quadrupedally normally, while walking upright bipedally on a few occasions. *He is virtually human* (*huwa ka-l insān*). An ape will see that he gets married and will be protectively jealous of his female [counterparts] and these are two qualities which humans take pride in.
> (al-Damīrī 2005: 3:476, *emphasis added*)

Al-Damīrī's rich entry contains several observations on the ape that deserve mention. He is eminently aware of the similarities between the ape and humans. These similarities start with the nature and degree of their intelligence. Apes have wit, a natural intelligence, and importantly can learn crafts. They were a natural part of human society as caretakers and thieves. Al-Damīrī continues to document how apes physically resemble humans in their gait, manner of sitting, and social aspects such as laughter, joy, and imitation. Beyond this, there are psychological similarities in that apes marry and get jealous over their female counterparts. al-Damīrī's account does not limit similarity to morphological, rather he extends it to the social, psychological, and intellectual. For al-Damīrī, the similarities are so striking that the ape is virtually a human.

Of interest is what is missing here in al-Damīrī entry. Despite the degree of similarities between humans and apes, al-Damīrī does not conclude from his observations that there is a genealogical relationship between humans and apes. The conclusion that the observed similarities, or even some of them, are due to common ancestry, as Elliot Sober (2011) writes, is one of the two central concepts in Darwin's theory; it is, however, absent from al-Damīrī's analysis. For the purposes of scholarly engagement, it is thus important to note that there is no logical necessity between the observation and the posited conclusion.

So, how did he and other scholars, particularly Sunnī *kalām* theologians, understand similarities and their causal origins? In other words, how do Sunnī *kalām* theologians conceive of similarities and how do they account for how these similarities come about.[20] I will turn to answer this question in what follows.

Sunnī *kalām* theologians hold that similarity or commonality (*ishtirāk*) between two things is also seen as two things being "one" (*waḥda*) in as much as it relates to the thing shared. Based on a substance-accident distinction along with an essentialist metaphysics, similarities are classified as either being related to the thing's essence (per se; *dhātī*), or accidentally related (per accidens; *ʿaraḍī*) to the thing. Essential similarities can be when two things or more have the same species (*nawʿ*) in common. For example, John and Fred are similar due to them both having the essence "human" in common. Commonality in essence is termed (1) *mumāthila*, the two things are analogous due to likeness. Essential similarity can also be when two things are (2) congeneric or have the same genus in common (*mujānisa*). This can be seen in the case of a horse, a dog, and a human, which all have in common the genus of "animal." Accidental similarities are in quantity (*kammīyya*), such as two animals having the same height or weight. This similarity is called (3) equality or *musāwa*, the two things have equal quantities of the same accident. Accidental similarity in quality (*kayfīyya*), such as a horse, dog, and human are all brown, or have the same skin texture. This kind of commonality is termed (4) resemblance or *mushābaha*. They can also be similarities in relation (*nisba*), such as the relation of the soul to the body is similar to the relation of a king to a city. Such a similarity is called (5) analogous or relational similarity (*munāsaba*). Similarity in form is when two things have shape in common, such as drops of water and the earth having sphericity in common. Such similarities are called (6) conformity or *mushākala*. Similarities in points or edges are called (7) correspondence (*muṭābaqa*), whereas similarities in position (*waḍʿ*), such as different parts having the same position in relation to each other, is termed (8) parallelism (*muwāzā*) (al-Taftāzānī 1887: 1:137–138).

Similarities, in this essentialist framework, are either by virtue of the essence of the thing or accidental. In light of this, humans and apes will have similarities by virtue of both belonging to the same genus in that they are both animals. Examples of this would be that they are both sentient beings, have mass, volition, and other qualities that all animals exhibit by virtue of their essence "animal." These similarities are

[20] This account is by no means of the Sunnī *kalām* schools only. It is also largely held by the Peripatetic school (Ibn Sīnā 1985) and by Shīʿī theologians (al-Ḥillī 1988).

of the *mujānisa* category. While some of the similarities identified by al-Damīrī fall into this category, the other similarities he identified do not belong to the categories of *mujānisa* or *mumāthila* as they are neither by virtue of the genus "animal" nor of the essence "human." Rather, they would fall under those similarities that are accidental; many of those listed such as the ability to learn crafts and being jealous fall under the *mushābaha* category of similarities, meaning that they exist neither by virtue of the essence "human" nor the essence "ape" but are accidental qualities to them. Hence, when al-Damīrī speaks of similarity between apes and humans, he uses the term resembles (*shabīhun*). This implies that the similarities are not in the category of essential similarities and accidental in nature.

Describing the current attitudes of evolutionary biologists toward similar traits between living things, philosopher of biology, Elliot Sober notes that: "For most evolutionists, the similarities that different species share make it obvious that they have common ancestors, and there is no reason to puzzle further over the question. The kind of genealogical question that attracts far more attention in evolutionary biology concerns how various species are related to each other, not whether they are" (Sober 2008: 264). Yet, it is important to note that evolutionary biologists do not classify all similarities between living things as being due to common ancestry. Some traits are, while others are not. Similar traits due to common ancestry are described as homologous traits. Examples of such traits according to evolutionary biologists are the forelimbs of vertebrates. "A homoplasy," Sober notes, "in contrast, is a similarity that is not due to inheritance from a common ancestor but instead arose because independent origination events occurred in separate lineages; the wings of birds and bats are an example. So defined, the concept of homology already has built into it the claim of common ancestry" (Sober 2008: 286). The alternative hypothesis to common ancestry to consider when it comes to explaining how a similar trait arose is that it is due to separate ancestry. For an evolutionary biologist, both are possibilities, and the question becomes which of the two has stronger evidence in favor for it. There are many things to be appreciated in Sober's approach to science, but what is significance for our purpose is the probabilistic epistemology of scientific findings that he highlights. While some empirical findings may be known with certainty, most scientific hypotheses, which are inferences based on observations, are affirmed probabilistically or presumptively. This is also the case with common ancestry as understood by evolutionary biologists (Sober 2024: 77–108). Sunnī *kalām* theologians would agree with Sober that certainty is not a requirement for inquiries to take place in the natural sciences. And while this is not significant within the field of natural science, it is very significant when interpreting revelatory transmitted knowledge in light of current positions of evolutionary biologists as explained in Chapter 2 of this book.

In light of the metaphysical commitments of the Sunnī *kalām* theologians outlined above, it seems that affirming essences for living things generates some related observations for the common ancestry hypothesis. One observation is that "[s]imilarity, ergo common ancestry" (Sober 2008: 265) is taken for granted and this implies a logical problem of *petitio principii*. Holding that similarity implies a genealogical relationship also seems like a category mistake in that similarities are conceived as a genealogical relationship, whether in a one-to-one relationship or otherwise. There are many cases of actual genealogical relationship resulting in vast amount of differences and also of remote genealogical relationship resulting in a vast amount of similarities. Another observation is that the evolutionary approach leaves much to be desired in that the metaphysical status of each trait is not investigated, thereby resulting in a flattening of each trait and a collapse of any distinctions between traits.

More importantly, in the evolutionary framework, the essence or "whatness" of a living thing is now solely determined by its genealogical ancestors. And since all living things have a common ancestor stemming from a universal common ancestor, all living things differ only in degrees and the idea of a living thing having an essence is jettisoned and plays no explanatory role as to the "whatness" of a living thing and consequently why it possesses its traits. This is completely in line with Darwin's metaphysics as he stated that, "the difference in mind between man and the higher animals, great as it is, is certainly one of degree and not of kind" (Darwin 1981: 105). This is also confirmed by contemporary evolutionary biologists Roger Lewin and Robert Foley as they state, confirming Darwin's thesis: "Differences between humans and other primates are ones of degree, not of kind" and although the uniqueness of human beings is acknowledged, it is "the product of evolutionary processes" (Foley and Lewin 2004: 104). Yet, this sits in great tension with the essentialist metaphysics of Sunnī *kalām*. For them, living things have an essence which constitutes their "whatness" and along with the substance-accident distinction, the traits observed, similar or dissimilar, are understood in relation to that essence. Summarizing the impact of this on humans and apes, the Ottoman Ashᶜarī theologian, Muṣṭafā Ṣabrī (d. 1954) said, "It is faulty logic to conclude that based solely on resemblance (*tashābuh*) in form between human body parts and any other animal that the human species and apes are descended from a common ancestor. For the difference between humans and other living animals is a difference resulting from difference in essences and are not differences in degree" (Ṣabrī 1981: 2:280). From the point of view of theologians like Ṣabrī, the question of common ancestry is a philosophical or metaphysical question, not an empirical question.

Conclusion

Conducting an inquiry from the starting points of the doctrinal classification scheme, the doctrine of scholarly consensus (*ijmāᶜ*), and the metaphysics of similarity exemplifies what Alasdair MacIntyre (1988) called tradition-constituted reasoning. MacIntyre argues that rational inquiry is always embedded within historically developed traditions, which define the standards of reasoning, justification, and metaphysical commitments informing any rational inquiry within that tradition. Within the Sunnī legal and *kalām* tradition, the classification of theological doctrines, the authority of scholarly consensus, and the metaphysical foundations of similarity (*tashābuh*) form key elements of such tradition-constituted reasoning. Sunnī inquiries into human origins—whether in relation to creation, the nature of humanity, or the reconciliation of transmitted proofs and rational proofs—are thus embedded within an established epistemological framework informed by the theological and metaphysical commitments of the tradition.

In the Sunnī tradition, the doctrine of scholarly consensus plays a central role, serving as a hermeneutical guide for interpreting transmitted knowledge. When evaluating whether Adamic exceptionalism is a hermeneutically possible reading of the Qurʾan and *ḥadīth*, this chapter demonstrates that the scholarly consensus regarding the Prophet Adam as the first human and progenitor of the rest of humankind precludes Adam exceptionalism an interpretive possibility. This implies that interpretations of data invoked on the question of human origins and inferences made from such data must account for this commitment, exemplifying MacIntyre's tradition-constituted reasoning. Along with this understanding of the Qurʾan and *ḥadīth*, the interpretation of human evolution and its similarities with other species would shape how such data is explained and understood. The overall aim of this chapter was to incorporate the role of scholarly consensus into the epistemological framework presented in Chapter 2, and to engage with current Muslim scholarship on the issue of human origins. Finally, I critically examined the role that similarity between living things plays in evolutionary biology, supporting one of its central ideas: common ancestry. I identified several logical difficulties and tensions that homology creates with the Sunnī *kalām* tradition, and offered alternative ways of understanding these similarities with different metaphysical commitments.

References

Abu Amina Elias. 2012. Musnad Ahmad no. 3600. https://www.abuaminaelias.com/dailyhadithonline/2012/12/11/urf-sahabah-custom-good-to-allah/ [Last accessed on 22 Dec, 2025].

Andani, Khalil. 2022. "Evolving Creation: An Ismaili Muslim Interpretation of Evolution." *Zygon: Journal of Religion and Science* 57 (2): 443–66. https://doi.org/10.1111/zygo.12774.

Bruggink, Paul. 2015. "Were Adam and Eve Historical? Yes, Maybe, No: A Brief Survey of Views." Last modified August 29, 2015. https://cdn.ymaws.com/network.asa3.org/resource/dynamic/forums/20150829_143039_10526.pdf.

Al-Būṭī, Muḥammad Saʿīd Ramaḍān. 2013. "*Naqḍ al-Naẓarīyya al-Taṭawwur.*" YouTube video, May 13. www.youtube.com/watch?v=byWBQT6zmuw.

Al-Būṭī, Muḥammad Saʿīd Ramaḍān. 2017. "*Kubrā al-Yaʿqīnīyyāt al-Kawnīyya*—25." YouTube video, December 27. www.youtube.com/watch?v=p5OIAbEW4aE&list=PL01kYZIehH-ps7QqYS0t0AFt5BWcRJwEP&index=31.

Al-Būṭī, Muḥammad Saʿīd Ramaḍān. 2020. *Kubrā al-Yaʿqīnīyyāt al-Kawnīyya*. Damascus: Dār al-Fikr.

Craig, William Lane. 2021. *In Quest of the Historical Adam: A Biblical and Scientific Exploration*. Grand Rapids, MI: William B. Eerdmans Publishing Co.

Al-Damīrī, Muḥammad ibn Mūsa. 2005. *Ḥayāt al-ḥayawān al-kubrā*. 4 vols. Edited by Ibrāhīm Ṣāliḥ. Damascus: Dār al-Bashāʾir.

Al-Daqar, ʿAbd al-Ghanī. 1986. *Muʿjam al-Qawāʿid al-ʿArabīyya fī al-Naḥw wa-l Taṣrīf*. Damascus: Dār al-Qalam.

Darwin, Charles. 1981. *The Descent of Man, and Selection in Relation to Sex*. Princeton, NJ: Princeton University Press.

Ennius. 2018. *Fragmentary Republican Latin, Volume II. Ennius: Dramatic Fragments. Minor Works*. Edited and translated by Sander M. Goldberg and Gesine Manuwald. Cambridge, MA: Harvard University Press.

Foley, Robert, and Roger Lewin. 2004. *Principles of Human Evolution*. 2nd ed. Malden, MA: Blackwell.

Farghalī, Muḥammad Maḥmūd. 1971. *Hujjīya al-Ijmāʿ*. Cairo: Dār al-Kitāb al-Jāmiʿī.

Al-Ghazālī, Muḥammad ibn Muḥammad. n.d. *Al-Mustaṣfā min ʿIlm al-Uṣūl*, 2 vols. Edited by Aḥmad Zakī Ḥammād. Riyadh: Dār al-Maymān.

Al-Ghursī, Muḥammad Ṣāliḥ. 2018. *Badr al-Tamām*. Jordan: Dār al-Fatḥ.

Gleave, Robert. 2018. "Imami Shiʿi Legal Theory: From Its Origins to the Early Twentieth Century." In *The Oxford Handbook of Islamic Law*, edited by Anver M. Emon and Rumee Ahmed. Oxford: Oxford University Press.

Guessoum, Nidhal. 2011. *Islam's Quantum Question: Reconciling Muslim Tradition and Modern Science*. London: I.B. Tauris.

Hallaq, Wael B. 1984. "Was the Gate of Ijtihad Closed?" *International Journal of Middle East Studies* 16 (1): 3–41. https://doi.org/10.1017/S0020743800027598.

Hameed, Salman. 2015. "Making Sense of Islamic Creationism in Europe." *Public Understanding of Science* 24 (4): 388–89.

Ḥasan, Aḥmad. 1978. *The Doctrine of Ijmāʿ in Islam*. Islamabad: Islamic Research Institute.

Al-Ḥillī, al-Ḥasan ibn Yūsuf ibn ʿAlī ibn al-Muṭahhar. 1988. *Kashf al-Murād fī Sharḥ Tajrīd al-Iʿtiqād*. Beirut: Muʾassasat al-Aʿlamī.

Al-Ḥillī, al-Ḥasan ibn Yūsuf ibn ʿAlī ibn al-Muṭahhar. 2016. *Mabādiʾ al-Wuṣūl Ilā ʿIlm al-Uṣūl*. UK: Shīʿah Institute Press.

Ibn al-Qaṭṭān, ʿAlī. 2011. *Al-Iqnāʿ fī Masāʾil al-Ijmāʿ* [The Sufficiency: On Issues Regarding Scholarly Consensus]. 4 vols. Qatar: Idāra al-Shuʾūn al-Islāmīyya.

Ibn Sīnā. 1985. *Kitāb al-Najāt*. Edited by Moḥammad Taqī Dāneshpajūh. Tehran: Enteshārāt-e Dāneshgāh.

Ibn Taymiyya, Aḥmad. 1995. *Majmūʿ al-Fatāwa*. 37 vols. Saudi Arabia: Wizāra al-Shuʾūn al-Islāmīyya wa-l Awqāf wa-l Daʿwa wa-l Irshād.

Inloes, Amina. 2023. "Other 'Adams': Twelver Shiʿism and Human Evolution." *Theology and Science* 21 (4): 708–31. https://doi.org/10.1080/14746700.2023.2255953.

Iqbal, Muzaffar. 2023. "Evolution in Light of the Qurʾān and Tradition: Seyyed Hossein Nasr in Conversation with Muzaffar Iqbal." *Islamic Sciences* 16 (1): 51–58. https://cis-ca.org/_media/pdf/2023/1/jis-16-1_nasr_iqbal.pdf.

Jalajel, David Solomon. 2009. *Islam and Biological Evolution: Exploring Classical Sources and Methodologies*. Bellville: University of the Western Cape.

Jalajel, David Solomon. 2018. "Tawaqquf and Acceptance of Human Evolution." Yaqeen Institute for Research. https://yaqeeninstitute.org/wp-content/uploads/2018/08/CORRECT-VERSION-Tawaqquf-and-Acceptance-of-Human-Evolution-2-David-Jalajel.pdf.

Kampourakis, Kostas, and Kevin McCain. 2020. *Uncertainty: How It Makes Science Advance*. New York: Oxford University Press.

Al-Kātibī, Najm al-Dīn. 2024. *Najm al-Dīn al-Kātibī's al-Risālah al-Shamsiyyah*. Translated by Tony Street. New York: New York University Press.

Kocsenda, Karim Gabor. 2022. "Shīʿī Readings of Human Evolution: Ṭabāṭabāʾī to Ḥaydarī." *Zygon: Journal of Religion and Science* 57 (2): 418–42. https://doi.org/10.1111/zygo.12781.

Al-Lucknawī, ʿAbd al-ʿlī. 2002. *Fawātiḥ al-Raḥamūt Bi Sharḥ Musallam al-Thubūt*. 2 vols. Lebanon: Dār al-Kutub al-ʿIlmīyya.

MacIntyre, Alasdair. 1988. *Whose Justice? Which Rationality?* Notre Dame, IN: University of Notre Dame Press.

Al-Maḥallī, Muḥammad ibn Aḥmad. 2005. *Al-Badr al-Ṭāliʿ fī Ḥall Jamʿ al-Jawāmiʿ*. Edited by Murtaḍā ʿAlī al-Daġistānī. Beirut: Muʾassasa al-Risāla Nāshirūn.

Malik, Shoaib Ahmed. 2021. *Islam and Evolution: Al-Ghazālī and the Modern Evolutionary Paradigm*. New York: Routledge.

Malik, Shoaib Ahmed. 2023. "Adam, Eve, and Human Evolution Is There a Conflict?" In *Islamic Philosophy of Religion: Essays from Analytic Perspectives*, edited by Mohammed Saleh Zarepour. New York: Routledge.

Malik, Shoaib Ahmed, and Elvira Kulieva. 2020. "Does Belief in Human Evolution Entail Kufr (Disbelief)? Evaluating the Concerns of a Muslim Theologian." *Zygon: Journal of Religion and Science* 55 (3): 638–62. https://doi.org/10.1111/zygo.12620.

Mansur, Ali. 2023. "The Human Being as the Mystery of *Kun Fa Kān*: An Engagement with Shoaib Ahmed Malik's *Islam and Evolution: Al-Ghazālī and the Modern Evolutionary Paradigm*." *Theology and Science* 21 (4): 732–44. https://doi.org/10.1080/14746700.2023.2255954.

Al-Mardāwī, ʿAlī ibn Sulaimān. 2000. *Al-Taḥbīr Sharḥ al-Taḥrīr*. 9 vols. Riyadh: Maktaba al-Rushd.

Pierret, Thomas. 2015. "Al-Būṭī, Muḥammad Saʿīd Ramaḍān." In *Encyclopaedia of Islam Three Online*, edited by K. Fleet, G. Krämer, D. Matringe, J. Nawas and D.J. Stewart. Leiden: Brill.

Qureshi, Omar, Afifi al-Akiti, and Aasim I. Padela. 2022. "Islam and Science: Reorienting the Discourse." In *Islam and Biomedicine*, edited by Afifi al-Akiti and Aasim I. Padela, 245–54. Cham: Springer.

Rajani, Kumail, and Robert Gleave, eds. 2023. *Shiʾite Legal Theory: Sources and Commentaries*. Edinburgh: Edinburgh University Press.

Ṣabrī, Muṣṭafā. 1981. *Mawqif al-ʿAql wa-l ʿIlm wa-l ʿĀlam Min Rabbil ʿĀlamīn.* Beirut: Dār Iḥyāʾ al-Turāth al-ʿArabī.

Sober, Elliott. 2008. *Evidence and Evolution: The Logic Behind the Science.* Cambridge: Cambridge University Press.

Sober, Elliott. 2011. *Did Darwin Write the* Origins *Backwards? Philosophical Essays on Darwin's Theory.* Amherst, NY: Prometheus Books.

Sober, Elliott. 2024. *The Philosophy of Evolutionary Theory: Concepts, Inferences, and Probabilities.* Cambridge: Cambridge University Press.

Somogyi, Joseph de. 1950. "Ad-Damīrī's Ḥayāt al-Ḥayawān: An Arabic Zoological Lexicon." *Osiris* 9: 33–43. https://www.journals.uchicago.edu/doi/abs/10.1086/368522.

Swamidass, Joshua. 2019. *The Genealogical Adam and Eve: The Surprising Science of Universal Ancestry.* Downers Grove, IL: InterVarsity Press.

Syed, Mairaj. 2015. "*Ijmaʿ*." In *The Oxford Handbook of Islamic Law*, edited by Anver M. Emon and Rumee Ahmed. Oxford: Oxford University Press.

Shihadeh, Ayman. 2012. Classical Ash'ari anthropology: Body, life and spirit. *The Muslim World* 102 (3–4): 433–477.

Al-Taftāzānī, Saʿd al-Dīn. 1887. *Sharḥ al-Maqāṣid al-Ṭālibīn fī ʿIlm Uṣūl al-Dīn.* 2 vols. Istanbul: Maṭbaʿa al-Ḥājj Muḥrim Efendi al-Bosnawī.

Al-Taftāzānī, Saʿd al-Dīn. 2020. *Sharḥ al-ʿAqāʾid al-Nasafiyya.* Edited by Anas Muḥammad ʿAdnān al-Sharfāwī. Damascus: Dār al-taqwā.

Al-Tirmidhī, Abū ʿĪsā Muḥammad b. ʿĪsā. n.d.*Jamiʿ al-Tirmidhī*, no. 2176. https://sunnah.com/tirmidhi:2167.

Al-Ṭūsī, Muḥammad ibn al-Ḥasan. 1997. *Al-ʿUdda fī Usūl al-Fiqh.* 2 vols. Edited by Muḥammad Riḍā al-Ansārī. Qumm: Sitārah.

Usmānī, Muḥammad Rafiʿ. 1983. *Fiqh May Ijmāʿ Kā Maqām.* Karachi: Idāra al-Maʿārif.

Zaidān, Muḥammad al-Amīn ibn Aḥmad. 1993. *Marāqī al-Sūʿūd Ilā Marāqī al-Sūʿūd.* Cairo: Maktaba Ibn Taymiyya.

CHAPTER 5

Are We Programmed or Do We Have Free Will? Discussions on Fate and Genetic Determinism in Islam and Bioscience

Mohd Yusof Mohamad and Abdurezak Abdulahi Hashi

Introduction

Queries about human fate, particularly if mankind's actions and behaviors are destined or if humans are the architects of their actions, have been made repeatedly. The answer to these queries, whether arguing for predetermination or free will, has a significant impact on life, hence this topic cannot be easily ignored. As we observe our surroundings and our lives, we develop a competing, and sometimes conflicting, sense about the fate of man. For instance, if we look into the functions of the body systems, one's sex, place, and time of birth, as well as the impact of natural laws such as the laws of motion and gravitational force on our bodies, one feels that our lives are predetermined and that we have no free will. But if we

M. Y. Mohamad · A. A. Hashi (✉)
International Islamic University Malaysia, Pahang, Malaysia
e-mail: hashi@iium.edu.my

M. Y. Mohamad
e-mail: yusofkahs@iium.edu.my

© The Author(s), under exclusive license to Springer Nature Switzerland AG 2026

A. I. Padela (ed.), *Bridging Bioscience and Islam*, Islam & Science,
https://doi.org/10.1007/978-3-032-09996-9_5

consider the choices we make every day and the observable achievements made by man, one concludes that human beings have choice and free will. Free will is the power of acting without the limitation of necessity, while choice is "the ability to freely decide between alternatives. Choice is a corollary of the traditional notion of free will" ("Choice," n.d.). Hence, one of the major puzzles to the human mind is the question of free will. Are human beings free to choose their actions in life? Or is human life fated and predetermined, programmed to act according to predestined path of life? This would mean that human beings are fated creatures, programmed entities, without free will. If so, the question of morality is then raised. This is because moral accountability is compatible with the notion of human free will, but it is difficult to reconcile morality with the notion of a predestined human nature.

This chapter addresses the topics of human free will and predestination within the scope of biosciences and Islamic theology. Theology teaches that human beings are created with the faculty of intellect that guides them to make rational decisions in life. It also recognizes that human intellect is not static and is influenced by various internal and external factors that lead to human actions. The theory of genetic determinism also posits that environmental factors impact gene expression and that human behaviors are modifiable. This "uncertainty" provides space for the moral culpability of human actions. Similarly, we believe that theological doctrines on fate and free will allow human choice such that culpability for actions can be assigned. In addressing this issue, this chapter describes an Islamic view on determinism and free will. In doing so, it explains the basic dimensions of human life. The fated aspects of human life and the dimensions of life in which human beings are given the choice to choose freely their fate are illustrated. This study attempts to identify the meaning of human accountability (*mukallaf*) and its implications. Other important topics of this chapter are the elaboration of the empirical evidence for free will, the description of gene interactions and human behavior, and discussions of the types of genetic determinism and their relations to human behavior. A hypothetical case of criminal behavior is used to elucidate the strength of evidence and possible conflicts between the domains of bioscience and theology with reference to the epistemological framework presented in Chapter 2.

Concepts Related to Free Will and Determinism

Human "destiny," which is also sometimes described in theology as "*fate*," "*Divine decree*," "*predestination*," or "*Divine will*," refers to lifetime events or circumstances that happen to a person from birth until death. Sometimes the word is explained as "the future destined for a person or thing," or "the predetermined or inevitable course of events," as well as "the ultimate power or agency that predetermines the course of events" (see "Destiny," n.d. versus "Fate," n.d.). As mentioned earlier, free will is understood as the ability to act based on one's own choice without constraints or controls. The term carries the sense of "independent choice," "voluntary decision," and "the ability to make a choice without coercion." This term signifies the freedom "to make choices that are not determined by prior causes or by divine intervention" ("Free Will," n.d.). In philosophy, the term reflects "the doctrine that the conduct of human beings expresses personal choice and is not simply determined by physical or divine forces" (Dictionary.com 2023).

The existence of genuine free will, independent of external factors, is associated with libertarianism. Advocates of this philosophy argue a level of indeterminism in the universe which is not determined and unpredictable. For example, the random movements of subatomic particles in quantum physics suggest that, at the most basic level, the universe is unpredictable. Furthermore, the ability to act upon alternative possible actions signifies freedom of choice. Nevertheless, no one is able to claim full freedom as our actions are governed by external factors. For example, we may have the freedom to choose what food we order in a restaurant but that is restricted by the items on menu. Thus, compatibilism takes a middle stance by compromising between libertarianism and hard determinism. This position, which is sometimes described as soft determinism, asserts that free will and moral responsibility are mutually compatible with determinism.

In contrast, proponents of hard determinism, including supporters of contemporary psychological behaviorism theory, deny free will. Hard determinist positions hold that free will is an illusion wherein human actions and choices are causally determined by forces and influences that are beyond human control (Honderich, n.d.). Just as natural events in nature are caused by antecedent factors, human actions are also the consequences of prior forces. In this view, the causation process in natural phenomena (cause and effect) also exists in human actions and behavior.

Human beings are seen as biological entities that function according to the natural laws of living things.

In relation to human behavior there are those who maintain the deterministic view that "all events, physical or mental, including human behaviour, are the necessary results of antecedent causes or other forces" ("Genetic Determinism" 2018). In this view, not only are the current actions and behaviors of human beings determined by prior causes, but also "the past and the future are fixed [thus]… all human behaviours result from specific, efficient causal antecedents, such as biological structures or processes, environmental conditions, or past experience" ("Genetic Determinism" 2018). According to this view, just as causes and effects in natural incidents are found, human behavior and its antecedents are also identifiable.

Biological determinism argues that "psychological and behavioural characteristics are entirely the result of constitutional and biological factors. Environmental conditions serve only as occasions for the manifestation of such characteristics" ("Genetic Determinism" 2018). Unlike biological determinism, which focuses on the biological factors, environmental determinism focuses on the environmental factors and argues that "psychological and behavioural characteristics are largely or completely the result of environmental conditions" ("Genetic Determinism" 2018). It gives less importance to biological factors compared to environmental influences. Part of biological determinism is the theory of genetic determinism, which hypothesizes that the actions of living things, including that of mankind, are determined. It gives no attention to social and environmental influences on human behavior. It holds that "human and nonhuman animal behaviour and mental activity are largely (or completely) controlled by the genetic constitution of the individual and that responses to environmental influences are for the most part innately determined" ("Genetic Determinism" 2018).

In biological determinism, human free will must be understood through and within the laws of genes, cells, and tissues, the functions of which are governed by natural laws. This theory contends that "there is a real external world out there containing rocks, atoms, cells, animals, plants, etc., and that this material world is all there is" (McFarlane 2001). It adds that "the cosmos has evolved over billions of years to form galaxies, suns, and planets, and—on our planet—an incredible variety of complex biological organisms" (McFarlane 2001). The theory does not recognize psychological dimensions of human experience, hence it argues that "mental experiences, feelings, thoughts, hopes, dreams, and human consciousness itself, are nothing but the activity of neurons in the brain, and thus these things are not more than states of the mind" (Hashi 2016: 29).

Based on these definitions and concepts of human behavior, particularly free will and determinism, one ponders the perspective of Islamic theology toward human behavior. Do Islamic teachings support deterministic views or free will? Is it true that Islamic theology is inclined toward compatibilism (i.e., between hard determinism and total libertarianism)?

Islamic Theology Concerning Free Will and Determinism

As stated in the texts of the Qurʾan, the Divine scripture of Islam, and in the Sunnah, the way of life of Prophet Muhammad, one of the six tenets of Islamic faith is the belief in *al-qaḍāʾ wa-l-qadar* (i.e., divine destiny and decree). Along with the other five pillars of faith, namely, belief in the existence and oneness of God, belief in the angels, belief in the Divine books of revelation, belief in the prophets, and belief in the Day of Judgment, a Muslim is required to believe in "*Divine destiny.*" *Qaḍāʾ* is God's knowledge of what happens in the future, while *qadar* (measurement) is God's creation of the world according to His knowledge and will (al-Ṭaḥāwī and ʿAlī Ibn Abī al-ʿIzz 1997: 1:79). Though belief in Divine destiny is fundamental to the Islamic faith, this concept and its implications in life have been a major concern and sometimes a subject of bitter theological debate among Islamic theologians.

For God knows everything in the past, present, and future, events happen according to His knowledge and will. Do humans have any degree of free will to change events in life? If we assume that all events, including human actions and behavior, are predestined, and thus humans have no choice in life, then moral responsibility and the belief in retribution (i.e., reward and punishment) would be meaningless.

In contrast, the view that humans have free will and are therefore accountable for their choices in life suggests that events occur without Divine will, a position which is theologically not acceptable. Another area of major concern is the nature of human beings, particularly whether the choices made by humans are a result of free choice or an outcome of biological function? The Qur'an states that God has created everything in the universe including the deeds of man, in due proportion:

> Such is God, your Lord, The Creator of all things. There is no god but He: Then how ye are deluded Away from the Truth! (Q.40:62)

But it also states that man has the ability to choose his direction in life:

> Say, 'The Truth is From your Lord': Let him who will, Believe, and let him Who will, reject (it). (Q.18:29)

With a firm belief that the meaning of these two verses is not contradictory, the question is then how to understand the belief in the existence of the All-Knowing God and human beings with free will. The challenge is how to understand the degree of "human free will" given the Creator is all-knowing (*ʿalīm*) with absolute power (*qādir*) and has destined everything for eternity.

What is the basis of the belief in human destiny and fate in Islamic theology? In what sense does man have free will? What is the role of prophethood and revelation in guiding the choices of man in life? Is human free will and determinism a matter of consensus among Islamic schools of theology?

Understanding *Qaḍāʾ* and *Qadar*

The word *qaḍāʾ* is often translated into English in various terms such as "fate," "divine decree," "predestination," "decree of Allah," or "preordainment," the core meaning of which is "divine destiny" (Wehr 1974: 771). The term *qadar* signifies "measuring out," or "being able and having power," which together connote "the power of Allah," or "the divine measurement" (Wehr 1974: 771). When discussing the Islamic belief in predestination, these two terms are often used together as *al-qaḍāʾ wa-l-qadar* (i.e., the Divine decree and predestination). Though these terms are used interchangeably, Islamic theologians agree that the two terms address two levels of Divine destiny. For instance, Ibn Ḥajar al-ʿAsqalānī (d.1449 CE), a leading figure of the Sunnī Shāfiʿī School of Law, mentions that: (i) the Divine decree or *qaḍāʾ* refers to the knowledge of God, whose knowledge covers the past, present, and future (Ibn Ḥajar al-ʿAsqalānī 1959: 11: 477). Since He is All-Knowing, the events and conditions of the entire universe are comprehensively covered by His eternal knowledge. In this aspect, God knows what will happen in life, including when and how it will happen, and (ii) the Divine measurement or *qadar* reflects another level of Divine destiny which allows particular events and conditions to happen or take place in their fixed time and place.

Said another way, the former term (*qaḍāʾ*) signifies God's knowledge of all the events of life (what happened before, now, and in the future), while the latter (*qadar*) refers to causing events to occur in their specific condition, time, and place. This means that the latter *qadar* (i.e., the Divine measurement) is based on the former (i.e., *qaḍāʾ*, or the Divine decree).

The Qurʾan addresses Divine decree in various ways. For instance, it states that: "*With Him, everything is by measure*" (Q.13:8), while also stating that: "*it is He Who created All things, and ordered them in due proportions*" (Q.25:2). A closely related meaning is found in another verse, which states: "*We only send down Thereof in due and ascertainable measures*" (Q.15:21). As stated by Shaykh Muṣṭafā al-Marāghī (1881–1945), these and similar verses of the Qurʾan indicate:

> everything that happens in our lives in this world is predetermined by Allah and its formation is in accordance with the determination of His wisdom the All-Knowing and His law is comprehensive and suitable with the *sunnah* He has set on His creations.
> (Al-Mīrghanī 1900)

In these verses, everything that Allah creates has its own set of laws governing it. In contrast, there are verses in the Qurʾan that suggest the sense of having free will and the ability to make choices in life. For instance, the Qurʾan states that mankind has been shown the way (Q.76:3), but mankind has a choice to either gratefully follow the right way or deny it. The Qurʾan (Q.17:29) also states that everyone is working according to his or her disposition and choice. These and similar verses of the Qurʾan state that man has the ability to freely and consciously make choices that are self-motivated. This is to say that each person acts in accordance with what is appropriate for him, though Allah knows best who is most guided to the truth (*al-Tafsīr al-Muyassar* 1988). This means that humans have the capacity to be rational agents, choosing a course of action from among various alternatives.

Belief in Divine Decree

As noted above, the belief in Divine decree is a fundamental pillar of Islamic faith. Theologians often mention four essential elements of this belief, namely, (i) Divine knowledge, (ii) Divine writing, (iii) Divine will, and (iv) creation (al-Ashqar 2005: 29). If all or one of these elements is missing, then the belief in Divine decree is incomplete (*al-Tafsīr al-Muyassar* 1988).

The first belief is that Allah is omniscient. The Qur'an affirms that God knows what humans conceal within themselves as well as what they disclose (Q.2:33). It then expands outward, declaring that His knowledge encompasses everything in the heavens and the Earth (Q.22:70). From there, it emphasizes universality, stating that God possesses complete knowledge of all things (Q.29:62). Finally, it underscores the precision of this knowledge, noting that He takes account of every single matter in existence (Q.72:28). Islamic theologians explain that nothing escapes Him whether great or small, visible or hidden. Secrets and public affairs are alike to Him, and even the smallest detail, down to an atom's weight, is encompassed by His perfect knowledge (al-Ṭabarī 2013).

Second, God has recorded His eternal knowledge in a book. This record is sometimes described as the Preserved Tablet (*al-Lawḥ al-Maḥfūẓ*). The Qur'an states that everything, regardless of how big or small, has been recorded in a "*Clear Record*" (Q.6:59), while it also confirms that misfortunes and calamities in life are all "*recorded in a decree*" (Q.57:22), long before such misfortunes occur. In explanation of these and similar verses of the Qur'an, the Prophet Muhammad once said: "Allah had written down the destinies of the creatures fifty thousand years before He created the heavens and earth."[1] In another narration the Prophet Muhammad affirms that, Allah recorded the destiny of each and every one of us long before we were born, and no one can change this record.[2] The reason according to the Prophet Muhammad is that "the pens have been lifted and the pages have dried."

Islamic theologians debated the meaning of these texts, as well as other similar ones, particularly whether such texts convey a sense of fatalism or not. The question is, if the Allah has recorded everything in a *Preserved Tablet*, and the pen has been lifted, can there be room for human free

[1] Reported by Imam Muslim, Ḥadīth no. 2653.

[2] Reported by Imam al-Tirmizī, Ḥadīth no. 2516.

will? The majority of the Islamic theologians, particularly those from Sunnī and Muʿtazilī schools hold that, these texts are not a proof for fatalism (Ibn Abi al-ʿIzz 1997). Rather, these verses imply recording God's all-encompassing knowledge. He knows what is going to happen before it happens, hence He recorded all that is, all that was, and all that will be. All things are set in the eternal knowledge of Allah, yet man and woman are encouraged to have faith, seek goodness, and fulfill their religious obligations (Ibn Abi al-ʿIzz 1997).

Third, Allah is the True Creator Who created everything. We are asked, is there "*a Creator, other than God?*" (Q.35:3). Later, the Qurʾan asks whether it could be that man created "*the heavens and the earth?*" (Q.52:36). The universe has no creator other than Allah. Humans are not the creators of heaven and earth, rather, it is the Lord, as stated in the Qurʾan: Who is "*the Creator of all things*" (Q.40:62). A closely related message states: "*it is He Who created all things, and ordered them in due proportions*" (Q.25:2). Not only has Allah created living and non-living things, but "*God has created you and your handiwork!*" (Q.37:96). According to Muslim exegetes like Imam al-Ṭabarī, the phrase "and your handiwork!" has two possible meanings. It could be in the sense of "God created you and your work, i.e., your actions" or it could be in the sense that: "God created you and what you do: that is, what you make idols from, which is wood, copper and the things from which they used to carve their idols" (al-Ṭabarī 2010). The former implies that God is the creator of everything including human actions.

The fourth element of Divine decree is to believe that God's will is all-encompassing; there is nothing that falls outside His will. His will is perfect; whatever He wills happens, and whatever He does not will does not happen. This is to say that nothing happens, whatsoever that might be, without His will. Likewise, nothing takes place against His will. In addition, His will is effective; whatever He wills happens, and whatever He does not will to happen, will not happen. His will is the first cause of everything that exists; without His will, nothing exists. The Qurʾan presents different descriptions of Divine will. For instance:

a. In the Qurʾan (Q.22:14) and (Q.85:16), it states that Allah is a doer of what He intends to do. His "Will is itself the Word and the Deed. There is no interval between them. He does not change His mind. But circumstance whatever can come between His Will and the execution thereof" (Ali 1997: 1629).

b. In the Qur'an (Q.16:40) as well as in (Q.36:82) and elsewhere in the Qur'an, it is stated that, if Allah wills something to happen, or something to exist, He commands "Be" and it is.
c. In the Qur'an (Q.10:99), 11:118, 16:93 and (Q.32:13), and elsewhere in the Qur'an, it is stated that God, through his will, could have forced everyone to believe. Because He desires to grant limited free will to man, He let man choose faith. In fact, "His omnipotence could have made all mankind alike: all would then have had a Faith, but that Faith would have reflected no merit on them" (Ali 1997: 505).

Human Nature in Islam

Islam describes various dimensions of human nature, including physiological and psychological characteristics, feelings, and behavioral traits that are shared by all humans. Human life is influenced by various factors, including natural laws, natural instincts, and social and psychological factors. First, the physiological and biological dimensions of human life are governed by the laws of nature including gravity, Newtonian physics, and the natural systems and functions of the human body such as the immune system, cardiovascular system, respiratory system, and the like. These systems function according to the laws of nature and are beyond the will of human beings. Human beings have no power to create these systems, nor do they function according to human will. In this aspect, mankind has no free will because these areas of human life function according to natural and universal laws of existence in which man has no free choice. Events and occurrences in this area of life take place beyond human will, hence humans are not accountable for events and occurrences in this area of life.

"Natural instincts" (*fiṭra*) constitute the second area of human nature which are manifested in the involuntary drive toward certain activities. These are "primordial, structural elements of the human psyche" (Luton, n.d.), which are essentially ways of behaving, thinking, or feeling that are not learned and not inherited but are inborn and common to all. These are natural desires or natural tendencies that motivate an individual's conduct or desire for certain things. These are inborn impulses or motivations to action typically performed in response to specific external stimuli.

> The only innate behaviors in humans are reflexes. A reflex is a response that always occurs when a certain stimulus is present. For example, a human infant will grasp an object, such as a finger, that is placed in its palm. The infant has no control over this reaction because it is innate.
> ("Innate Behavior of Animals," n.d.)

There are a variety of lists of human natural instincts. Some psychologists like Jung "identified five prominent groups of instinctive factors: creativity, reflection, activity, sexuality, and hunger" (Luton 2022–2023). While hunger is a primary instinct of self-preservation, sexual desire helps to divert biological energy into other channels. Activity is manifested in travel, love of change, restlessness, and play, while the instinct of reflection refers to the search of meaning in life. These instincts are innate in the sense of being: (a) unlearned; (b) genetically inherited; (c) internally patterned and controlled; and (d) possessed by all members of at least one sex of a species. Though innate instincts are inborn, not acquired, the fulfilment of innate instincts is subject to your choice. Hence, humans can direct and guide their desires. Since natural instincts are part of the primordial nature of mankind, which cannot be separated from human nature, mankind is not accountable for having such instincts, rather we are accountable only for directing them in the right way and through the right means.

The third dimension of human life is manifested in the faculty of intellect and human rationality, in which man and woman are gifted with the ability of thinking, and thus making choices (*khiyārāt muktasaba*, *ikhtiyārāt*). Since man is equipped with the ability of making choices, he is therefore accountable for the choices he makes. As stated in Qur'an—(Q.76:3), (Q.90:10), (Q.91:8), and (Q.41:40)—it is God's will that He created human beings with limited free will, in which man and woman can achieve their goals and aims in life, and are given the ability to acquire and achieve these intended goals and aims. But nothing happens beyond His will (Q.76:30), human will cannot go against or beyond His Divine will.

a. The presence of this faculty gives man and woman the special power of making choices in life, which is known as *mukallaf*.
b. The absence of power (*ikrāh*) means the absence of *taklīf*.
c. Though human choices in given deeds and actions are all created by Allah, yet human beings have the ability of acquisition (*kasb*).

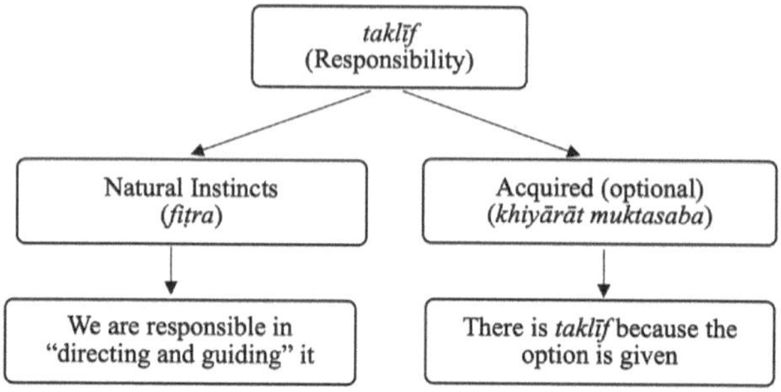

Fig. 5.1 Human nature and *taklīf*

Human nature and *taklīf* are elucidated clearly in Fig. 5.1.

The Role of Guidance (*Hidāya*)

Part of the Islamic teaching on human free will (*ikhtiyār*) and accountability (*taklīf*) is the concept of Divine guidance (*hidāya*) provided to humankind. Guidance means advice or information provided for the sake of resolving a problem or difficulty, particularly advice or information provided by someone with authority. Reading passages of the Qurʾan and Sunnah, one finds two levels of guidance: rational guidance and revelational.

First, rational guidance is achieved through the intellect in terms of reflection and thinking. The ability to reason and understand guides a person not only to know what is right or wrong in life but to overcome problems and thus succeed in achieving life goals. This faculty is manifested in the capacity for logical, rational, and analytic thought, and embraces comprehending, evaluating, and drawing conclusions. With this faculty, an individual is identified as *mukallaf*, which means mature and mentally sound and able to make a choice when given options. *Mukallaf* indicates accountability for one's actions. In the Qurʾan, Allah says "*We showed*" man and woman "*the Way*" (Q.76:3), but it is up to them to be grateful or otherwise. A similar message is found where Allah shows mankind the "*two highways*" (Q.90:10), so that man can make a choice from the given options in life: good or bad. Since the ability to choose between good and bad is given to mankind, as stated in the Qurʾan

(Q.91:8), success in life—both now and in the hereafter—depends on the choices made in this life.

Second, in Islamic teachings, prophethood and revelation are described as a guidance to mankind. It is Allah, who by his eternal will and mercy, decreed to send the prophets with Divine revelation as guidance for mankind. As stated in the Qur᾿an (Q.13:7), to each community, a guide has been sent, to convey the message of Allah. The Qur᾿an states that, "*verily, you (Prophet Muhammad) guide mankind to the right path*" (Q.42:52). Similarly, as reflected in the Qur᾿an (Q.41:17), the role of prophethood and the Divine revelation is to show the right way (*hidāya al-dilāla*), but in the absence of empowering guidance (*hidāya al-tamkīn*) to succeed in following the right way, a community or moral agent may not follow the right way, but rather choose the wrong way. In essence, there are two types of guidance:

1. *Hidāya al-dilāla*: guidance in the sense of showing the direction and the right way, but the actual details of achieving it, is left to the person. The function of this guidance is to show the path, not the detailed steps and processes which are left to the moral agent. As stated in the Qur᾿an (Q.28:56), prophets are tasked to convey the Divine message to mankind, yet the actual power to guide belongs to Him alone.
2. *Hidāya al-tamkīn*: guidance that is empowering in nature, which is sometimes known as the "guidance of success." This type of guidance empowers the heart with guidance that propels man to action. The power to generate guidance in the heart of the person belongs to Allah, the Creator and Master of the heart. At this level of guidance, not only is the person shown the right path, but the person is Divinely assisted in the practical details and steps of achieving it. In the Qur᾿an (Q.24:46), and elsewhere in the Qur᾿an, it states that "*God guides whom He wills to a way that is straight*" (Q.24:46). According to al-Ṭabarī: "God guides whomever He wills of His servants to succeed in life" (al-Ṭabarī 2010). Though God guides whom He wills, this type of guidance is given by God, not sought by man.

Based on the above discussions, a number of conclusions are worth mentioning:

1. Divine decree is one of the pillars of faith in Islam. In fact, a person's belief is incomplete without the belief in Divine decree.
2. The belief in Divine destiny includes belief in the Divine knowledge, Divine writing, Divine will, and God's act of creation.
3. The belief in the Divine decree is based on authentic revealed texts/ transmitted proofs.
4. In certain areas of life, mankind has a choice between alternatives, but there are aspects of life that are predetermined.
5. Divine guidance is provided as an aid to man for making correct choices in life.

Islamic Theological Views

The debate on whether mankind's actions are a result of free will or predestination was among the earliest theological debates in Islamic history. This debate resulted in the appearance of various theological views about human free will. Some Islamic theologians argue that man's actions are predestined, while others uphold the notion of free will. Yet another group advocates for compatibilism.

First, the supporters of determinism take the position that a deed cannot be "attributed to mankind," so they "ascribe it to God" (Shahrastānī 1984: 72). A person does not have free will, rather all acts are predestined in all respects. The theology of this group holds that it is Allah, who gives us the ability to believe or to disbelieve and compels us to act. In other words, actions are done or performed under God's compulsion. In this understanding, every action or deed is actuated by God. This means that mankind is doomed to his or her fate. Based on this deterministic view, no one is accountable for his or her sins. However, this view is often classified as deviant by most Islamic schools of theology (Shahrastānī 1984: 72).

Second, in contrast, the Muʿtazilīs posit that God bestows *free will* upon man so that man deserves reward or punishment for his deeds, hence "one cannot ascribe to God evil and injustice or an act of unbelief and sin, because if He creates an injustice he would be unjust; likewise if he created justice he would be just" (Shahrastānī 1984: 42). The Muʿtazilīs assert that God does not create actions; rather, mankind is the creator of his or her actions. For this school of Islamic theology, God does not interfere in the actions of the humans. Humans are fully free to choose their actions and behavior.

Third, both moral positions were denied by the Sunnī theologians of Ashʿarī and Māturīdīs, as well as those of Atharīs, who present a moderate position of ***compatibilism*** as the nature of human will and freedom. The Ashʿarī school of theology argues that God is "the sole, true author of every action," including human actions (Imam al-Bajuri 2002: 188). Mankind has the ability of "taking possession by means of acquisition (***kasb***) of the actions created by God" (Imam al-Bajuri 2002: 188). Hence, man is accountable for acquired actions (bad or good). This means that humans are not creators of their actions, rather they have the ability to acquire actions that are created by God. The Māturīdī school of theology's position is closely related to that of the Ashʿarī school. According to the Māturīdī belief, all active possibilities (***mumkināt***) are created by God. Humans act according to their free intention and are able to choose which action to follow. That is intention precedes the created action and the capacity by which the actions are acquired (***kasb***) (al-Māturīdī 1970). Islamic theologians address human free will through the contents of the Divine texts (***waḥy***), they differ, however, on the role of rationality (***ʿaql***), in understanding these texts. The Muʿtazilī school of theology favors rationality in understanding this issue, while Atharīs favor a more conservative and traditional interpretation. Ashʿarīs and Māturīdīs seek a middle ground in using revelation and reason to understand human free will. Islamic theologians' views on human free will vary, ranging from libertarianism to compatibilism to determinism. This diversity of views arises because the revelational texts concerning nature and types of predetermination (***qadar***) are open to various interpretations. As a result, the conclusions drawn from them are often speculative (*ẓannī*) rather than definitive (*qaṭʿī*).

It is worth mentioning that both theology and the biological sciences address the subject of human free will. The questions are: What is the scientific explanation of human free will? and, Does the theory of biological determinism offer a conclusive position about human nature (determinism versus free will)? These and related questions will be addressed in the following sections.

The Science of Free Will

Free will is a complex concept that is encompassed by many fields including neuroscience and philosophy. While philosophy ignites deep questions about moral implication, neuroscience attempts to link our free will to the circuitry of the brain. For instance, is flexing our wrist consciously due to our volition or is it predetermined by brain activity? The latter is dependent on the physical, chemical, and biological functioning of the brain. The former is subject to our own life experiences and consciousness as we feel we have freedom to act, although we may be unaware of the cause determining those actions (Adomaitis and Grinbaum 2024).

Neuroscientists have conducted many experiments to investigate the nature of free will and brain activity. Readiness potential (RP or Bereitschaftspotential), initially discovered by Hans Helmut Kornhuber and Lüde Deecke in 1964, offers a new paradigm in understanding the timing of voluntary actions. Using an electroencephalogram (EEG), participants' electrical signals were measured before and during lifting their fingers. An electrical potential was found on the scalp preceding the voluntary actions by approximately 1.5 seconds (Schmidt et al. 2016). Libet's experiment in 1983 added an additional element to this experiment by adding an oscilloscope with a dot on a clock face rotating at a speed of 2.5 seconds (Libet et al. 1983). The participants were asked to lift their finger spontaneously and note the position of the dot on the clock face at the moment of the voluntary action (Schmidt et al. 2024). The presence of RP was recorded 550 milliseconds prior to the action. The data suggests the presence of a gradual build-up of electrical impulses known as RP in the motor cortex even before the person is aware of his intention and consciously decides to make a voluntary movement (The Information Philosopher, n.d.). These results, taken together with similar experiments are proffered as evidence by free will deniers that free will is an illusion. This is because if neuroscientists can predict the timing of a particular voluntary movement prior to the conscious decision, then people might be controlled by unconscious neural processes (Buccella and Dominik 2024).

Haggard and Eimer expanded on Libet's findings by introducing fixed-movement and free-choice conditions. In the fixed-movement condition, participants consistently used the same hand, while in the free-choice condition, they selected which hand to move in each trial. This design allowed for the measurement of the lateralized readiness potential (LRP), which reflects hand-specific motor preparation. Their findings showed that LRP correlated with the timing of conscious intention (W), whereas the readiness potential (RP) did not show this correlation and occurred earlier, suggesting LRP may be more directly linked to the experience of intention than RP (Haggard and Eimer 1999). Astoundingly, the use of fMRI revealed a more extended and distributed preparatory process in the brain. In the study conducted by John-Dylan Haynes (2011), participants were shown rapidly changing letters on a screen and instructed to press a button with either hand whenever they felt the urge, while also reporting the letter visible at the moment they made their decision. The results demonstrated that brain activity in the frontopolar cortex and posterior cingulate cortex could predict the participants' choices up to seven seconds before they became consciously aware of them (Haynes 2011).

Other regions of the brain, such as the supplementary motor area (SMA), pre-SMA, and anterior cingulate cortex (ACC), showed that brain activity could predict both the timing of the urge to move and the hand that would be used. Itzhak Fried and his team demonstrated that in epilepsy patients with electrodes implanted in these areas, decisions could be predicted up to 500 milliseconds before the reported "W time." Participants were instructed to press a key at any moment of their choosing while observing an analog clock, and to report the exact moment they felt the urge to act (Fried et al. 2011).

Nevertheless, Libet experiments have been challenged and reinterpreted. For instance, Libet himself in other experiments demonstrated that an RP is recorded even without voluntary action. The initiation of movement starts unconsciously but can be inhibited consciously regardless of RP measurements in what is famously known as "free won't" (Libet 1999). The timing between RP and initiation of movement is not correlated as temporal gap between LRP and W was smaller than that between RP and W, as demonstrated by Haggard and Eimer (1999). Another study by Schlegel and colleagues showed neither RP nor LRP is causally related to conscious will (Schlegel et al. 2013). RP is not fixed in timing, is more complex in nature, and can include anticipation or spontaneous random fluctuation. As with Haynes' findings mentioned earlier, the process of willing an action begins much earlier—up to 7 seconds beforehand. Nevertheless, the reliability and robustness of the results can be questioned as the prediction accuracy reached only 60%. In terms of regions of brain involved and the associated brain activity, SMA signifies higher prediction accuracy, but it is not clear whether it is the place of origin for movement initiation (Soon et al. 2008).

These counter-interpretations and criticisms of Libet experiments suggest that the evidence showing free will is an illusion are not certain. Objective analysis of empirical data indicates the experiments are more concerned with arbitrary motor decisions rather than deliberate purposeful decisions. In other words, it may not fully capture the nuanced nature of decisions made by humans in context-rich situations (Nichelli and Grafman 2024).

Genetic Determinism and Human Behavior

Human behavior is intricately influenced by a multitude of interconnected factors. Biological determinism posits that factors such as genetics, hormones, and brain activity determine human behavior and traits. The term "genetic determinism" emerged in the scientific literature championing the role of genetics in determining human behavior.[3] Indeed, the sequence of nucleotides in a particular gene, which codes for a specific protein synthesis, serves as a blueprint for trait development (Pathak and Bordoni 2023).[4] The idea of genetically controlled behavior can lead to fatalism. Genetic fatalism views genes as overwhelmingly influencing human behavior and thus dictating one's fate; this belief undermines the role of the environment and free will. The classical example of fatalism is the myth of Oedipus from Greek mythology. In this famous myth, he is prophesized to kill his father in order to marry his mother. Soon after birth, he was given to a shepherd to be killed because his parents wanted to avoid destiny. But the shepherd was unable to complete the task and presented him to another king. He was raised with both intellect and bravery. On one occasion, he killed a caravan of men which, unknown to him, included his own father. He earned a kingship and later married his own mother whom he had never met. He was living in disgrace, unable to defy destiny. Applied to the current context, those who believe everything is fated in the genes without any other factors, resemble Oedipus. This

[3] The word genetic determinism is sometimes used interchangeably with biological determinism; see Allen (n.d.).

[4] The number of base pairs is huge on each chromosome. For instance, chromosome 1 is formed by approximately 249 million base pairs of nucleotides. Even the smallest chromosome (no. 22) has about 50 million nucleotides.

idea impacts people's perceptions in different ways. Either they become pessimistic attributing all negative actions to genes or become narcissists claiming superiority over others (Heine et al. 2017).

In the late nineteenth and early twentieth centuries, genetic determinism laid the foundation for the rise of the eugenics movement. Rooted in a misinterpretation of Mendelian genetics, eugenics advocated for the selective breeding and sterilization of individuals deemed "unfit" in an effort to improve the human population. Adam Rutherford, in *Control: The Dark History and Troubling Present of Eugenics*, traces the movement's origins to figures such as Francis Galton and Charles Davenport, and highlights how American eugenic policies directly shaped Nazi initiatives, including the Hereditary Health Courts and the Action T4 euthanasia program. By the onset of World War II, approximately 60,000 people had been forcibly sterilized in the United States, with justifications shifting toward the supposed protection of vulnerable women from unwanted pregnancies (Reilly 2015).

Advances in genetic studies have transformed the deterministic worldview of genes-to-behavior to probabilistic in nature. The availability of massive genetic databases and advancements in technology has enabled an understanding of the role of genetics in behavior. Molecular genetics permits direct assessment of DNA variability which contributes to heritability of behavioral traits. A person's entire genome can be read to identify genetic variants typically known as single nucleotide polymorphism (SNPs). Genomic-wide association studies (GWAS)[5] utilize complex statistical methods to determine the frequency of SNPs found in individuals with specific traits to compare to those without. In addition to GWAS, numerical polygenic scores may be assigned to individuals affected by many genetic variants. The score for each person indicates the genetic risk of a predisposition for a certain trait. Nevertheless, findings from genome-wide association studies (GWAS) on behavior must be interpreted with caution and situated within broader social, environmental, and developmental frameworks. GWAS faces several limitations, including the issue of missing heritability, population specificity, the polygenic nature of complex traits, limited understanding of gene function,

[5] A genome-wide association study (GWAS) is an approach to compare the genomes from many different people to find genetic markers associated with a particular phenotype or risk of disease; see National Cancer Institute (n.d.).

and ethical concerns surrounding data use and interpretation. Methodological challenges such as the need for large sample sizes, the risk of multiple testing errors, spurious associations, and difficulties in identifying causal variants further complicate interpretation (Omidiran et al. 2024). Polygenic scores derived from GWAS account for only a small fraction of variance in behavioral traits. For example, a GWAS involving nearly 150,000 participants identified 128 genomic regions associated with schizophrenia, yet these variants explained merely 7% of the observed phenotypic variation (Akbay 2025). Such findings underscore the probabilistic, rather than deterministic, nature of genetic contributions to behavior.

Nature Versus Nurture

The debate between the roles of nurture versus nature continues in the bioscientific community. Twin studies are used to assess the influence of genes in the environment. Comparing identical and fraternal twins enables us to observe the effect of nature on behavior. The trajectory of the lives of twin babies may verify or refute inheritability by determining whether the twins turn out to behave in a similar way though separated and exposed to different social and environmental cues. Twin studies started as early as 1875 in a study by Sir Francis Galton (Burbridge 2001). Later, in the 1920s, the research began to delineate the differences between monozygotic (MZ) who share nearly 100% of their genes with dizygotic (DZ) twins, who share approximately 50%. Contemporary research continues to affirm the value of this approach. For example, a study by Willoughby et al. (2023) reaffirmed twin studies in demonstrating genetic contributions to behavioral and cognitive traits ranging from 53 to 73% of heritability degree. Similarly, Clemmensen et al. (2025) employed a population-based twin sample to show that tattooing behavior displays modest heritability, illustrating how twin designs can capture complex social traits. Moreover, Smatti et al. (2025) emphasized the unique contribution of MZ/DZ comparisons in understanding host responses to infectious diseases, extending twin methodologies into immunogenetics. Additionally, Portugal et al. (2025) used twin cohorts to examine genetic influences on infant eye movements, offering early insights into attentional development. These studies collectively underscore the enduring value of twin design, especially when integrated with modern statistical modeling and genomic technologies. Taken together, such findings suggest that while genetic predispositions significantly shape human behavior and traits, they do not wholly eliminate the possibility of free will.

Genetic and environmental factors that work together to influence human behavior are referred to as the gene-environment correlation. In simple words, this is the inherited genotype of a particular person along with the environment in which they are raised. This can be further classified as active, passive, or evocative. An active gene-environment correlation indicates that individuals seek out an environment that supports their genetic tendencies. For example, a child predisposed with greater cognitive abilities might opt for more challenging questions. An evocative or a reactive gene-environment correlation on the other hand refers to an individual whose genetic make-up elicits responses from others in their environment. For instance, a child predisposed to friendliness may elicit positive interactions from others. A passive genetic-environment correlation refers to an individual whose genetic profile inherited from the parents matches the environment provided by the parents. This is demonstrated in a study by UK Biobank involving 6311 adoptees, the genetic influences on education are mediated through the home environment. Nevertheless, the study did not characterize the home environment and the mechanism of transmission of educational attainment needs further investigation (Cheesman et al. 2020). With regard to psychiatric disorders, genetic influences are generally considered more significant than environmental factors. Plomin and Vassos (2025) highlight that, on average, genetic factors account for approximately 50% of the variation observed in psychiatric conditions. Stronger genetic contributions have been observed in disorders such as schizophrenia and bipolar disorder. The shared family environment appears to have minimal impact on mental health similarities among siblings.

The nature versus nurture debate is further complicated by changes occurring beyond the genetic sequence. Epigenetics refers to the modification of DNA that is independent of the genetic sequence (Li 2021). Gene expression is altered indirectly causing changes in protein synthesis. Epigenetic modification includes DNA methylation, histone modification, and non-coding RNAs. These alterations occur outside the genetic sequence and are derived from the environment and behavior. Stress response, behavior, appearance, and even longevity are some of the factors influencing epigenetics (Tammen et al. 2013). The mechanism for epigenesis may be direct or indirect. Direct effect refers to the changes that happen in the individual's lifetime. Indirect effect is related to changes during gestation or changes to the individual's predecessors (Lacal and Ventura 2018). Countless epigenetic changes occur during a person's lifetime due to daily encounters with the environment. But can epigenetics be inherited by subsequent offspring? Interestingly, current studies support the idea of epigenetics inheritability (Lind and Spagopoulou 2018). For example, a painful experience may cause an epigenetic mark that is passed to the offspring and lasts for several generations. Yehuda and Lehrner (2018) identified two epigenetic mechanisms involved in intergenerational transmission of trauma effects. The first factor is the influence on offspring during early environmental exposure (e.g., maternal stress during pregnancy). The second factor refers to alteration in gene expression due to trauma experienced by a parent before conception affecting the germline.

Deterministic to Probabilistic: Assessing the Evidence of Genetic Determinism

Evaluating the evidence for genetic-behavior interaction shows it is more probabilistic than deterministic. As mentioned earlier, the complexity of gene interactions, gene-environment interplay, and epigenetic modifications offer a nuanced perspective on genetic influences of traits and behaviors. Genetic determinism warrants understanding a one-to-one relationship between genes and behavior. In reality, gene interactions are diverse and complex. As seen in Fig. 5.2, interactions can be between one gene associated with one trait, one gene associated with several traits, several genes associated with one trait, many genes associated with many traits, or one or more genes along with environmental and epigenetic factors (Carver et al. 2017).

Consequently, pinpointing specific behavior to specific genes is difficult. For instance, no single gene can be attributed to intelligence. Even a combination of genes may not yield consistent behavioral outcomes. Certain genes may increase or decrease the tendency to act or adopt certain behaviors (Walinga 2014). Even in controlled environments with known genes, we do not yet have a clear understanding of how environmental variations interact with genes. Traditional GWAS may not be sufficient to determine the complex interactions between genes and behavior (Robinson et al. 2024). In the past decade, polygenic scores have been used to predict common disorders and complex traits. For

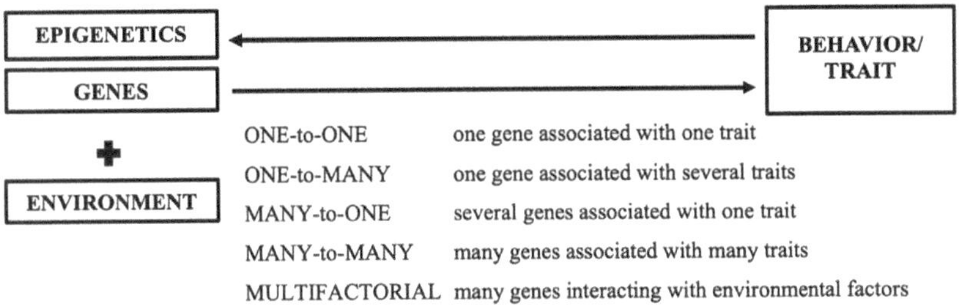

Fig. 5.2 Complex interaction between genes and behavior

example, the polygenic score for intelligence is 7%, 11% for educational attainment, and 15% for school performance at 16 years old. Nevertheless, these scores will never be able to predict complex traits because heritability is about 50% for most behavioral traits (Plomin and Von Stumm 2022). The evidence suggests that the probabilistic nature of gene-behavior interaction overrules the concept of determinism.

Recent advances in gene-editing technologies, particularly CRISPR-Cas9, have significantly challenged the traditional concept of strict genetic determinism. If humans now possess the capability to modify DNA sequences, it raises fundamental questions about the validity of viewing the genome as a fixed, predetermined blueprint. Does this technological capacity imply that we can engineer specific behaviors by altering genetic code? While such prospects are theoretically intriguing, they remain limited by the complexity of gene-behavior relationships and the profound influence of environmental, epigenetic, and developmental factors. Currently, CRISPR being applied in human medicine to treat certain genetic disorders, though not complex traits such as behavior. A notable advancement is Casgevy, a CRISPR-based therapy recently approved for the treatment of sickle cell disease and beta-thalassemia. In this treatment, a patient's stem cells are extracted, the defective gene is corrected using CRISPR, and the modified cells are then reintroduced into the body. Ongoing clinical trials are also exploring the use of CRISPR for treating cancers and rare protein disorders. While these developments highlight the therapeutic promise of gene editing, applying CRISPR to modify behavioral traits remains a possibility, constrained by both ethical concerns and the scientific complexity of behavior (Balch 2021).

Case Study: Assessing Criminal Actions from the Perspectives of Bioscience and Islam

In order to demonstrate the interplay between the bioscience and Islamic perspectives of free will and determinism, consider the hypothetical case study below:

> Mr. A, a 28-year-old man, was recently arrested for armed robbery at a convenience store. During interrogation, he attributed his actions to God and his genetic predisposition.

Important questions raised based on bioscience and Islamic theology are:

1. Can criminal action be attributed solely to our genes?
2. Can someone blame God for his evil action? Is he culpable?

In responding to the first question, certainly a person's genetic make-up can influence his/her behavior and traits. Studies have shown that two genes are associated with violent behavior: monoamine oxidases (MAOs) and cadherin 13 (CDH13). MAOs have gained significant attention over the past 20 years beginning in 1993 when Brunner et al. reported aggression in a Dutch family with an X-linked mutation in the MAO-A gene (Brunner et al. 1993). Piton et al. (2014) identified a second case of the mutation in a small family, followed by a third report involving two Australian families (Mentis et al. 2021). The MAO genes code for the enzyme monoamine oxidase A, which is vital for regulating the amount of serotonin and dopamine in the brain. More recently, Mustafayeva et al. (2024) reaffirm that males who have only one X chromosome are more susceptible to the behavioral effects of MAO-A mutations due to the gene's X-linked inheritance pattern. Under normal circumstances, MAO-A maintains neurotransmitter balance; however, a low-activity variant can lead to excessive serotonin levels in early development, which has paradoxically been associated with increased impulsivity and aggression (Godar et al. 2016).

However, the association of MAO with aggressive behavior is disputed. A large-scale meta-analyses by Vassos, Collier, and Fazel found no significant relationship even after reviewing 1331 potential cases, from 185 studies encompassing 277 independent associations with 31 genes including MAO (Vassos et al. 2014). This suggests mutation of MOA and other genes alone cannot be the cause of aggressive behavior. It appears that other environmental stressors such as abuse must be present at early stages of development in order for a particular person to be predisposed to commit crime (Mentis et al. 2021).

Another gene candidate is CDH13, which is expressed in amygdala mediating cell signaling in the brain. Increases in CDH13 lead to violent behavior. It has been implicated in impulse control and neurodevelopmental disorders, with one Finnish study estimating that up to 10% of all severe violent crimes may be linked to CDH13 variants (Tiihonen et al. 2015). Recent literature further underscores its potential role. A 2023 study in the *Journal of Psychology & Clinical Psychiatry* reviewed cases of extreme violence, noting significant overlaps between CDH13 and MAO-A genotypes, suggesting a polygenic influence rather than a single-gene determinant (Oliveira do Nascimento and Nascimento Silva Leandro 2023). However, CDH13's behavioral effects in human populations remain under-researched, and experts caution against over-interpreting early findings without replication In sum, while genetic predispositions may increase an individual's susceptibility to antisocial or aggressive behavior, they do not determine behavior in isolation. Environmental influences, including parenting, socioeconomic status, trauma, and community context, play an equally vital role. Recent findings reiterate that criminal behavior results from the dynamic interplay between genes and life experiences, not from genetic programming alone.

In addressing the second question, Islamic theology attributes moral responsibility for the evil committed by a person. One cannot simply blame God for his own evil actions. He is given the capacity to act and react accordingly based on his choice. Criminal behavior is considered a sin and is punishable in Islamic law. With any voluntary action, culpability for the action is attributed to the perpetrator alone. Culpability is related to the state of mind, intention, and the circumstances of the offence performed as governed by the *sharīʿa*. Intentional crime is considered purposeful with rebellious intent. There is a difference between crimes performed with purpose as opposed to in the state of negligence (Khan et al. 2022). This is based on the maxim "*actions are to be judged by intention and everyone shall have what he intends.*" The absence of coercion is another important criterion in establishing culpability.

Evaluating this case study based on key assumptions of Sunnī *kalām* discourse (A1–A6) and the steps of al-Tahānwī's framework delineated in

Chapter 2, the claim of Mr. A is not supported empirically or theologically. Empirically, Mr. A has no conclusive evidence to claim his action is fully due to genetic predisposition. Recent studies estimated only up to 60% of heritability (Koyama et al. 2024). Scientific findings on heritability offer probabilistic conclusions placing them in the realm of speculative (*ẓannī*) knowledge rather than definitive (*qaṭʿī*) certainty. On theological evidence, no scriptural proof conclusively (*ẓannī al-dalāla*) separates Mr. A from his crime action. He bears responsibility when he possesses the capacity to act with a sound mind, has knowledge of the act, and is not acting under coercion.

Weighing both domains of transmitted and rational proof it seems no apparent conflict exists. Reconciliation between domains is necessary when one form of proof contradicts another. As cited by Omar Qureshi in Chapter 2, drawing on al-Taḥānwī, "two proofs are said to be in conflict when 'the truth of one of two judgments entails the falseness of the other'." In the present case, however, no such conflict arises. Both the transmitted (theological) and (scientific) proofs point to the same conclusion: criminal actions are not entirely determined by genetic predisposition, and the individual retains the capacity for choice and moral responsibility in his actions. Thus, there is no urgency to harmonize or prioritize either domain over the other since the propositions of bioscience and theology are not contradictory.

Conclusion

No doubt, the discourse on free will and determinism is intricate, raising many intriguing questions. Current research acknowledges the significant influence of genetic factors yet also emphasizes the powerful role of environmental conditions and epigenetic modifications. These findings support a probabilistic, rather than deterministic, model of human behavior, one that does not eliminate the possibility of agency or conscious decision-making. Thus, empirical evidence from bioscience does not necessarily refute the existence of free will; instead, it reveals the complexity of the mechanisms that underlie human action. In terms of theology, the Big Questions on whether human actions are fated has long been debated. The scriptural proof of free will and determinism is conclusive, yet the interpretation is presumptive. Islamic theology takes a middle stance by accommodating both deterministic and free will paradigms.

The concept of moral responsibility (*taklīf*) in Islam is closely tied to the presence of volitional control, freedom from coercion, knowledge of the act, and sound mental capacity. A person is only accountable when these conditions are met, which aligns with many contemporary discussions in neuroethics and behavioral science concerning agency and culpability.

In a nutshell, the perceived tension between fate and freedom, or between theology and bioscience, need not be seen as a contradiction. Rather, when properly contextualized, both disciplines offer complementary insights. The biosciences provide nuanced understandings of the biological underpinnings of behavior without negating human agency, while Islamic theology affirms a framework in which Divine will and human choice coexist.

References

Adomaitis, Laurynas, and Grinbaum, Alexei. 2024. "Neurotechnologies, Ethics, and the Limits of Free Will." *Integrative Psychological and Behavioral Science*: 1–14. https://doi.org/10.1007/s12124-024-09830-2.

Akbay, Gökhan. 2025. "Defining Genetic Determinism: Three Old Concepts, One New Problem and a Modest Proposal." *Teorie vědy/Theory of Science*. https://doi.org/10.46938/tv.2025.649.

Allen, Garland Edward. n.d. "Biological Determinism." Britannica. https://www.britannica.com/topic/biological-determinism.

Ali, Abdullah Yusuf. 1997. *The Meaning of the Quran*. Brentwood, MD: Amana Publications.

Al-Ashqar, ʿUmar b. Sulaymān b. ʿAbd Allāh al-ʿUtaybī. 2005. *Al-Qaḍāʾ wa-l-Qadar*. Amman: Dār al-Nafāʾis.

Imam al-Bajuri, Burhan al-Din Ibrahim. 2002. *Hashiyat al-Imam al-Bajiri ʿAla Jawharah al-Tawhid*. Edited by ʿAlī Jumʾah Mohd al-Shafiʾiyyi. Cairo: Dar al-Salam Li al-Tabaʾah wa al-Nashar wa al-Taqziʾ wa al-Tarjamah.

Balch, Bridget. 2021. "The Future of CRISPR Is Now." Association of American Medical Colleges, December 2. https://www.aamc.org/news/future-crispr-now.

Brunner, H. G., M. Nelen, X. O. Breakefield, H. H. Ropers, and B. A. van Oost. 1993. "Abnormal Behavior Associated with a Point Mutation in the Structural Gene for Monoamine Oxidase A." *Science* 262 (5133): 578–580. https://doi.org/10.1126/science.8211186.

Buccella, Alessandra, and Dominik, Tomáš. 2024. "Free Will Is Only an Illusion If You Are, Too." *Scientific American*, February 20. https://www.scientificamerican.com/article/free-will-is-only-an-illusion-if-you-are-too/.

Burbridge, David. 2001. "Francis Galton on Twins, Heredity and Social Class." *The British Journal for the History of Science* 34 (3): 323–340. https://doi.org/10.1017/S0007087401004332.

Carver, Rebecca Bruu, Jérémy Castéra, Niklas Gericke, Neima Alice Menezes Evangelista, and Charbel N. El-Hani. 2017. "Young Adults' Belief in Genetic Determinism, and Knowledge and Attitudes Towards Modern Genetics and Genomics: The PUGGS Questionnaire." *PloS one* 12 (1): e0169808. https://doi.org/10.1371/journal.pone.0169808.

Cheesman, Rosa, Avina Hunjan, Jonathan R. I. Coleman, Yasmin Ahmadzadeh, Robert Plomin, Tom A. McAdams, Thalia C. Eley, and Gerome Breen. 2020. "Comparison of Adopted and Nonadopted Individuals Reveals Gene–Environment Interplay for Education in the UK Biobank." *Psychological Science* 31 (5): 582–591. https://doi.org/10.1177/0956797620904450.

"Choice". n.d. *Encyclopaedia Britannica.* https://www.britannica.com/topic/choice. Accessed July 6, 2024.

Clemmensen, Signe B., Jonas Mengel-From, Jaakko Kaprio, Jennifer R. Harris, Henrik Frederiksen, and Jacob von Bornemann Hjelmborg. 2025. "Tattooing Is Mainly Cultural: A Representative Twin Study of Tattooing Determinants." *Behavior Genetics* 55: 114–123. https://doi.org/10.1007/s10519-025-10215-3.

"Destiny." n.d. *Collins English Dictionary.* HarperCollins Publishers. https://www.collinsdictionary.com/dictionary/english/destiny. Accessed February 5, 2023.

Dictionary.com. 2023. "Free Will." Based on The Random House Unabridged Dictionary, Random House. https://www.dictionary.com/browse/free will. Accessed February 5, 2023.

"Fate." n.d. *Merriam-Webster.com Dictionary.* https://www.merriam-webster.com/dictionary/fate. Accessed February 5, 2023.

"Free Will." n.d. *Merriam-Webster.com Dictionary.* https://www.merriam-webster.com/dictionary/free will. Accessed February 5, 2023.

Fried, Itzhak, Roy Mukamel, and Gabriel Kreiman. 2011. "Internally Generated Preactivation of Single Neurons in Human Medial Frontal Cortex Predicts Volition." *Neuron* 69 (3): 548–562. https://doi.org/10.1016/j.neuron.2010.11.045.

"Genetic Determinism." 2018. American Psychology Association Dictionary. https://dictionary.apa.org/genetic-determinism. Accessed February 6, 2023.

Godar, Sean C., Paula J. Fite, Kenneth M. McFarlin, and Marco Bortolato. 2016. "The Role of Monoamine Oxidase A in Aggression: Current Translational Developments and Future Challenges." *Progress in Neuro-Psychopharmacology and Biological Psychiatry* 69: 90–100. https://doi.org/10.1016/j.pnpbp.2016.01.001.

Haggard, Patrick, and Martin Eimer. 1999. "On the Relation Between Brain Potentials and the Awareness of Voluntary Movements." *Experimental Brain Research* 126 (1): 128–133. https://doi.org/10.1007/s002210050722.

Hashi, Abdurezak Abdulahi. 2016. *Worldviews and Belief Systems*. Kuala Lumpur: Fajar Ulung.

Haynes, John-Dylan. 2011. "Decoding and Predicting Intentions." *Annals of the New York Academy of Sciences* 1224 (1): 9–21. https://doi.org/10.1111/j.1749-6632.2011.05994.x.

Heine, Steven J., Ilan Dar-Nimrod, Benjamin Y. Cheung, and Travis Proulx. 2017. "Essentially Biased: Why People Are Fatalistic About Genes." *Advances in Experimental Social Psychology* 55: 137–192. https://doi.org/10.1016/bs.aesp.2016.10.003.

Honderich, Ted, ed. n.d. *The Determinism and Freedom Philosophy Website*. https://www.ucl.ac.uk/~uctytho/dfwIntroIndex.htm. Accessed July 4, 2024.

Ibn Abi al-ʿIzz, ʿAli ibn ʿAli. 1997. *Sharh al-ʿAqidah al-Ṭaḥāwiyyah*. Beirut: Muʾassasat al-Risala.

Ibn Ḥajar al-ʿAsqalānī, Aḥmad ibn ʿAlī. 1959. *Fatḥ al-Bārī bi-Sharḥ al-Bukhārī*. Vol. 11. Beirut: Dār al-Maʿrifah

Ibn Kathīr, Ismāʿīl Ibn Omar. 2000. *Tafsīr Ibn Kathīr*. Riyadh: Darussalam.

The Information Philosopher. n.d. "Libet Experiments." https://www.informationphilosopher.com/freedom/libet_experiments.html.

"Innate Behavior of Animals." n.d. In *Introductory Biology*. Biology Libre Texts. https://bio.libretexts.org/Bookshelves/Introductory_and_General_Biology/Book%3A_Introductory_Biology_(CK-12)/10%3A_Animals/10.04%3A_Innate_Behavior_of_Animals#:~:text=The%20only%20innate%20behaviors%20in,reaction%20because%20it%20is%20innate. Accessed on February 11, 2022.

Khan, Asif, Kashif Javed, Abdus Samad Khan, and Aatir Rizwi. 2022. "Aggression and Individual Criminal Responsibility in the Perspective of Islamic Law." *Competitive Social Science Research Journal* 3 (1): 35–48.

Koyama, Emiko, Tuana Kant, Atsushi Takata, James L. Kennedy, and Clement C. Zai. 2024. "Genetics of Child Aggression: A Systematic Review." *Translational Psychiatry* 14 (252). https://doi.org/10.1038/s41398-024-02870-7.

Lacal, Irene, and Rossella Ventura. 2018. "Epigenetic Inheritance: Concepts, Mechanisms and Perspectives." *Frontiers in Molecular Neuroscience* 11: 292. https://doi.org/10.3389/fnmol.2018.00292.

Li, Yuanyuan. 2021. "Modern Epigenetics Methods in Biological Research." *Methods* 187: 104–113. https://doi.org/10.1016/j.ymeth.2020.06.022.

Libet, Benjamin W. 1999. "Do We have Free Will?" *Journal of Consciousness Studies* 6 (8–9): 47–57.

Libet, Benjamin, Curtis A. Gleason, Elwood W. Wright, and Dennis K. Pearl. 1983. "Time of Conscious Intention to Act in Relation to Onset of Cerebral Activity (Readiness-Potential) the Unconscious Initiation of a Freely Voluntary

Act." *Neurophysiology of Consciousness* 106 (3): 249–268. https://doi.org/10.1093/brain/106.3.623.
Lind, Martin I., and Foteine Spagopoulou. 2018. "Evolutionary Consequences of Epigenetic Inheritance." *Heredity* 121 (3): 205–209. https://doi.org/10.1038/s41437-018-0113-y.
Luton, Frith. 2022–2023. "Instinct, Consciousness, and Five Instinctive Factors." FrithLuton.com. https://frithluton.com/articles/instinct/#:~:text=Jung%20identified%20five%20prominent%20groups,%2C%20activity%2C%20sexuality%20and%20hunger. Accessed February 11, 2023.
Luton, Frith. n.d. "What Is an Archetype?—Working with Universal Archetypes." *Jungian Dream Analysis and Psychotherapy*, FrithLuton.com. https://frithluton.com/articles/what-is-an-archetype/. Accessed February 11, 2023.
McFarlane, Thomas J. 2001. "Questioning the Scientific Worldview." Center for Sacred Sciences. http://www.centerforsacredsciences.org/index.php/Articles/questioning-the-scientific-worldview.html. Accessed September 2, 2015.
Al-Māturīdī, Abu Manṣūr Muḥammad. 1970. *Kitāb at-Tawḥīd*. Beirut: Dar El-Machreq.
Al-Mīrghanī, Muḥammad ʿUthmān. 1900. *Tāj al-Tāfasīr li-Kalām al-Malik al-Kabīr*. Beirut: Dar al-Fikr.
Mentis, Alexios-Fotios A., Efthimios Dardiotis, Eleni Katsouni, and George P. Chrousos. 2021. "From Warrior Genes to Translational Solutions: Novel Insights into Monoamine Oxidases (MAOs) and Aggression." *Translational Psychiatry* 11 (1): 130. https://doi.org/10.1038/s41398-021-01257-2.
Merriam-Webster.com Dictionary, s.v. "Free Will." https://www.merriam-webster.com/dictionary/free will. Accessed February 5, 2023.
Muhtaroglu, Nazif. 2024. "Akkirmânî's Occasionalist Approach to the Neuroscientific Research on the Human Will." *Religions* 15 (9): 1134. https://doi.org/10.3390/rel15091134.
Mustafayeva, Ulviyya, Burcin Yuksel, and Ozlem Akkan. 2024. "Usability of MAO-A Gene in the Judicial Process." *Medicine Science* 13 (1): 227–233. https://doi.org/10.5455/medscience.2024.01.09.
National Cancer Institute. n.d. "Genome-Wide Association Studies." Division of Cancer Epidemiology & Genetics at the National. https://dceg.cancer.gov/research/how-we-study/genomic-studies/gwas-overview#:~:text=A%20genome%2Dwide%20association%20study,phenotype%20or%20risk%20of%20disease.
Nichelli, Paolo F., and Jordan Grafman. 2024. "The Place of Free Will: The Freedom of the Prisoner." *Neurological Sciences* 45 (3): 861–871. https://doi.org/10.1007/s10072-023-07138-4.
Oliveira do Nascimento, Julia, and Maria K. do Nascimento Silva Leandro. 2023. "Serial Killers: A Review About the Genetic Influence on Violent

Behavior." *Psychology and Clinical Psychiatry* 14 (5): 148–150. https://doi.org/10.15406/jpcpy.2023.14.00744.

Omidiran, Oluwaferanmi, Aashna Patel, Sarah Usman, Ishani Mhatre, Habiba Abdelhalim, William DeGroat, Rishabh Narayanan, Kritika Singh, Dinesh Mendhe, and Zeeshan Ahmed. 2024. "GWAS Advancements to Investigate Disease Associations and Biological Mechanisms." *Clinical and Translational Discovery* 4 (3): e296. https://doi.org/10.1002/ctd2.296.

Pathak, Isha, and Bruno Bordoni. 2023. "Genetics, Chromosomes." In *StatPearls [Internet]*. Treasure Island, FL: StatPearls Publishing.

Piton, Amélie, Hélène Poquet, Claire Redin, Alice Masurel, Julia Lauer, et al. 2014. "*20 ans après*: A Second Mutation in MAOA Identified by Targeted High-Throughput Sequencing in a Family with Altered Behavior and Cognition." *European Journal of Human Genetics* 22 (6): 776–783. https://doi.org/10.1038/ejhg.2013.243.

Plomin, Robert, and Sophie von Stumm. 2022. "Polygenic Scores: Prediction Versus Explanation." *Molecular Psychiatry* 27 (1): 49–52. https://doi.org/10.1038/s41380-021-01348-y.

Plomin, Robert, and Evangelos Vassos. 2025. "What Clinicians Should Know About the Contribution of Modern Behavioral Genetics to Psychiatric Problems." *Psychological Medicine* 55: e83. https://doi.org/10.1017/S0033291725000273.

Portugal, Ana Maria, Mark J. Taylor, Kristiina Tammimies, Angelica Ronald, and Terje Falck-Ytter. 2025. "Dissociable Genetic Influences on Eye Movements During Abstract Versus Naturalistic Social Scene Viewing in Infancy." *Scientific Reports* 15 (4100). https://doi.org/10.1038/s41598-024-83557-3.

Al-Qurṭubī, Muhammad Bin Ahmad. 2013. *Tafsīr al-Qurṭubī: Al-Jamiʿ li Ahkam al-Qurʾan al-Karīm*. Online Version by Markaz al-Turāth al-Birmajiyat.

Reilly, Philip R. 2015. "Eugenics and Involuntary Sterilization: 1907–2015." *Annual Review of Genomics and Human Genetics* 16: 351–368. https://doi.org/10.1146/annurev-genom-090314-024930.

Robinson, Gene E., Rina Bliss, and Matthew E. Hudson. 2024. "The Genomic Case Against Genetic Determinism." *PLoS Biology* 22 (2): e3002510. https://doi.org/10.1371/journal.pbio.3002510.

Rutherford, Adam. 2022. *Control: The Dark History and Troubling Present of Eugenics*. New York: W. W. Norton.

Schlegel, Alexander, Prescott Alexander, Walter Sinnott-Armstrong, Adina Roskies, Peter U. Tse, and Thalia Wheatley. 2013. "Barking Up the Wrong Free: Readiness Potentials Reflect Processes Independent of Conscious Will." *Experimental Brain Research* 229 (3): 329–335. https://doi.org/10.1007/s00221-013-3479-3.

Schmidt, Stefan, Han-Gue Jo, Marc Wittmann, and Thilo Hinterberger. 2016. "'Catching the Waves'–Slow Cortical Potentials as Moderator of Voluntary

Action." *Neuroscience & Biobehavioral Reviews* 68: 639–650. https://doi.org/10.1016/j.neubiorev.2016.06.023.

Schmidt, Stefan, Prisca R. Bauer, and Fynn-Mathis Trautwein. 2024. "Neurophenomenology in Action: Integrating the First-Person Perspective into the Libet Experiment." *Mindfulness*: 1–14. https://doi.org/10.1007/s12671-024-02368-9.

Shahrastānī, Muḥammad b. ʿAbd al-Karīm. 1984. *Muslin Sects and Divisions: The Section on Muslim Sects in Kitāb al-Milal wa ʾl-Niḥal by Muḥammad b. ʿAbd al-Karīm Shahrastānī (d. 1153)*. Translated by A. K. Kazi and J. G. Flynn. Abingdon: Routledge.

Smatti, Maria K., Hadi Mbarek, Hamdi M. Yassine, and Dorret I. Boomsma. 2025. "Understanding Heritable Variation Among Hosts in Infectious Diseases Through the Lens of Twin Studies." *Genes* 16 (2): 177. https://doi.org/10.3390/genes16020177.

Soon, Chun Siong, Marcel Brass, Hans-Jochen Heinze, and John-Dylan Haynes. 2008. "Unconscious Determinants of Free Decisions in the Human Brain." *Nature Neuroscience* 11 (5): 543–545. https://doi.org/10.1038/nn.2112.

Al-Ṭabarī, Abi Jaʾfar Mohd. 2010. *Tafsīr al-Ṭabarī*. Cairo: Dar al-Hadith.

Al-Ṭabarī, Mohd bin Jarir bin Yazid. 2013. *Tafsīr al-Ṭabarī: Jamiʿ al-Bayān ʿan Taʾwīl Ay al-Qurʾan*. Riyadh: Dar al-Nashr.

Al-Tafsīr al-Muyassar li-l-Qurʾān al-Karīm. 1988. Cairo: Maktabat al-Nahḍah al-Miṣriyyah.

Al-Ṭaḥāwī, Aḥmad ibn Muḥammad, and ʿAlī ibn ʿAlī Ibn Abī al-ʿIzz. 1997. *Sharḥ al-ʿAqīdah al-Ṭaḥāwīyyah*. Beirut: Muʾassasat al-Risālah.

Tammen, Stephanie A., Simonetta Friso, and Sang-Woon Choi. 2013. "Epigenetics: The Link Between Nature and Nurture." *Molecular Aspects of Medicine* 34 (4): 753–764. https://doi.org/10.1016/j.mam.2012.07.018.

Tiihonen, J., M.-R. Rautiainen, H. M. Ollila, E. Repo-Tiihonen, M. Virkkunen, et al. 2015. "Genetic Background of Extreme Violent Behavior." *Molecular Psychiatry* 20 (6): 786–792. https://doi.org/10.1038/mp.2014.130.

Vassos, E., D. A. Collier, and S. Fazel. 2014. "Systematic Meta-Analyses and Field Synopsis of Genetic Association Studies of Violence and Aggression." *Molecular Psychiatry* 19 (4): 471–477. https://doi.org/10.1038/mp.2013.31.

Walinga, Jennifer. 2014. "Is Personality More Nature or More Nurture? Behavioural and Molecular Genetics." In *Introduction to Psychology*, 1st Canadian ed., by Charles Stangor and Jennifer Walinga, 557–565. Victoria, BC: Pressbooks. Accessed from https://opentextbc.ca/introductiontopsychology/chapter/11-3-is-personality-more-nature-or-more-nurture-behavioral-and-molecular-genetics/.

Wehr, Hans. 1974. *A Dictionary of Modern Written Arabic–English*. Edited by J. Milton Cowan. Beirut: Maktabah Lubnān.

Willoughby, Emily A., Tinca J. C. Polderman, and Brian B. Boutwell. 2023. "Behavioural Genetics Methods." *Nature Reviews Methods Primers* 3 (10): 32. https://doi.org/10.1038/s43586-022-00191-x.

Yehuda, Rachel, and Amy Lehrner. 2018. "Intergenerational Transmission of Trauma Effects: Putative Role of Epigenetic Mechanisms." *World Psychiatry* 17 (3): 243–257. https://doi.org/10.1002/wps.20568.

CHAPTER 6

Assessing Evidence for Extraterrestrial Intelligent Life and Implications for Human Uniqueness

Bilal Ali Ansari and Ahsan Arozullah

Introduction: Extraterrestrial Intelligent Life and Human Uniqueness

The question of human uniqueness has long occupied a central place in philosophical, theological, and scientific discourse and has gained renewed significance in light of contemporary debates about the possibility of observable extraterrestrial intelligent life.[1] Within the Islamic intellectual tradition, this question was not approached in isolation from the broader

[1] Editor's Note: This chapter divides sources of knowledge and truth claims into revelatory, rational, and empirical categories. The authors hold that this division allows for precisely assessing and resolving apparent conflicts. Although slightly different terms are utilized, this schema largely aligns with Chapter 2's epistemic framework.

B. A. Ansari · A. Arozullah (✉)
Darul Qasim College, Glendale Heights, IL, USA
e-mail: ahsan.arozullah@darulqasim.org

B. A. Ansari
e-mail: bilal.ali@darulqasim.org

© The Author(s), under exclusive license to Springer Nature Switzerland AG 2026

A. I. Padela (ed.), *Bridging Bioscience and Islam*, Islam & Science,
https://doi.org/10.1007/978-3-032-09996-9_6

epistemological framework developed by Islamic discursive theologians (*mutakallimūn*) who identified empirical evidence, reason, and revelation as harmonious sources of knowledge, all capable of producing certainty. This chapter applies the broad epistemological concepts of certainty and uncertainty to revelatory, rational, and empirical knowledge that informs our understanding of the captivating question: Are humans the sole bearers of intelligence in the universe? In this chapter, we use the following terms to describe three sources of knowledge as described in Chapter 2: (1) "revelatory" to refer to testimonial knowledge in the form of true reports (*al-khabar al-ṣādiq*); (2) "rational" to refer to rational knowledge from the intellect (*ʿaql*); and (3) "empirical" to refer to knowledge from the five external senses which provide humans sensory knowledge. We use astrobiology (also referred to as exobiology or xenobiology) to refer to a growing multidisciplinary field exploring the nature, existence, and search for extraterrestrial life. Extraterrestrial life has been defined as life that may exist or may have existed in the universe outside of our Earth (Margulis et al. 2024). Additionally, extraterrestrial intelligent life is defined as hypothetical extraterrestrial life capable of thinking and purposeful activity (Shostak 2023). Lastly, in the context of this chapter, "observable" is used to distinguish the kind of extraterrestrial intelligent life sought by modern astrobiologists, life forms whose existence could, in principle, be empirically detected through observation, measurement, or rational inference, from beings such as angels and *jinn*.

Discovery of extraterrestrial intelligent life could significantly challenge notions of human uniqueness, a concept that acknowledges the uniqueness of humans and asserts that humans, as a species, possess certain characteristics or attributes that set them apart from other living organisms. Both theological and empirical approaches to human uniqueness agree that humans have significantly advanced cognitive and intellectual abilities compared to other species. This includes complex thinking, problem-solving, language, and the development of advanced cultures and societies. In addition, humans can develop moral and ethical reasoning, which has led to complex ethical systems and the ability to ponder questions of right and wrong. Humans demonstrate a high level of self-awareness and consciousness, enabling them to reflect on their own existence, the nature of reality, and their place in the world. Humans also have a unique capacity for religious and spiritual beliefs, often resulting in the search for meaning and purpose in life.

An Islamic theological perspective on human uniqueness asserts that humans have been favored with Divine guidance through human messengers who receive *waḥy* (Divine revelation) and then distribute it. These human messengers exemplify specific characteristics such as honesty, integrity, patience, generosity, and trustworthiness that distinguish them as exceptional individuals. Islamic teachings assert that messengers are protected by Allah from committing sins, known as *ʿiṣma* (Divine protection), ensuring that they maintain the highest level of integrity and are reliable conveyors of Divine revelation. Allah chooses these human messengers to fulfill the critical role of guiding humanity and serving as examples for all people to follow. Their lives, teachings, and the Divine support they received are central to Islamic belief and serve as a foundation for the moral and spiritual guidance of Muslims.

In terms of Islam's perspective on Prophetic exceptionalism, Islamic scholars unanimously agree that the Prophet Muhammad is the greatest and most exceptional of creation in status before Allah.[2] His supreme uniqueness among creation is supported by the following honors that Allah granted him: Allah took him as a close friend (*khalīl*); He made him the Seal of His Messengers; He revealed the Qurʾan to him; He caused miracles to happen at his hand (e.g., the splitting of the moon) and made his message the truth that applies until the end of the world. On the Day of Resurrection, the Messenger of Allah will be the first one for whom the grave will be opened; the first one to intercede and the first one whose intercession will be accepted by Allah; the first one to cross the *ṣirāṭ* (the bridge that crosses Hell into the Garden); the first one to knock at the gates of the Garden, and the first human to enter the Garden. The reality that the Messenger of Allah is a human being gives further support for the uniqueness of human beings among all of Allah's creation.

In addition to the key elements described above for messengers, the Qurʾan states that humans in general are favored above many other creatures: "*Indeed, We have honored the children of Adam, carried them on land and sea, granted them good and lawful provisions, and privileged them far above many of Our creatures*" (Q.17:70). Allah created the human being in the best form, proportionately designed, in perfect measures: "*We have created humans in the best of forms*" (Q.95:4). This best form places human beings on an exceptional level in comparison

[2] See al-Nawawī (1972: *sub ḥadīth* 4223); Ibn Ḥajar (1970: *sub ḥadīth* 7404); and al-Ālūsī (n.d.: 4:284).

to other living beings. The ability of humans to communicate in a linguistic manner, vocally as well as through literature, is another exceptional human quality: "*He has created man. He has taught him eloquence (how to express himself).*" (Q.55:2–3).

In summary, this chapter applies the epistemological categories of certainty and uncertainty to revelatory, rational, and empirical knowledge concerning extraterrestrial intelligent life. Importantly, the discovery of extraterrestrial intelligent life could challenge notions of human uniqueness. From believers, it invites reflection on how humanity's distinct status among God's creation would be reinterpreted, and whether other intelligent beings would receive divine guidance analogous to that conveyed through human messengers. For scientists, it poses the challenge of reassessing claims about humanity's biological and evolutionary exceptionalism. By situating these questions within the integrated framework of Islamic theology, the chapter offers a perspective in which the search for extraterrestrial intelligence is not a challenge to faith but an extension of long-standing reflections on creation, guidance, and the distinctive role of the human being in the cosmos.

Epistemological Frameworks for the Search of Extraterrestrial Intelligent Life

It is instructive at this stage to briefly return to the discussion of knowledge sources outlined in Chapter 2, where an Islamic epistemological framework was introduced in detail. Given the intervening chapters, a concise review is warranted in order to reorient the reader to the categories of certainty and uncertainty and the three recognized sources of knowledge: reason, sense perception, and reliable testimony. Revisiting these concepts here also serves as a case study that applies the epistemological framework to a stimulating and fascinating question, namely, the possibility of observable extraterrestrial intelligent life. In doing so, it demonstrates the utility of the framework for addressing contemporary issues and brings earlier theoretical discussions into direct engagement with a live theological and scientific debate.

The foundational principle in Islamic epistemology is that all knowledge in existence, in both the physical and metaphysical realms, originates in a single, Divine source. Due to this synthesis of origin, all knowledge is essentially "religious," and the scientific study of the natural world cannot be divorced from the sciences that study the Divine, scripture, law, ethics, and so on. At the same time, Divinely stipulated knowledge that has been conveyed with conclusive certainty and whose meaning is conclusively certain holds the highest rank of epistemological strength; it is understood, interpreted, and corroborated using knowledge from sensory and rational sources (al-Taftāzānī 2020: 54; al-Shāmī 2021: 60–61).

In contrast to this approach, a purely empirical/rational approach to the search for extraterrestrial intelligent life can be traced back to Epicurus (341–270 BCE). He believed that the goal of human life was happiness, defined as the absence of physical pain and mental disturbance. From Epicurus' perspective, which included the absence of an afterlife, discovering extraterrestrial intelligent life would diminish the belief in human uniqueness and would challenge the superiority of transcendent entities that generate fears, such as punishment after death (Konstan 2022). Eliminating those fears would free people to pursue the pleasures, both physical and mental, to which they are naturally drawn and achieve satisfaction and tranquility. In support of this philosophy, Epicurus and his followers pursued the understanding of nature based on atomic materialism through an empiricist approach to knowledge in which sensations, together with the perceptions of pleasure and pain, are infallible criteria. This investigative approach is far more limited than the Muslim approach.

According to the broad epistemological framework articulated by Islamic discursive theologians (*mutakallimūn*), all knowledge is classified as either pre-eternal (*qadīm*) or temporal (*ḥādith*). The former pertains exclusively to the Creator, whereas human knowledge belongs to the latter category and is further divisible into what is necessary (*ḍarūrī*) or what is acquired (*naẓarī*). Necessary knowledge is created by God in the knower without volition or deliberation; examples include knowledge of one's own existence or of one's hunger or pain. Acquired knowledge, although likewise created by God, is obtained through human effort and volition. As outlined in Chapter 2, Sunnī *kalām* theologians agree that there are three sources of acquired knowledge, each independently capable of yielding certainty: (1) the intellect (*ʿaql*), which provides humans with rational knowledge; (2) the five external senses, which produce sensory knowledge; and (3) testimonial knowledge in the form of true reports (*al-khabar al-ṣādiq*), the most authoritative and truest form of which is Divine revelation (al-Ṣābūnī 1440/2018: 11). These three sources are conventionally designated as (1) sensory (*ḥissī*), (2) rational (*ʿāqlī*), and (3) transmitted (*naqlī*) or testimonial.

This simple yet broadly accommodating epistemological framework enables the appropriation and acceptance of empirical evidence, rational understandings, and revelatory claims, while challenging the false dichotomy between natural science and sacred tradition. While all three knowledge sources are equally capable of providing certainty, they are not equal in terms of sacredness. And each must be judged in accordance with the strength of its evidence, as to whether they yield certainty (*qaṭʿīyya*) or presumptive likelihood.

In this epistemological framework, no two objective knowledge claims can truly contradict one another. In the presence of an apparent contradiction, a process of reconciliation involving various steps outlined below is applied to demonstrate an actual cohesion between knowledge sources. This is because Islamic theologians agree that while empirical data gathered from the senses can establish conclusive truths (al-Nasafī 1990: 1:9), the singularity of the origin of knowledge necessitates a harmony of empirical data with sound rational or revelatory claims.

A relatively basic scheme to help navigate apparent conflicts between knowledge claims provides four basic scenarios for evidentiary resolution. For the sake of simplicity, we specifically present here apparent conflict scenarios between revelatory evidence on one hand and empirical or rational evidence on the other. In each scenario, testimonial evidence may be either epistemically conclusive (*qaṭʿī*) or presumptive (*ẓannī*) in both its establishment (*thubūt*), meaning how the proof is conveyed, and in its meaning (*dalāla*). For example, *ḥadīth* reports may be conclusively conveyed from the Prophet Muhammad but only probabilistic in their significance and meaning. Similarly, a report may be only presumptively attributable to the Prophet, but conclusively certain in its meaning. Empirical or rational evidence may likewise be either conclusively certain or presumptively likely in establishment and meaning. Each of the three knowledge sources thus consists of four levels of evidentiary strength, reaching a total of twelve possibilities. In only the first of the four levels is the evidence conclusive overall, as in each of the other three cases any probability, in either establishment or indication, renders the evidence probabilistic as a whole (see Table 6.1).

In the first of four conflict scenarios, an individual perceives a conflict between a conclusive, revelatory testimonial argument and a conclusive empirical/rational argument. Such a contradiction is impossible in an Islamic epistemological framework. In the last of the four scenarios, both revelatory and rational or empirical evidence are probabilistic. In

Table 6.1 Epistemic value of testimonial evidence

	Establishment (thubūt)	*Signification* (dalāla)	*Overall epistemic value*
1	Conclusively certain	Conclusively certain	Conclusively certain
2	Conclusively certain	Presumptive	Presumptively authoritative
3	Presumptively	Conclusively certain	Presumptively authoritative
4	Presumptive	Presumptive	Presumptively authoritative

this scenario, one reconciles the two arguments by either avoiding the literal sense of the arguments and preferring some form of interpretation of the sources considering one another, or they may simply give preference to the literal understanding of revelation and dismiss the rational/empirical argument. In a third scenario, when the revelatory evidence is conclusively certain and the rational/empirical is presumptively authoritative, the revelatory argument is always given precedence. However, in the final scenario, when the rational or empirical argument is conclusively certain and the revelatory argument is presumptively authoritative, either in respect to its establishment or its significance, the former argument is given precedence, and the revelatory report is interpreted in a non-literal sense (see Table 6.2) (al-Rāzī 2001; Thanvī 1992: 25–26).

Table 6.2 Relationship between revelatory and empirical/rational evidence based on level of certainty

	Revelatory evidence	*Empirical/rational evidence*	*Contradiction*
1	Conclusively certain	Conclusively certain	Impossible
2	Conclusively certain	Presumptively authoritative	Revelation is understood literally
3	Presumptively authoritative	Conclusively certain	Revelation is interpreted
4	Presumptively authoritative	Presumptively authoritative	Revelation is understood literally first, but interpretation is possible

Rational and Empirical Evidence for Extraterrestrial Intelligent Life

The broad rational metaphysical possibility of extraterrestrial intelligent life has been explored in Hamza Karamali's "Theological Information on the Existence of Intelligent Life Outside Our Solar System: Metaphysics, Scripture, and Science" (Karamali 2024). In sum, Karamali argues that the late *kalām* tradition[3] contains rational methods that allow us to gain insight into the potential or actual existence of something, including extraterrestrial intelligent life. Rationally speaking, there is no evidence that would lead us to conclude that extraterrestrial intelligent life exists or does not exist. Extraterrestrial intelligent life is neither rationally impossible (*mustaḥīl*) nor rationally necessary (*wājib*). It is rationally possible (*jāʾiz*). The *kalām* tradition's metaphysical reasoning, which is built on the foundation of revelatory evidence, does in fact grant us conclusive evidence of a certain type of extraterrestrial intelligent life that is usually beyond the interest of modern scientists: angels and *jinn*, both of which are non-human intelligent beings frequently mentioned in the Qurʾan and which Sunnī creed necessitates belief in. Modern reductionist natural scientists are disinterested in such beings because, though they are non-human and intelligent, they are also normally unobservable through empirical methods. Revelation informs us that such life exists beyond our Earth, evidenced by verses in the Qurʾan such as Q.35:1, Q.42:5, and Q.70:4. Still, this knowledge of angels and *jinn* beyond our Earth is based on revelatory sources and not on pure reason. Nevertheless, a sound process of rational investigation into the possibility of empirically observable extraterrestrial intelligent life will lead us to conclude that it is possible for it to exist, just as it is possible for it not to exist.

For scientists, the search for ETI is based on empirical and rational evidence. Scientists utilize tools, such as the Drake Equation, to organize their search for key evidence supporting the existence of ETI and to estimate the likelihood that ETI will be detectable by us on Earth. The Drake Equation is named after the famous astronomer, Frank Drake, who created the formula to organize an inaugural meeting of experts held in West Virginia in 1961. The formula estimates N, defined as the number

[3] By late *kalām* tradition, Karamali refers broadly to the works, figures, and learning centers of the pre-modern Sunnī scholarly tradition, without demarcating a specific point that differentiates the early and late scholarly traditions.

of civilizations in the Milky Way galaxy whose electromagnetic emissions are detectable:

$$N = R* \times f_p \times n_e \times f_l \times f_i \times f_c \times L$$

(Shostak 2021)

The terms in the Drake Equation are defined as follows:

R_* The rate of formation of stars suitable for the development of intelligent life (number per year).
f_p The fraction of those stars with planetary systems.
n_e The number of planets, per solar system, with an environment suitable for life.
f_l The fraction of suitable planets on which life actually appears.
f_i The fraction of life bearing planets on which intelligent life emerges.
f_c The fraction of civilizations that develop a technology that produces detectable signs of their existence.
L The average length of time such civilizations produce such signs (years).
(Shostak 2021)

From an epistemological view, advances in technology such as newer telescopes, exploratory probes, and increased computing capacity have improved the estimated values (level of certainty) for variables such as f_p (fraction of stars with planetary systems) and n_e (number of planets, per solar system, with an environment suitable for life). From an empirical perspective, research using the Kepler space telescope estimated that there could be as many as 300 billion potentially habitable planets within our galaxy (SETI Institute 2020). Of these 300 billion planets, approximately 10 percent exist in a habitable zone defined as its relative location to its star being like the Earth's location relative to the sun (NASA n.d.). Therefore, empirical evidence suggests that there may be 10 billion habitable planets in the universe.

However, one major hurdle in applying the Drake Equation is that a variable such as L (average length of time such civilizations produce detectable signs) is extremely challenging to quantify through biology or astronomy, but rather depends on alien behavior. Furthermore, Noor (2022) notes that one cannot assume that ETI will follow the patterns of life that we observe on Earth. If life exists on other planets, does such life follow hereditary patterns like species on Earth and does evolution function differently? How do these factors affect the length of time such alien civilizations could produce detectable signs? From a rational perspective, one can surmise that ETI may exist completely independently of life on Earth. This would include aspects of life that we may mistakenly assume are universal (e.g., lifespan, heredity, modes of communication) because our conception of life itself is limited by our observations of life on Earth.

One area of technological advances in detecting ETI is related to detecting and decoding extraterrestrial radio signals. These advances can improve the estimate of f_c (fraction of civilizations that develop a technology that produces detectable signs of their existence) as these civilizations' signals can now be more easily detected. Zhang et al. (2023) provide an overview of real-time fast radio burst searching. The first Fast Radio Burst (FRB) was discovered in 2007 and FRBs are currently the brightest bursts in the radio band detectable in the universe. Thousands of burst pulses reach the Earth each day, with the nearly 1,000 FRBs that have been discovered, dozens of these have been identified as repeated bursts of FRBs. Another area of rapid advancement in the search for ETI is the application of artificial intelligence (AI) to the increasingly massive amounts of data retrieved by telescopes and probes. Saphalya Peta describes several possible applications of AI for analyzing the growing amounts of data available to find exoplanets and gain new insights. The search for ETI utilizes machine learning to find artificially created electromagnetic waves in data sources, including other techno signatures (Peta 2023). However, based on the ongoing activities and discoveries, we currently have no conclusive knowledge regarding the origin or physical mechanism of the FRBs. Therefore, ongoing research in this field is focusing on optimizing the FRB origin positioning accuracy and recording the data generated when the pulse bursts (Zhang et al. 2023).

In summary, over the past 60 years since the Drake Equation was formulated, there has been significant progress in refining the estimates for some variables based on empirical evidence using advanced technology. For example, evidence from cutting-edge telescopes has refined the fraction of new stars with planetary systems and the number of planets per solar system that are habitable for complex life. However, due to significant uncertainties in other variables, estimates for N have ranged from 1, a scenario where Earth houses the only transmitting galactic society, to several million. Drake himself suggested that $N = 10{,}000$ based on the assumption that new transmitting societies are produced at intervals of one per year, with an average lifetime of 10,000 years. The Drake

Equation retains considerable utility for organizing research and discussions about ETI, even though it cannot be conclusively "solved" or even accurately estimated. In conclusion, until we establish two-way communication or a face-to-face interaction with ETI, the existence of ETI remains inconclusive and speculative based on the available empirical and rational evidence.

Exploring Revelatory Evidence for ETI

It is instructive to conceive of Islamic theologians as explorers navigating an expansive cosmos of revelatory evidence. Analogous to astronomers who survey the skies for phenomena both visible and imperceptible to the naked eye, scholars of Islam's textual sources, principally the Qurʾan and *ḥadīth*,[4] scan the vast corpus of transmitted reports to formulate observations concerning the created order. In this capacity, they explore and investigate the Divine word with methodological rigor to deepen their understanding of the self, the cosmos, and the Divine within Islam's integrated *tawḥīdic* ontology.

These explorers actually discovered revelatory evidence for extraterrestrial intelligent life in a number of verses that speak to the existence of angels and *jinn* in the skies. However, it is also known through revelatory evidence and the general absence of empirical evidence that such beings are not generally observable to humans.[5] Thus, while revelatory evidence provides a quick answer to the question of ETI in the form of angels and *jinn* that occupy the cosmos, the question of normally observable ETI beyond our Earth requires a deeper investigation of revelation.

It is here that Muslim scholars find possible evidence of not only potentially observable extraterrestrial intelligent life but also multiple earths. Two Qurʾanic verses and several explanatory *ḥadīth* reports of these verses are the focus of our attention:

[4] A *ḥadīth* is generally defined as any report containing a statement, action, or tacit approval attributed to the Prophet, or, by extension, his Companions or their Followers. A *ḥadīth* consists of two components: (1) the action or statement being transmitted (*matn*); and (2) the source of the action/statement, which is represented by a chain of reporters who transmitted it (*sanad*).

[5] See, for instance, the implications of the Qurʾanic verses: Q.25:22, Q.17:94–95, Q.3:164, Q.6:9, and so on.

(1) Q.65:12, in which the existence of multiple earths is suggested, and
(2) Q.42:29, in which the scattering of walking, or mobile, creatures throughout the skies and the earth is mentioned.

The Multiple Earths Verse

In their study of Q.65:12, Qurʾanic interpreters as early as the Companions of the Prophet entertained a series of inquiries about the existence of multiple earths, over a millennium before modern astronomers.

> *It is Allah who created the seven skies and of the earth their like, the command descending down through all of them, so that you may know that Allah has power over all things and that Allah encompasses all things in knowledge.* (Q.65:12)

Two interpretations are offered to explain the "likeness" described in the expression: *and of the earth their like*. In the first interpretation, likeness simply refers to a similarity in complexity and not in number. There is a single earth (*arḍ*) that is similarly complex to the skies in that it is composed of a variety of elements: earth, air, water, fire, and so on. However, the overwhelming majority of Qurʾanic interpreters posit that the verse is likening the earth to the skies in number and that the verse is proving a multiplicity of earths like the well-established notion of a multiplicity of the skies (al-Qurṭubī 1433/2012: sub Q.65:12; see also al-Rāzī 2001; Ibn Kathīr 2013; and al-Bayḍāwī 2001: sub Q.65:12). According to a claim cited by the Persian polymath and classical Qurʾanic exegete Jār Allah al-Zamakhsharī (d. 538/1143), no other verse in the Qurʾan suggests that there are seven earths (al-Zamakhsharī 1433/2012: sub Q.65:12).[6]

After asserting that the scholarly majority agrees that the verse proves that there are "seven" earths, the Andalusian polymath Abū ʿAbd Allah

[6] Al-Zamakhsharī says that these multiple earths are similar to the multiple skies in that "the command is descending down through all of them," meaning that the Divine command is enforced throughout them without exception (al-Zamakhsharī 1433/2012: sub Q.65:12). He quotes the early Basran exegete and traditionist Qatāda (d. 117/735) as stating that each sky and earth contains a form of creation as well as the Divine command and decree.

al-Qurṭubī (d. 671/1273) describes these earths as layered one over another, separated by the distance of what lies between the earth and sky, with each earth inhabited by various creations of Allah. The early Qurʾanic exegete al-Ḍaḥḥāk ibn Muzāḥim (d. 102/721) similarly suggests the idea of seven earths but disagrees as to the nature of their arrangement, stating that they are not layered one above another unlike the skies (al-Qurṭubī 1433/2012 sub Q.65:12).

The majority interpretation that suggests multiple earths is considered the preponderant view due to the corroboration of several *ḥadīth* reports (Ibn Kathīr 2013: 1:34–41). One such report is a Prophetic supplication found in the rigorously authentic *ḥadīth* collections which reads: "O Allah, Lord of the seven skies and what they shelter, Lord of the seven earths and what they carry..." (al-Tirmidhī 1998: *Daʿawāt*, *bāb minhu* no. 3523, 5:424).[7] More explicitly and significantly, an exegetical statement of the Companion and cousin of the Prophet, ʿAbd Allah Ibn ʿAbbās (d. 68/687) in commentary of this verse (Q.65:12) suggests that the seven earths are not only inhabited by intelligent life but also by beings who are analogous to this Earth's prophets: "In each earth there is the like of Abraham" (al-Ṭabarī 1420/2000: 23:469 sub Q.65:12). In another version of the report, Ibn ʿAbbās states: "[There are] seven earths, on each earth there is a prophet like your prophet, an Adam like your Adam, a Noah like your Noah, an Abraham like your Abraham, and Jesus like your Jesus." (al-Ḥākim 1411/1990: *ḥadīth* no. 3822, 3833, 3869; al-Bayhaqī 1413/1993: *ḥadīth* 831 and 832, 2:267–268). If the *ḥadīth* critical methodology can determine the authenticity of Ibn ʿAbbās's report, testimonial evidence would certainly support at least the possibility or probability of ETI.

Sourcing the Seven-Earths Report

The critical methodological apparatus employed by traditionists (*muḥaddith*) to determine the soundness of such a report begins with sourcing. Ibn ʿAbbās's above report is sourced in a variety of works, ranging from *tafsīr* (Qurʾanic exegesis) works to *ḥadīth* collections and theological monographs. The earliest of such works is the Persian

[7] Al-Ḥākim (1411/1990: 1:614) affirms that its chain is sound while noting that neither al-Bukhārī nor Muslim transmit it (see also Ibn Khuzayma 1400/1980: 4:150; and Ibn Ḥibbān 1432/2011: 6:426).

polymath and exegete Ibn Jarīr al-Ṭabarī's (d. 310/923) *Jāmiʿ al-Bayān fī Tafsīr Āyi al-Qurʾān*, in which the brief version of the report reads:

> ʿAmr b. ʿAlī and Muḥammad b. al-Muthannā both state that Muḥammad b. Jaʿfar narrated to us that Shuʿba narrated to us from ʿAmr b. Murra from Abū al-Ḍuḥā from Ibn ʿAbbās that he said about the verse: "*Allah is the one who created the seven skies and of the earth their like*: ʿAmr states that (Ibn ʿAbbās) states: 'On each earth is the like of Ibrāhīm and that which resembles the creation on this earth'."
>
> (al-Ṭabarī 1422/2001: 23:77, sub Q.65:12)

The early *ḥadīth* scholar and critic al-Ḥākim al-Nīsāpūrī (d. 405/1014) cites two fully connected chains to Ibn ʿAbbās in his *al-Mustadrak* (al-Ḥākim 1411/1990: *ḥadīth* no. 3822, 3833, 3869) with more detailed wording, and judges the report to be sound.[8] His version of the report reads: "[There are] seven earths; in each earth there is a prophet like your prophet, and an Adam like your Adam, and a Noah like your Noah, and an Abraham like your Abraham, and Jesus like your Jesus."[9] See Fig. 6.1 for a visual representation of the chains in al-Ḥākim's *al-Mustadrak* and Ibn Jarīr al-Ṭabarī's *al-Tafsīr*.

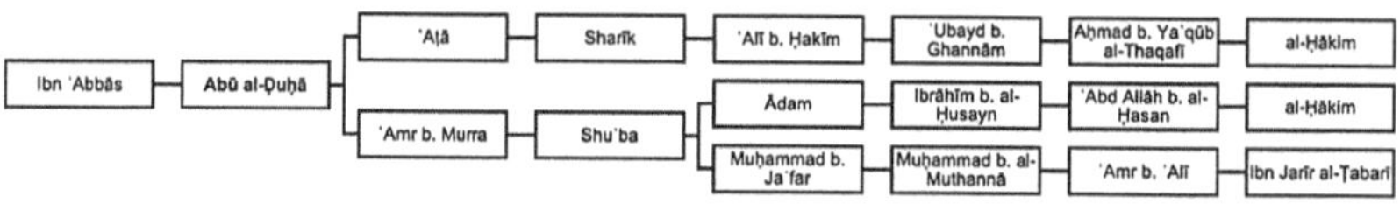

Fig. 6.1 All the known chains of transmission for the Ibn ʿAbbās exegetical report on the seven earths

[8] Ibn ʿAbbās's report is also transmitted by Abū Bakr al-Bayhaqī (d. 458/1066) in his *al-Asmāʾ wa-l-Ṣifāt* using two chains (see al-Bayhaqī 1413/1993: *ḥadīth* 831 and 832, 2:267–268). However, both chains run directly through those of his teacher, al-Ḥākim al-Nīsāpūrī, and provide us no additional information.

[9] Ibn Abī Ḥātim al-Rāzī (d. 327/938) transmits the same text in his *Tafsīr Ibn Abī Ḥātim*, (1419/1998: sub Q.65:12, 10:3361 *ḥadīth* no. 18,918), but with only a partial chain.

Chain Analysis of the Report

It should be noted that the report is explicitly traceable only to the Companion Ibn ʿAbbās (may Allah be pleased with him), and therefore cannot be categorically or explicitly raised (*marfūʿ*) to the Prophet (pbuh). While it may be inferred that the statement implicitly originates in Prophetic teaching, an analysis of this particular report's chain to determine its reliability ultimately ends with a conclusion that it is more accurately attributable to Ibn ʿAbbās.[10] To determine even this, the *ḥadīth* critical apparatus employed by traditionists considers two main aspects of the *sanad*: (1) the credibility of each reporter (*rāwī*) in respect to their accuracy (*ḍabṭ*) as well as their probity (*ʿadāla*); and (2) the manner of their transmission of the report in order to rule out the possibility of a disconnected chain. Beyond these two considerations, traditionists search both the text (*matn*) and chain for anomalies (*shudhūdh*) and subtle defects identifiable only to experts. Further analysis of the report's text and its various implications is entertained once the sourcing of the report is properly authenticated.[11]

The authenticity of Ibn ʿAbbās's seven-earths report has long been a subject of some debate. Expert traditionists like Abū Bakr al-Bayhaqī (d. 458/1066) and Shams al-Dīn al-Dhahabī (d. 748/1348) admitted that its chain appears sound (*ṣaḥīḥ*), that is, it meets the requirements of narrator accuracy, probity, and chain connection, yet criticized the report as being wholly anomalous (*shādhdh bi-marra*), and note that only one student of Ibn ʿAbbās, namely, Abū al-Ḍuḥā (d. 100/719), transmits it (al-Bayhaqī 1413/1993: 2:2). Al-Dhahabī nevertheless asserts about its chain that despite the relative laxity (*līn*) in two of its narrators, namely, Sharīk and ʿAṭāʾ, it does not reach a degree that would justify rejecting their reports. About the report's meaning and import, al-Dhahabī calls

[10] Editor's note: This analysis pertains to the type of *ḥadīth* based on the chain. With regard the Companion reporting something he heard directly from the Prophet Muhammad, further discussion on the matter follows below in the origin of the report section.

[11] The *muḥaddith* leverages an extensive knowledge of narrator biographies and historical timelines to conduct a rigorous evaluation of the chain. Five main considerations assist in determining the soundness of a report, two related to the reporters themselves, namely, their probity and accuracy, one related to the connectedness of the chain of reporters, and two auxiliary considerations: one to ensure an absence of anomalies; and the second to ensure the absence of hidden defects.

it a "perplexing problem that leaves the listener bewildered... it falls under the category of: 'Listen and be silent.'" (al-Dhahabī 1416/1995: 75).[12] Other *ḥadīth* authorities, such as Ibn Kathīr (d. 774/1373) and Jalāl al-Dīn al-Suyūṭī (d. 911/1505) agree with this criticism, the former suggesting that if the report is ascribable to Ibn ʿAbbās, it must have originated in Judeo-Christian sources (*isrāʾiliyyāt*), and the latter stating that they find al-Ḥākim's authentication of the report odd (Ibn Kathīr 1999: 8:157, 2013: 1:39, sub Q.65:12; al-Suyūṭī 1437/2016: 3:274–277).[13] Their declaration of the report as anomalous is likely due to a combination of several factors: (1) that only one student of Ibn ʿAbbās narrates the report; (2) that the idea of beings similar to our prophets inhabiting other earths is not found in other sound *ḥadīth*; and (3) that a report of such potential significant theological importance should be attested to by a larger body of independent reporters.[14]

According to the seven-earth report's critics, who assert that since it is: (1) not Prophetic; (2) singularly transmitted through Ibn ʿAbbās; and (3) odd in its meaning, the report provides no clear evidence on the question of extraterrestrial intelligent life. Later scholars like the post-Mughal-era Indian theologians and contemporaries Muḥammad Qāsim al-Nānawtawī (d. 1297/1880) and ʿAbd al-Ḥayy al-Laknawī (d. 1304/

[12] The Qurʾanic exegete, Abū Ḥayyān, strangely declares that the report is an unquestionable forgery, a transmission of the liar al-Wāqidī (see Abū Ḥayyān 1441/2020: 20:382–383 sub Q.65:12). Al-Laknawī responds by noting that the paths of the report transmitted by al-Ḥākim and al-Bayhaqī, and which are ruled to have sound chains by them, do not contain al-Wāqidī at all, so his presence in other paths is not problematic here. Further, al-Wāqidī is not universally asperse, as some investigative traditionists, such as Yaʿqūb b. Abī Shayba, Musʿab al-Zubayrī, Yazīd b. Hārun, and others, consider him to be trustworthy (*thiqa*) (see al-Laknawī 1419 AH: 403; Ibn Sayyid al-Nās n.d.: 68–70).

[13] Al-Laknawī is critical of this claim, stating that no one made such a claim before Ibn Kathīr and those after him simply imitated his supposition, which is refuted by "clear texts and shining proofs" (see al-Laknawī 1419 AH: 414).

[14] Ibn Ḥajar al-Haytamī notes that al-Bayhaqī's criticism of the *ḥadīth* is one of anomaly in the text and not the chain. He subsequently cites al-Suyūṭī's strong endorsement of this criticism, since the soundness of a report's chain does not necessitate the soundness of its meaning. Furthermore, once it becomes apparent that a *ḥadīth* is weak (in meaning), one is absolved of the need to interpret it, since with this type of subject matter weak *ḥadīth* are unacceptable. That said, Ibn Ḥajar grants the possibility of interpreting the report to explain away these other "prophets" as simply warners from the *jinn* who convey the messages of human prophets to these earths. It is not far-fetched to imagine that they are referred to by the names of the prophets whose messages they convey (see al-Haytamī 1989: 165).

1886), while admitting the at-best probabilistic nature of the report as evidence, warn against dismissing the evidentiary value of the report. As al-Laknawī explains in one of three treatises that he devotes to the analysis of the report, *Zajr al-Nās ʿalā Inkar Athar Ibn ʿAbbās*:

> I adopt a moderate path in this, avoiding the rejectionist position (*ifrāṭ*) of one group and the excess (*tafrīṭ*) of another, as one group seeks to negate the cited report and deem it weak, forged, and so on, while another group inclines to strengthen it from the perspective of the chain and explain it in such a way as to lead to ruin. I am neither content with one approach nor the other.
> (al-Laknawī 1419 AH: 1:401 [9])

Al-Laknawī's argument for why the outright dismissal of the report is unjustified revolves around the following points: (1) all the report's narrators are trustworthy; (2) a number of masters of report authentication (*arbāb al-taṣḥīḥ*), such as al-Ḥākim, al-Bayhaqī, al-Dhahabī, and Ibn Ḥajar al-ʿAsqalānī (d. 852/1449), admit that its chain is sound, while other scholars skilled in judging the preponderance of reports (*arbāb al-tarjīḥ*), such as al-Zurqānī (d. 1122/1710) and al-Shiblī (d. 1332/1914), are silent of any criticism (al-Laknawī 1419 AH: 1:401 [9]). Further, Ibn Ḥajar al-ʿAsqalānī, al-Shiblī, and al-Suyūṭī cite the *ḥadīth* in an evidentiary context, meaning that they were not even dismissive of the report's implications.

Regarding the declaration of the report as anomalous, or *shādhdh*, al-Laknawī notes that anomalous are of two types, the acceptable and the unacceptable, the former which does not negate a report's soundness (*ṣiḥḥa*). This report is of the former category, as it is at worst isolated but not contradictory to what others transmit. Al-Laknawī emphasizes that although Abū al-Ḍuḥā may be alone in narrating the report from Ibn ʿAbbās, he is himself trustworthy and his report does not conflict with the narrations of those stronger than him.[15] He concludes that "in sum, the

[15] Al-Laknawī addresses and refutes six main objections to the report in *Zajr al-Nās*, noting: (1) that al-Ḥākim is an overly lenient *ḥadīth* transmitter; (2) that one of the narrators in the chain is dismissible as confused; (3) that some of the paths of the report's transmission have been criticized; and (4) are anomalous; (5) the unreliability of certain paths of transmission for Ibn ʿAbbās's exegetical comments; and (6) the report's absence in the six major *ḥadīth* collections. The other objections he addresses largely revolve around responding to weak claims that the report's implications are in conflict with other

chain of this report is undoubtedly *ṣaḥīḥ* (sound) or *ḥasan* (fair), even if it is anomalous" (al-Laknawī 1419 AH: 1:406). Nevertheless, as evidentiary knowledge, the report as a transmitted proof at best renders presumptive authenticity (*ẓannī al-thubūt*) and meaning (*ẓannī al-dalāla*).

Origin of the Report

Further, although he grants that the report is not explicitly Prophetic, as it is an exegetical statement of the Companion Ibn ʿAbbās, he cites the well-established principle in *ḥadīth* critical methodology that statements of Companions that are not derivable through independent analytical reasoning must be assumed to be implicitly Prophetic in nature. In other words, since Ibn ʿAbbās could not have reached the conclusion that analogous prophets exist on other earths through independent reasoning, it can be safely assumed that he received this information from the Prophet.[16]

Proliferation of the Report

Defenders of the legitimacy of the report also conceded that the seven-earths *ḥadīth* is relatively singular in its chains and did not widely proliferate in even the first three generations of Islam. The absence of a massively diffused and congruent set of chains (*tawātur*) places the report well into the category of *āḥād* (singular, or non-*mutawātir*) reports that do not yield certain knowledge. Nevertheless, proponents argue that

revelatory evidence and to claims that the report is pointless due to its ambiguity in explaining the manner in which each prophet is being compared to another (al-Laknawī 1419 AH: 1:401–415).

[16] Ibn Ḥajar al-ʿAsqalānī asserts that:

> when a Companion makes a statement in which there is no scope for independent reasoning (*ijtihād*), such as transmission of information about past events, such as the genesis of creation or the stories of prophets, or about future events, such as the trials and circumstances preceding the Day of Resurrection, the statement is considered equivalent in ruling to a Prophetic report (*marfūʿ*), since the relaying of such information presupposes that it was received by them through transmission as well, since there is no scope for *ijtihād* therein.
> (Ibn Ḥajar al-ʿAsqalānī 1413/1996: 107).

since the meanings of the report are not in conflict with certainly established aspects of the faith, at worst the singularity of the report yields probabilistic knowledge and maintains the revelatory possibility of ETI.

Meaning of the Report

The possibility of multiple earths is already established by the Qurʾanic verse Q.65:12 and further corroborated by reports such as the seven-earths report. The new information Ibn ʿAbbās's report provides is not only that of the possibility of ETI, but also that the life on such worlds may include more than just angels. Several Muslim scholars express no doubt that creatures inhabit each of the earths, or layers of the single earth, even if their detailed nature is unknown to anyone but Allah. This group of scholars, according to al-Laknawī, includes al-Khafājī, Sayyid al-Jurjānī, and Ibn Abī Jamra (al-Khafājī n.d.; al-Jurjānī 1419/1988; Ibn Abī Jamra 1444/2022). Some scholars, such as al-Shiblī, al-Zurqānī, al-Ḥalabī, and al-Qastallānī, proposed that these earths are inhabited by the *jinn* due to what is mentioned in certain reports. Some even claim that each consists of unique species (al-Laknawī 1419 AH: 416).[17] Al-Laknawī thus concludes that:

[17] Here, al-Laknawī is citing *Badāʾiʿ al-Duhūr* and "other works." Al-Laknawī quotes Wahb b. Munabbih as stating:

> When Allah created the earth, it was one layer, and He split it and transformed it into seven, as He did with the skies, and made the distance between one layer and the next layer five hundred years. The name of the highest layer is Adīm, the second Basīṭ, the third Thaqīl, the fourth Baṭīḥ, the fifth Janb, the sixth Māsika, and the seventh al-Thurā. The inhabitants of the second earth are peoples who are called al-Ṭamas. Their food is of their own flesh, and their drink is of their own blood. The inhabitants of the third are peoples whose faces are like the faces of the children of Adam. Their mouths are similar to the mouths of dogs. Their hands are similar to the hands of the children of Adam. Their legs are similar to the legs of cows. Hair like sheep wool covers their bodies which serves as their clothing. The inhabitants of the fourth are peoples called Al-Ḥulhām. They have no eyes or feet, though they have wings. The fifth of them are people like mules. They have tails, each tail around three hundred cubits [in length]. The sixth of them are peoples called al-Ḥuthūm. They have black bodies and claws like those of predatory animals. The seventh contains the abode of Iblīs and his armies. (al-Laknawī 1419 AH: 416–417)

> there can be no doubt about the existence of creatures in all the layers as testified to by the narrations. As for their specific description, to which species they belong, their form and shape, its knowledge is consigned to Allah the Exalted. Ibn ʿAbbās's report proves that there are Adams like our Adam and prophets like our prophet therein.
> (al-Laknawī 1419 AH: 417)

Given that al-Laknawī's discussion resides within the discourse of solitary reports (*al-khabar al-wāḥid*), it is fair to interpret his claim of "no doubt" to signify a denial that the report's implications are unlikely (*wahmī*) or doubtful (*shakkī*), not that he is claiming certainty.

Some ṣūfīs understood the seven-earths report to be referring to the World of Image-Exemplars (*ʿālam al-mithāl*), an isthmus realm between the two worlds of the Unseen (*ghayb*) and the Observable (*shahāda*) (al-Laknawī 1419AH: 417).[18] A second explanation, provided by the likes of al-Suyūṭī, al-Qasṭallānī, al-Ḥalabī, and al-Zurqānī, interprets each layer of the seven earthly layers as possessing a guide designated with the name of a prophet in this layer. They are not prophets in reality; rather they take the laws from the prophets in this layer and they convey it to the people in their layer (al-Laknawī 1419 AH: 418).[19] Al-Suyūṭī, for example, states that it is possible to understand these other "prophets" as "warners" who convey the message of the prophets of this layer to the *jinn* of other worlds. Al-Suyūṭī states:

> It is not farfetched to refer to each of them by the name of the prophet whose message they convey. Thus, our Prophet (pbuh) would have a messenger from the *jinn* whose name is the same as his.
> (al-Suyūṭī 1424/2004: *al-Ḥāwī*: 1:462; al-Laknawī 1419 AH: 418)[20]

[18] In this passage, al-Laknawī cites Ḥusayn al-Diyārbakrī's *Kitāb al-Khamīs fī Aḥwāl Anfas Nafī*, which in turn draws on Ibn ʿArabī's *al-Futūḥāt al-Makkiyya* in several places where he quotes a report of *Misbāḥ al-Ẓalām fī Dhikri Mazāl al-Ḥukamā al-Aʿlām*: "Know that numerous reports substantiate the [World of] Image-Exemplars. Among these reports is the *ḥadīth* narrated from Ibn ʿAbbās that reads: "The Kaʿba is among fourteen houses. In each of the seven earths, there exists a creation akin to ours, complete with an Ibn ʿAbbās resembling me." (see Ibn ʿArabī 1405/1985: 2:257).

[19] Here, al-Laknawī is citing al-Ḥalabī, *Insān al-ʿUyūn*.

[20] Al-Laknawī rejects this interpretation as unnecessary since an apparent, literal meaning of the report is not impossible.

A third interpretation, offered by al-Shiblī and preferred by al-Laknawī, posits that Allah sent actual prophets to the creatures of each earthly layer who resemble the Prophets Adam, or Nūḥ, or even the last prophet, Muhammad (pbuh) of this earthly layer. However, the specifics of these prophets and the creatures to whom they were sent are known only to Allah.

Epistemic Value of Report

In sum, the seven-earths report is neither dismissible nor certainly true. On the subject of extraterrestrial intelligent life, it does not independently provide us with conclusive evidence for or against its existence. Using the fivefold taxonomy of theological issues provided in Chapter 1 (see also al-Ghursī 1437/2016: 36–38), the theological evidence for the existence of observable extraterrestrial intelligent life falls into the fifth category, under such issues that are not evidenced by revelatory proofs that are certain in their establishment (*thubūt*) or meaning (*dalāla*), nor established by sound Prophetic reports.[21] At best, the possibility of observable extraterrestrial intelligent life here is proven by contentious, albeit not dismissible, reports. The report's detractors thus appear to consider its meanings to fall into the realm of doubtfulness (*shakk*), or at best unlikelihood or implausibility (*wahm*), while its defenders would place it at highest in the realm of presumptive likelihood (*ẓānn*). While such issues have a role in theological conversations, no general ruling can be offered on their basis and the issue remains of an *ijtihādī* nature in which ignorance is pardonable. In sum, all the above revelatory evidence discussed so far suggests that observable, non-human extraterrestrial intelligent life is neither impossible nor necessary. Like conclusions based on empirical and rational evidence, the Qurʾanic verse Q.65:12 and the seven-earths report suggest that it is possible (al-Ālūsī n.d.: 28:143 sub Q.65:12).

The Dābba Verse

The next set of revelatory evidence that speaks to the issue of observable, non-human ETI are the Qurʾanic verses: Q.42:29 and Q.16:8, as explained by classical exegetes. The nineteenth-century Iraqi exegete

[21] Additionally, it is significant that the idea of ETI is not generally found among the generations of early Muslims (*salaf*) outside of the isolated Ibn ʿAbbās report.

Maḥmūd b. ʿAbd Allah al-Ālūsī claims that the position of the scholars is that beyond the Earth, the skies are also inhabited by creatures, angels, and other living beings, just as they occupy the Earth. This is evidenced by this Qurʾanic verse:

> *And among His signs is the creation of the skies and the earth and the walking-creatures (*dābba*) He has scattered in both of them [i.e. the skies and the earth].* (Q.42:29)

According to many commentators, the wording in the Qurʾanic verse Q.42:29: "*in both of them*" suggests that walking, living creatures (*dābba*) exist both on Earth and in the skies. This is because, as al-Ālūsī explains, the word *dābba* lexically refers to any living creature (*ḥayawān*) that possesses a soul, be it male or female, intelligent or otherwise. The word derives from *dabīb*, which means "to walk lightly," and only in customary usage refers to that which possesses four legs. In this verse, the lexical meaning is intended (al-Ālūsī n.d.: 28:143 sub Q.65:12).

Now, the existence of walking creatures on the Earth is evident. The existence of walking creatures in the skies is not. Among the famous exegetes, al-Rāzī claims that angels are celestial beings and not walking creatures. Nevertheless, it is possible that the verse is informing us that they do walk somewhere beyond our earth. He asserts that it is also possible that Allah created different types of living beings in the skies that walk as humans walk on Earth (al-Rāzī 2001: 9:599).[22] Muḥammad Bakhīt al-Muṭīʿī similarly claims nothing precludes that *dābba* is general and includes angels, *jinn*, and living beings such as those on the Earth (al-Muṭīʿī 1437/2016: 323). Abū al-Suʿūd Effendi adds that the verse, … *and He creates what you don't know* (Q.16:8), may also be hinting at the existence of such extraterrestrial life (Abū al-Suʿūd n.d.: 8:32). Still, he provides explanations for the verse that negate any implication of extraterrestrial intelligent life, by either interpreting *dābba* as any "living thing" and referring only to angels in respect to the skies, or by interpreting the statement "*and the walking-creatures He has scattered in both of them [i.e. the skies and the earth*]" as "the heavens and the earth as a whole," which would be true as long as walking-creatures exist on Earth, even

22 Ibn al-Munayyir objects to the inclusion of humans (or human-like beings) and angels in the meaning of *dābba*. He believes that the term is specific to four-legged animals on Earth (see al-Ālūsī n.d.: sub 42:29).

if they don't exist outside of it (Abū al-Suʿūd n.d.: 8:32). Nevertheless, scholars like al-Ālūsī conclude that sufficient evidence exists for the presence of *dawābb* (pl. *dābba*) in the skies just as they are present on the Earth (al-Ālūsī n.d.: 25:40).

Legitimacy of Theological and Metaphysical Inquiries

The challenges that revelatory evidence for the existence of non-human extraterrestrial intelligent life pose to incontrovertible Islamic understandings include the ideas of Prophetic superiority as well as human exceptionality. After all, Allah states in the Qurʾan: "*We have created humans in the best of forms and still reduced them to the lowest of the low*" (Q.94:4–5). It is also considered a matter of scholarly consensus among Muslims that "the best of all creation without exception is our prophet" (al-Laqānī 2009: 1:747 and 2:782–784).

Addressing these concerns, al-Laknawī states that the comparison between prophets across the earths or its layers does not imply an exact likeness. He refutes the notion that this comparison implies the existence of a prophet exactly like the Prophet Muhammad, as it is understood with certainty in the Sunnī tradition that no one can possess all of his special qualities. Comparison does not require exact likeness in all qualities; even sharing a single attribute is sufficient. Even if the comparison is of being the last prophet relative to that earth, the finality of the Chief of the Prophets, Muhammad, is not challenged by these evidences.

Al-Laknawī further addresses the theological legitimacy of the report in light of the concern that if there are multiple prophets with the title of "seal of the prophets," it may suggest that the Prophet Muhammad is not the ultimate Seal of the Prophets as is traditionally understood. Al-Laknawī posits that the term "seal" denotes the completion and perfection of prophethood, indicating that the last prophet in each earth or layer fulfills that role for their respective people. The comparison between different seals is only in terms of being the first or last in their respective abode, not in other qualities. Thus, while there may be multiple seals in different levels of existence, the Prophet Muhammad's sealship applies universally, ensuring that there will be no prophet after him in any level of existence. Therefore, the existence of multiple relative seals does not undermine the Prophet's unique status as the ultimate Seal of the Prophets nor the universality of his law and message. Any "counterpart" on other earths would necessarily have to chronologically precede him and be only final relative to their earth.[23]

Mawlānā Muḥammad Qāsim al-Nānawtawī (Nānotwī) resolves the uniqueness of the Prophet Muhammad with Ibn ʿAbbās's report in various ways. First, he distinguishes between non-derivative and intrinsic (*dhātī*) attributes and derivative attributes (*ʿaraḍī*), the former being like the non-derivate light of the sun, while the latter being like moonlight, or the light of planets. Prophethood, he argued, may be similar, in that the prophethood of Muhammad is non-derivate and that of other prophets is derivative (Nānotwī 1413 AH: 179).

In *Thahiri al-Nās*, Nānotwī extends this to argue that the report of Ibn ʿAbbās (Allah be pleased with him) enhances the supremacy and greatness of the Prophet Muhammad and does not in any way impinge on it (Nānotwī 1438/2017: 63–74). How so? Because all prophets, whether of this world or any other, ultimately derive their prophethood from his. Hence, the more prophets that are subordinate to his ultimate prophethood, the greater his status (Nānotwī 1438/2017: 74).

Nānotwī argued that the Qurʾanic idea of *khātm al-nubuwwa* (sealship of prophethood) referred to in the verse: "*Muḥammad is not the father of any of your men, but the messenger of Allah and the seal of prophets (khātam al-nabiyyīn)*" (al-Aḥzāb 33:40) contains two layers of meaning: (1) an exoteric and outer sealship (*ẓahr*), being chronological finality, or being a last of the prophets in time; and (2) an inner and esoteric meaning of sealship. The esoteric meaning of the title "Seal of the Prophets" is that prophethood is not possible without Muhammad. This is akin to the way the moon and planets receive their light from the sun, and not from each other. In this analogy, the sun represents Muhammad, and its unique position as the source of light mirrors his unique position as the final prophet. Just as the sun does not receive its light from any other

[23] Al-Laknawī concludes the discussion, and his monograph with an exploration of the idea of multiple seals of prophets by addressing three possibilities: (1) that the other seals existed during the time of Prophet Muhammad; (2) that they exist after him; or (3) that they existed before him (al-Laknawī 1419 AH: 420–421). He refutes the second possibility, as it contradicts the concept of Muhammad being the ultimate Seal of the Prophets, while acknowledging the possibility of the first and third, suggesting that if other seals existed before or alongside Muhammad, it wouldn't diminish his status. All prophets, if they lived during Muhammad's time, would have followed him. His message was inclusive across all time periods and his prophethood the most comprehensive and significant.

celestial body in the physical world, Muhammad's prophethood is not derived from any predecessor and the prophethood of earlier prophets is derived from him (Nānotwī 1402/1981: 245).

Therefore, Mawlānā Qāsim Nānotwī interpreted the title "Khātam al-Nabiyyīn" to encompass, even in its deeper, esoteric interpretation as the "pinnacle of prophethood," the notion of being the last in time as an inherent or suggested meaning, which he referred to as "an implicative meaning" (*dalāla iltizāmiyya*). He further articulated that, in his view, the phrase "Khātam al-Nabiyyīn" carries several meanings simultaneously, one key aspect of which is the concept of chronological finality. In *Tanwīr al-Nibrās*, Nānotwī clarifies that any counterparts of the Prophet on other, lower earths are relative seals (*khātam iḍāfī*), while the Muhammad is the true seal (*khātam ḥaqīqī*) (Nānotwī 1437/2016: 127).

The second challenge that the existence of ETI may pose to revelatory evidence is to the idea of human uniqueness. While the Sunnī scholarly tradition affirms that revelation is open to the possible existence of intelligent extraterrestrials, it also affirms that human beings are the most special of Allah's creation. This position is based on a central Qurʾanic theme, that Allah uniquely honors human beings among his creation. In Qurʾanic verse Q.2:29, for example, Allah tells human beings that it is He who created for you all that the earth contains. Similarly, He reminds man that everything in the universe was made for the advantage of humans in Q.31:20: Have you not seen that Allah has subjugated for you what is in the heavens and what is on the earth, and has perfected His blessings on you, both outward and inward? Elsewhere in the Qurʾan, Allah relates that he made the first human being "*with my own two hands*" (Q.38:75), which is understood to figuratively denote Allah's "complete concern for his creation... signifying that he should be venerated and revered" (Abū al-Suʿūd n.d.: 7:236). Further reason for human uniqueness is in the fact that the first human was created and given form in Paradise (Q.2:35) and was given the freedom to delight in its pleasures, unlike angels whose purpose is constant worship (Q.21:20). Additionally, Allah instructed the angels to bow down to the Prophet Adam, the first human, demonstrating His absolute honoring of humanity (Q.2:34). Sunnī scholars have interpreted these verses as a clear demonstration of humanity's elevated status over all other beings (Karamali 2024: 37). In order for the rhetorical purpose of the verses to be fulfilled, which is for man to be grateful for having been so honored, it follows that man must necessarily be more honored than all other forms of creation, including any possible intelligent extraterrestrial life (Karamali 2024: 37). In other words, if extraterrestrial intelligent life exists, it must be less honored by Allah than human beings.

Summary Findings About Extraterrestrial Intelligent Life

In sum, revelatory and empirical evidence prove that the universe is vast. Revelatory evidence tells us of the multiplicity of the skies and, presumptively, earth realms as well. Lastly, revelation also conclusively establishes the existence of non-human intelligent life, namely, angels and *jinn*, on Earth and presumptively throughout the entire known universe.

It is thus empirically, rationally, and revelatory possible for observable extraterrestrial intelligent life to exist and to not exist, although there may be a disagreement if the probability falls within the realm of presumptive likelihood (*ẓann*), unlikeliness (*wahm*), or doubtfulness (*shakk*), with some later scholars such as al-Laknawī inclining toward likelihood. Presuming such life exists, revelatory evidence necessitates that it will not contradict the creed of the finality of prophethood nor the idea of human uniqueness. Current empirical evidence neither confirms nor denies this necessity, nor the presence or absence of intelligent extraterrestrial beings. Should such evidence emerge in the future, it may offer valuable theological insights into revelatory evidence such as those found in the multiple earths verse, the seven-earths *ḥadīth*, and the *dābba* verses. Therefore, the scientific pursuit of empirical evidence for the existence or non-existence of intelligent extraterrestrials is potentially beneficial for theological understanding.

In sum, the available sources of knowledge exhibit a perfect cohesion in what they collectively demonstrate about the current evidence for or against the existence of observable extraterrestrial intelligent life.

On Speculative Inquiry in the Islamic Intellectual Tradition

This section addresses a common concern among Muslim scientists who wonder whether the Islamic intellectual tradition supports or opposes speculative and hypothetical inquiries that form the basis for scientific innovation in the modern age. Although the Islamic intellectual tradition generally welcomed hypothetical inquiries, including questions about the existence of extraterrestrial intelligent life, there is evidence of resistance to speculation among the early generations of Muslims. This resistance highlights an early appreciation for concerns regarding speculation in the absence of an ethical framework that guides and disciplines such questioning.

Much of the evidence pointing to a limitation on speculative inquiry relates to legal discourses, despite law being a field grounded in practical application. One of the earliest instances of opposition can be traced back to the era of the Prophet's Companions. For instance, when the Companion ᶜAbd Allah b. ᶜUmar was once consulted about a hypothetical situation, he responded: "Do not inquire about that which has not [yet] happened. I heard ᶜUmar b. al-Khaṭṭāb condemn (*yalᶜanu*) those who inquire about that which has not [yet] happened" (al-Dārimī 1412/2000: *al-ᶜIlm, bāb karāhiyyat al-futyā*, no. 128, 1:26).[24]

Similarly, the Companion, Zayd b. Thābit, when posed with hypothetical questions, would first ask: "Has this occurred?" If the response was yes, he would offer advice based on his knowledge and judgment. If the response was no, he advised "Let it be until it happens." (al-Dārimī 1412/2000: *al-ᶜIlm, bāb karāhiyyat al-futyā*, no. 128, 1:26).[25] Companions like ᶜAmmār b. Yāsir[26] and Ubayy b. Kaᶜb (al-Khaṭīb al-Baghdādī 1421/2000: 2:14, no. 262)[27] expressed similar sentiments upon being questioned.[28]

Al-Khaṭīb al-Baghdādī interprets the statements of these Companions as precautions about offering personal judgments due to the fear of making mistakes and out of respect for the potential dangers in making

[24] *Sunan al-Dārimī*'s commentator, Nabīl b. Muḥammad al-Ghamarī, in his *Fatḥ al-Mannān* (1999: 2:81), states that the chain for the report is acceptable (*maqbūl*).

[25] This report is transmitted through a disconnected chain between the narrator al-Zuhrī and Zayd b. Thābit (see al-Ghamarī 1999: 2:82).

[26] Al-Būṣīrī states in *Itḥāf al-Khiyara al-Mahara* (1419/1998: 1:301, no. 525), that the report is sound if al-Shaᶜbī indeed heard reports directly from ᶜAmmār, and its narrators are all reliable.

[27] Al-Khaṭīb transmits the report through Abū Khaythama but does not mention his chain to him, making the report disconnected (*munqaṭiᶜ*) and suspended (*muᶜallaq*) from its beginning end.

[28] Ibn al-Munayyir reports a similar habit of the *mujtahid* imam Mālik b. Anas, who would state that the one who responds to real questions is Divinely assisted, otherwise they are met with disappointment (see al-Ḥajawī 1416/1995: 2:420).

independent legal decisions (*ijtihād*). They opted for a cautious approach in situations that had not yet occurred, especially in unforeseen events. Their view was that should such situations arise, the need for their guidance would become apparent, and in those moments, Allah would assist those seeking to discern the truth. In the specific case of the Caliph ʿUmar b. al-Khaṭṭāb, other historical reports articulate his rationale for limiting speculative questioning: he simply prioritized the study of past events, emphasizing that Allah has already outlined what is to come (al-Dārimī 1412/2000: *muqaddima, bāb karāhiyyat al-futyā*, 1:242–244, no. 123–126; al-Khaṭīb al-Baghdādī 1421/2000: 2:12, no. 260).[29]

This context is crucial for understanding why, despite these reservations, a substantial body of early Islamic scholarship meticulously documented legal judgments covering a broad spectrum of real and speculative issues. This was done, according to some scholars, in order to guarantee that subsequent generations could easily access solutions to unforeseen future issues. These scholars understood the cautious warnings of certain Companions and their Followers as grounded in a combination of foresight and devoutness, rather than a blanket prohibition against speculative thought. Al-Bayhaqī, for instance, cites the fourth-century Shāfiʿī jurist, theologian, and traditionist Abū ʿAbd Allah al-Ḥalīmī (d. 403) as allowing students of law to explore hypothetical legal scenarios as a means to refine their skills in legal reasoning and independent judgment (al-Bayhaqī 1437/2016: *bāb man kariha al-masʿala ʿammā lam yakun*, 1:226).

[29] ʿAbd al-Fattāḥ Abū Ghudda notes that the chain of these reports are disconnected even if the narrators are all trustworthy in his *Manhaj al-Salaf fī al-Suʾāl ʿan al-ʿIlm* (Abū Ghudda 2008: 16). Muḥammad Taqī al-ʿUthmānī mentions that the prohibition of hypothetical inquiry may be evidenced by a *mursal* (disconnected chained) report through Abū Salama b. ʿAbd al-Raḥmān, that the Messenger of Allah (pbuh), said:

> Do not hasten affliction before it descends, for if you do not hasten it before its arrival, the Muslims will remain united. But when it does descend, among them are those who, if they speak, are guided and accurate. If you hasten it, inclinations will vary among you, leading you to take this or that path, and he gestured in front of him and to his right and his left.
> (al-Dārimī 1412/2000: *bāb al-tawarruʿ ʿan al-jawāb fīmā laysa fīhī kitāb wa lā sunna*, 1:238–239, no. 118).

Al-Khaṭīb al-Baghdādī, having relayed statements of restraint from several Companions and Followers, clarifies that the Prophet's aversion to certain types of queries stemmed from a desire to protect his community. This protective measure was born from concern that excessive speculation about a matter might inadvertently lead to its prohibition, despite its default lawful status. Essentially, the act of questioning could paradoxically result in forbidding something inherently permissible. This worry, according to al-Khaṭīb al-Baghdādī, ceased to be relevant following the Prophet's passing as such inquiries no longer have the power to affect the legal status of actions or practices (al-Khaṭīb 1421/2000: 2:30–31).

To illustrate the broad acceptance of speculative inquiry in the Islamic intellectual tradition, al-Khaṭīb al-Baghdādī references an occasion in which the Companion Rāfiᶜ b. Khadīj questioned the Prophet, saying: "Messenger of Allah! We anticipate confronting the enemy tomorrow, and we lack knives (implying that using their swords to slaughter animals might blunt them). May we slaughter using (sharpened) reeds instead?" The Messenger of Allah answered: "Anything that makes blood flow, and which is not teeth or nails, and upon which Allah's name is invoked, then eat (it)." (al-Bukhārī 1422/2001: *al-Zabāʾiḥ wa-l-Ṣayd, bāb al-tasmiya ᶜalā al-dhabīḥa wa-man taraka mutaᶜammidan*, no. 5498). Given that Rāfiᶜ b. Khadīj's inquiry concerned a future event, and the Prophet provided guidance without expressing any disapproval of the question, this incident underscores the acceptance of speculative questioning about future situations.

A clearer demonstration that speculative inquiry is generally acceptable within an Islamic ethical framework is found in an incident in which Salama b. Yazīd al-Juᶜfī asked the Prophet: "Messenger of Allah! What do you think if we have rulers over us who demand their rights from us but deny us our rights? Should we engage with them in combat?" Ashᶜath b. Qays stood up and objected: "You ask the Messenger of Allah about an issue that has not yet happened?!" The man retorted: "I will surely ask him until he forbids me" and proceeded to repeat the question. The Messenger of Allah responded, saying: "No. They are responsible for what they have been charged with, and you are responsible for what you have been charged with."[30] Al-Khaṭīb explains:

[30] Muslim 1433/2013: *al-Imārah, bāb fī ṭāᶜat al-umarāʾ*, no. 4745–4746; al-Tirmidhī 1998: Jāmiᶜ al-Tirmidhī, al-Fitan, *bāb mā jāʾa satakūnu fitan ka-qiṭaᶜ al-layl al-muẓlim*, no. 2199. Al-Tirmidhī grades the report as *ḥasan ṣaḥīḥ*.

> The Messenger of Allah (pbuh) did not forbid the man from asking this question, nor rebuke him for it. Instead, he replied without objection. Similar incidents in support of the aforementioned (idea of permissible speculative questioning) can be found in many historical reports. (al-Khaṭīb al-Baghdādī 1421/2000: 2:30–31).

Al-Khaṭīb al-Baghdādī then proceeds to state that it is reported that ʿUmar b. al-Khaṭṭāb, ʿAlī b. Abī Ṭālib, as well as other Companions would discuss the rules of scenarios before they had occurred and debated among themselves on issues of inheritance law. The scholars among the Followers[31] and then those after them throughout the Muslim world similarly engaged in discussions and debates on non-occurrences, indicating a type of consensus (*ijmāʿ*) on the allowance for some degree of speculative inquiry.

One clear example of such type of inquiry in the era of the Followers is a conversation that took place between the Qurʾanic scholar and traditionist Qatāda (d. 61/736) and the eponymous founder of the Ḥanafī school, Abū Ḥanīfa (d. 150) when the former came to Kufa and invited the residents to ask him legal questions. Abū Ḥanīfa stood up and asked:

> What do you say about a man who is absent from his wife (for a long time), and then her husband's death is announced (incorrectly). Subsequently, his wife marries another man and gives birth to children. But then her husband returns, and he says (to her): Adulteress! You married (another), while I am (still) your husband?!

At the same time, the other says to her: "Adulteress! You married (me) while you had a husband?!" Is the non-discretionary punishment (*ḥadd*) necessary, and to whom are the children attributed?" Qatāda remained in thought for a while and asked, "Has this situation actually occurred?" The people replied, "No." Qatāda exclaimed, "Then why do you ask me about what has not occurred?" Abū Ḥanīfa responded: "Scholars prepare for calamities before they strike, so if they do occur, they recognize them

[31] The Followers (*tābiʿūn*) are the generation of Muslims who followed the generation of the Companions of the Prophet, saw at least one of them, was rightly guided, and who died as rightly-guided believers.

well. They know how to enter into such situations and how to exit from them."[32]

Al-Khaṭīb al-Baghdādī's explanation of ʿUmar b. al-Khaṭṭāb's approach to non-occurrences helps us understand some of the epistemological parameters of scientific inquiry in the Islamic tradition. According to al-Khaṭīb al-Baghdādī, ʿUmar b. al-Khaṭṭāb's condemnation of speculative inquiries was likely aimed at questions posed out of stubbornness and deceit, rather than a genuine desire for understanding or seeking beneficial knowledge. This explains his actions against Ṣabīgh b. ʿAsal, whom he punished, exiled, and cut off financially for asking about complex verses of the Qurʾan. He was likely concerned that Ṣabīgh's questions were intended to undermine the faith of the less knowledgeable Muslims by introducing doubts and leading them astray with incorrect interpretations of the Qurʾan, away from its intended message and toward flawed understandings. Such acts were also forbidden and disapproved of by the Prophet, who condemned those who engaged in such trick questions, or *ughlūṭāt* (pl. *ughlūṭa*, meaning questions that are intended to baffle the scholars and cause them to err, giving rise to communal trouble and ruin) (Ibn Athīr 1399/1979: 3:378).[33] This is because such types of questions possess no religious utility (*ghayr nāfiʿa fī al-dīn*) and are usually posed in the realm of non-occurrences. This is likely the meaning of the Companion ʿAbd Allah b. Masʿūd's statement: "I warn you about complex discussions," which al-Awzāʿī explained as "complex questions" (Ibn ʿAbd al-Barr 1414/1994: *bāb mā jāʾa fī dhamm al-qawl fī dīn Allāh taʿālā*, 2:1227, no. 2038).

[32] This is a portion of the full incident, which can be found in full in: al-Ṣaymarī 1405/1985: 35–36; Ibn ʿAbd al-Barr 1417/1997: 304–305; al-Khaṭīb al-Baghdādī 1422/2002: 13:348–349; al-Muwaffaq 1321/1903: 1:102–103; and al-Kardarī 1321/1903: 1:158.

[33] About such types of trick questions, the Prophet Muhammad said: "There will be people from my nation who will (try to) cause their men of understanding to fall into error with perplexing (*ʿaḍl*) questions. They are the worst of my nation." See al-Khaṭīb 1421/2000: 2:30–31.

Conclusion

In summary, this chapter describes how one can apply the epistemological concepts of certainty and uncertainty in an integrative and coherent manner to assess revelatory, rational, and empirical knowledge regarding the existence of extraterrestrial intelligent life. While the discovery of extraterrestrial intelligent life could significantly challenge notions of human uniqueness, this did not preclude prior generations of Muslims from engaging in discourse regarding this theoretical question by exploring revelatory evidence in parallel with rational and empirical evidence. If we were to discover extraterrestrial intelligent life using empirical means, revelatory evidence suggests that Allah would guide those intelligent beings similarly to how humans are guided through messengers. For a Muslim scientist, revelatory evidence conclusively informs one's understanding of human uniqueness and uniqueness among all of God's creations even if extraterrestrial intelligent life were discovered. However, for a scientist using empirical and rational knowledge without revelatory knowledge, discovering extraterrestrial intelligent life could challenge the view that humans are biologically and evolutionarily exceptional.

References

Abū Ghudda, ʿAbd al-Fattāḥ. 2008. *Manhaj al-Salaf fī al-Suʾāl ʿan al-ʿIlm wa fī Taʿallum Mā Yaqaʿu wa Mā lam Yaqaʿ*. Beirut: Dār al-Bashāʾir al-Islāmiyya.

Abū Ḥayyān, Muḥammad b. Yūsuf. 1441/2020. *Al-Baḥr al-Muḥīṭ*, 22 vols. Beirut: al-Risāla al-ʿĀlamiyya.

Abū al-Suʿūd, Muḥammad b. Muḥammad. n.d. *Irshād al-ʿAql al-Salīm ilā Mazāyā al-Qurʾān al-Karīm*. 9 vols. Beirut: Dār Iḥyāʾ al-Turāth al-ʿArabī.

Al-Ālūsī, Abū al-Faḍl Maḥmūd. n.d. *Rūḥ al-Maʿānī*. 30 vols. Beirut: Idārat al-Ṭibāʿa al-Munīriyya.

Al-Bayḍāwī, Abū Saʿīd ʿAbd Allah b. ʿUmar. 2001. *Anwār al-Tanzīl wa Asrār al-Taʾwīl*, 2 vols. Beirut: Dar Ṣādir.

Al-Bayhaqī, Abū Bakr. 1413/1993. *Al-Asmāʾ wa-l-Ṣifāt*, 2 vols. Jeddah: Maktabat al-Sawādī.

Al-Bayhaqī, Abū Bakr. 1437/2016. *Al-Madkhal ilā ʿIlm al-Sunan*, 2 vols. Jeddah: Dār al-Minhāj.

Al-Bukhārī, Muḥammad b. Ismāʿīl. 1422/2001. *Ṣaḥīḥ al-Bukhārī*, or *al-Jāmiʿ al-Musnad al-Ṣaḥīḥ al-Mukhtaṣar min Umūr Rasūl Allāh ṣalla Allāh ʿalayhi wa sallam wa Sunanihi wa Ayyāmihi*, 9 vols. Beirut: Dār Ṭawq al-Najāt.

Al-Būṣīrī, Aḥmad b. Abī Bakr. 1419/1998. *Itḥāf al-Khiyara al-Mahara bi-Zawā᾽id al-Masānīd al-ᶜAshara*, 11 vols. Riyadh: Maktabat al-Rushd.

Al-Dārimī, Abū Muḥammad ᶜAbd Allāh b. ᶜAbd al-Raḥmān. 1412/2000. *Sunan al-Dārimī*, 4 vols. Riyadh: Dār al-Mughnī.

Al-Dhahabī, Shams al-Dīn. 1416/1995. *al-ᶜUluww li-l-ᶜAlī al-Ghaffār*. Riyadh: Maktaba Aḍwā᾽ al-Salaf.

Al-Ghamarī, Abū ᶜĀṣim Nabīl. 1999. *Fatḥ al-Mannān*, 10 vols. Beirut: Dār al-Bashā᾽ir al-Islāmiyya.

Al-Ghursī, Muḥammad Ṣāliḥ. 1437/2016. *Taḥqīq Masā᾽il Muhimmāt mi ᶜIlm al-Tawḥīd wa-l-Ṣifāt*. Amman: Dār al-Fatḥ.

Al-Ḥajawī, Muḥammad b. al-Ḥasan. 1416/1995. *Al-Fikr al-Sāmī fī Tārīkh al-Fiqh al-Islāmī*, 2 vols. Beirut: Dār al-Kutub al-Islāmiyya.

Al-Ḥākim al-Nīsāpūrī, Abū ᶜAbd Allāh Muḥammad b. ᶜAbd Allāh. 1411/1990. *Al-Mustadrak ᶜalā al-Ṣaḥīḥayn*, 4 vols. Beirut: Dār al-Kutub al-ᶜIlmiyya.

Al-Haytamī, Aḥmad b. Ḥajar. 1989. *Al-Fatāwā al-Ḥadīthiyya*. Cairo: Mustafā Bābī al-Ḥalabī.

Ibn ᶜAbd al-Barr, Abū ᶜUmar Yūsuf b. ᶜAbd Allāh al-Qurṭubī. 1414/1994. *Jāmiᶜ Bayān al-ᶜIlm wa Faḍlihi*, 2 vols. Dammam: Dār Ibn al-Jawzī.

Ibn ᶜAbd al-Barr. 1417/1997. *Al-Intiqā᾽ fī Faḍā᾽il al-A᾽imma al-Thalātha al-Fuqahā᾽*. Aleppo: Maktab al-Maṭbūᶜāt al-Islāmiyya.

Ibn Abī Ḥātim al-Rāzī, *Tafsīr al-Qur᾽ān al-ᶜAẓīm l-Ibn Abī Ḥātim*. 1419/1998. Mecca: Maktaba Nizār Muṣṭafā al-Bāz.

Ibn Abī Jamra, Abū Muḥammad ᶜAbd Allāh b. Saᶜd. 1444/2022. *Bahjat al-Nufūs Sharh Mukhtaṣar Ṣaḥīḥ al-Bukhārī*, 5 vols. Beirut: Dār al-Rayāḥīn.

Ibn Athīr, Abū al-Saᶜādāt al-Mubārak b. Muḥammad. 1399/1979. *Al-Nihāya fī Gharīb al-Athar*, 5 vols. Beirut: Dār al-Kutub al-ᶜIlmiyya.

Ibn Ḥajar, al-ᶜAsqalānī, Aḥmad ibn ᶜAlī. 1970. *Fatḥ al-Bārī bi Sharḥ al-Bukhārī*, 13 vols. Egypt: al-Maktaba al-Salafiyya.

Ibn Ḥajar al-ᶜAsqalānī, Aḥmad b. ᶜAlī. 1413/1996. *Sharḥ al-Nukhbah Nuzhat al-Naẓar fī Tawḍīḥ Nukhbat al-Fikar fī Muṣṭalāh al-Athar*. Damascus: Maṭbaᶜat al-Ṣabāḥ.

Ibn Ḥibbān, Muḥammad. 1432/2011. *Ṣaḥīḥ Ibn Ḥibbān bi-Tartīb Ibn Balbān*, 18 vols. Beirut: al-Risāla al-ᶜĀlamiyya.

Ibn Kathīr, Abū al-Fidā᾽ Ismāᶜīl. 1999. *Tafsīr al-Qur᾽ān al-ᶜAẓīm*, 8 vols. Riyad: Dār Ṭayba li-l-Nashr wa-l-Tawzīᶜ.

Ibn Kathīr, Abū al-Fidā᾽ Ismāᶜīl. 2013. *Al-Bidāya wa-l-Nihāya*, 20 vols. Damascus: Dār Ibn Kathīr.

Ibn Khuzayma, Abū Bakr Muḥammad b. Isḥāq. 1400/1980. *Ṣaḥīḥ Ibn Khuzayma*, 4 vols. Beirut: al-Maktab al-Islāmī.

Ibn Sayyid al-Nās, Abū al-Fatḥ Muḥammad b. Muḥammad. n.d. *ᶜUyūn al-Athar fī Funūn al-Maghāzī wa-l-Shamā᾽il wa-l-Siyar*. Damascus: Dār Ibn Kathīr.

Al-Jurjānī, al-Sayyid al-Sharīf ʿAlī b. Muḥammad. 1419/1988. *Sharḥ al-Mawāqif*, 8 vols. Beirut: Dār al-Kutub al-ʿIlmiyya.

Karamali, Hamza. 2024. "Theological Information on the Existence of Intelligent Life Outside Our Solar System: Metaphysics, Scripture, and Science." In *Islamic Theology and Extraterrestrial Life: New Frontiers in Science and Religion*. Edited by Shoaib Ahmed Malik and Jörg Matthias Determann, 25–41. London: I.B. Tauris.

Al-Kardarī, Ḥāfiẓ al-Dīn Ibn al-Bazzāz. 1321/1903. *Manāqib al-Imām al-Aʿẓam*, 2 vols. Hyderabad: Majlis Dāʾirat al-Maʿārif al-Niẓāmiyya.

Al-Khafājī, Shihāb al-Dīn Aḥmad b. Muḥammad. n.d. *Ḥāshiyat al-Shihāb ʿalā Tafsīr al-Bayḍāwī*, 8 vols. Beirut: Dār al-Ṣādir.

Al-Khaṭīb al-Baghdādī, Abū Bakr Aḥmad b. ʿAlī. 1421/2000. *Al-Faqīh wa-l-Mutafaqqih*, 2 vols. Dammam: Dār Ibn al-Jawzī.

Al-Khaṭīb al-Baghdādī. 1422/2002. *Tārīkh Madīnat al-Salām*, or *Tārīkh Baghdād*, 16 vols. Beirut: Dār al-Gharb al-Islāmī.

Konstan, David. 2022. "Epicurus." *The Stanford Encyclopedia of Philosophy*. Edited by Edward N. Zalta and Uri Nodelman.

Al-Laknawī, Muḥammad ʿAbd al-Ḥayy. 1419 AH. *Zajr al-Nās*. In vol. 1 of *Majmūʿ Rasāʾil al-Laknawī*, 5 vols. Karachi: Idārat al-Qurʾān wa-l-ʿUlūm al-Islāmiyya.

Al-Laqānī, Burhān al-Dīn Ibrāhīm. 2009. *Hidāyat al-Murīd li-Jawharat al-Tawḥīd*. Cairo: Dār al-Basāʾir.

Margulis, Lynn, Carl Sagan, and Dorion Sagan. 2024. "Extraterrestrial Life." *Encyclopedia Britannica*, June 3. https://www.britannica.com/science/extraterrestrial-life.

Muslim b. al-Ḥajjāj. 1433/2013. *Ṣaḥīḥ Muslim*, 8 vols. Jeddah: Dār al-Minhāj.

Al-Muṭīʿī, Muḥammad Bakhīt. 1437/2016. *Tawfīq al-Raḥmān li-l-Tawfīq bayna Mā Qālahu ʿUlamāʾ al-Hayʾa wa bayna Mā Jāʾa fī al-Aḥādīth al-Ṣaḥīḥa wa Āyāt al-Qurʾān*. Beirut: Dār al-Minhāj.

Al-Muwaffaq al-Makkī, Abū al-Muʾayyad. 1321/1903. *Manāqib al-Imām al-Aʿẓam Abī Ḥanīfa*, 2 vols. Hyderabad: Majlis Dāʾirat al-Maʿārif al-Niẓāmiyya.

Nānotwī, Muḥammad Qāṣim. 1413 AH. *Āb-i Ḥayāt*. Multan: Idārat Taʾlīfāt Ashrafiyya.

Nānotwī, Muḥammad Qāsim. 1402/1981. *Ajwiba Arbaʿīn Radd Rawāfiḍ*. Gujranwala: Idārat Nashr wa Ishāʿat Madrasa Nasrat al-ʿUlūm.

Nānotwī, Muḥammad Qāsim. 1438/2017. *Taḥdhīr al-Nās min Inkār Athar Ibn ʿAbbās*. Saharanpur: Ḥujjat al-Islām Academy.

Nānotwī, Muḥammad Qāsim. 1437/2016. *Tanwīr al-Nibrās ʿalā man Ankar Taḥdhīr al-Nās*. Karachi: Markaz Ahl al-Sunnat wa-l-Jamāʿat.

NASA. n.d. "Fast facts: What is the Habitable Zone?" *NASA: National Aeronautics and Space Administration*. https://science.nasa.gov/exoplanets/habitable-zone/.

Al-Nasafī, Abū al-Muʿīn. 1990. *Tabṣirat al-Adilla*. Edited by Ḥusayn Ātāy. Ankara: Riʾāsat al-Shuʾūn al-Dīniyya li-l-Jumhūriyya al-Turkiyya.

Al-Nawawī, Yaḥya ibn Sharaf. 1972. *al-Minhāj Sharḥ Ṣaḥīḥ Muslim ibn Ḥajjāj*, 9 vols. Beirut: Dār Iḥyāʾ al-Turāth al-ʿArabī.

Noor, Mohamed A.F. 2022. "Thinking Outside Earth's Box—How Might Heredity and Evolution Differ on Other Worlds?" *Evolution: Education and Outreach* 15 (13). https://doi.org/10.1186/s12052-022-00172-4.

Peta, Saphalya. 2023. "Journey of Artificial Intelligence Frontier: A Comprehensive Overview." *Global Journal of Computer Science and Technology* 23 (D2): 9–36. https://doi.org/10.34257/GJCSTDVOL23IS2PG9.

Al-Qurṭubī, Abū ʿAbd Allah Muḥammad b. Aḥmad. 1433/2012. *Al-Jāmiʿ li-Aḥkām al-Qurʾān wa-l-Mubayyin li-mā Taḍammanahu min al-Sunna wa-Āyī al-Qurʾān*, 24 vols. Beirut: al-Risāla al-ʿĀlamiyya.

Al-Rāzī, Fakhr al-Dīn. 2001. *Al-Tafsīr al-Kabīr li-l-Imām al-Fakhr al-Rāzī*, 11 vols. Beirut: Dār Iḥyāʾ al-Turāth al-ʿArabī.

Al-Rāzī, Fakhr al-Dīn. n.d. *Al-Maḥṣūl fī ʿIlm Uṣūl al-Fiqh*, 6 vols. Beirut: Muʾassasat al-Risāla.

Al-Ṣābūnī, Nūr al-Dīn. 1440/2018. *Al-Bidāyah fī Uṣūl al-Dīn*. Istanbul: al-Maktaba al-Ḥanīfiya.

Al-Ṣaymarī, Abū ʿAbd Allah Ḥusayn b. ʿAlī. 1405/1985. *Akhbār Abī Ḥanīfa wa Aṣḥābihi*. Cairo: ʿĀlam al-Kutub.

Seti Institute. 2020. "How Many Habitable Planets are Out There?" October 29. https://www.seti.org/press-release/how-many-habitable-planets-are-out-there.

Al-Shāmī, Muḥammad Amīn b. ʿĀbidīn b. 2021. *Nasamāt al-Asḥār ʿala Sharḥ al-Manār*. Beirut: Dār al-Kutub al-ʿIlmiyya.

Shostak, Seth. 2021. "Drake Equation." *SETI Research*, updated July 2021. https://www.seti.org/drake-equation-index.

Shostak, Seth. 2023. "Extraterrestrial Intelligence." *Encyclopedia Britannica*, September 14. https://www.britannica.com/science/extraterrestrial-intelligence.

Al-Suyūṭī, Jalāl al-Dīn. 1424/2004. *Al-Ḥāwī li-l-Fatāwī*, 2 vols. Beirut: Dār al-Fikr.

Al-Suyūṭī, Jalāl al-Dīn. 1437/2016. *Tadrīb al-Rāwī fī Sharḥ Taqrīb al-Nawāwī*, 5 vols. Jeddah: Dār al-Minhāj.

Al-Ṭabarī, Muḥammad ibn Jarīr. 1420/2000. *Jāmiʿ al-Bayān fī Tafsīr Āyi al-Qurʾān*, 24 vols. Beirut: Muʾassasat al-Risāla.

Al-Ṭabarī, Abū Jaʿfar Muḥammad b. Jarīr. 1422/2001. *Tafsīr al-Ṭabarī: Jāmiʿ al-Bayān ʿan Taʾwīl Āy al-Qurʾān*, 26 vols. Edited by ʿAbd Allāh b. ʿAbd al-Muḥsin al-Turkī et al. Cairo: Dār Hajar.

Al-Taftāzānī, Saʿd al-Dīn. 2020. *Sharḥ al-ʿAqāʾid al-Nasafiyya*. Damascus: Maktabat Dār al-Daqqāq.

Thanvī, Ashraf ʿAlī. 1992. *Answer to Modernism*. Karachi: Maktaba Darul Uloom Karachi.

Al-Tirmidhī, Abū ʿĪsā Muḥammad b. ʿĪsā. 1998. *Sunan al-Tirmidhī*, 6 vols. Beirut: Dār al-Gharb al-Islāmī.

Al-Zamakhsharī, Abū al-Qāsim Maḥmūd b. ʿUmar. 1433/2012. *Al-Kashshāf ʿan Ḥaqā ʾiq al-Tanzīl wa ʿUyūn al-Taʾwīl*, 4 vols. Cairo: Dār al-Ḥadīth.

Zhang, X. X., R. Duan, V. Gajjar, H. Y. Zhang, P. Wang, C. H. Niu, D. Werthimer, J. Cobb, S. Y. Li, X. Pei, Y. Zhu, and D. Li. 2023. "An Overview of FAST Real-Time Fast Radio Burst Searching System." *Research in Astronomy and Astrophysics* 23 (9): 095023. https://doi.org/10.1088/1674-4527/ace518.

CHAPTER 7

Is Human Height Changing, and Should Medicine Intervene? Transhumanism, Biomedicine, and Islamic Bioethics

Aasim I. Padela

Introduction

Recent decades have witnessed meteoric growth in biotechnological capacities and prowess. Breakthrough advancements in scientific knowledge and technology have yielded a host of remarkable therapeutics that have saved scores of lives. Feats such as the development of the "iron lung" staved off mortality from polio in the 1920s and was soon followed by novel vaccine-based preventive measures; the discovery of antibiotics heralded by penicillin in the 1960s led to an explosion of pharmaceutical research that has since revolutionized the treatment of infectious diseases; and discoveries in genetic science have yielded novel therapeutics marked most recently by CRISPR-Cas 9 and CAR T-cell immunotherapies for some cancers. Alongside these therapies that are externally applied

A. I. Padela (✉)
Medical College of Wisconsin, Milwaukee, WI, USA
e-mail: apadela@mcw.edu

The Initiative on Islam and Medicine, Glendale Heights, IL, USA

© The Author(s), under exclusive license to Springer Nature Switzerland AG 2026

A. I. Padela (ed.), *Bridging Bioscience and Islam*, Islam & Science,
https://doi.org/10.1007/978-3-032-09996-9_7

to the body or are temporarily in vivo, biotechnology also augments and supplants bodily functions. Illustratively, implantable devices such as cardiac defibrillators are used to prevent death from arrhythmias, prosthetics such as artificial knee and hip joints support physical mobility, bio-neural interfaces such as cochlear implants allow for hearing, and dialysis machines nearly fully replace the function of kidneys. These technologies are more internal to the body, and some meld mechanical devices with the human body's biological substrate.

As we are increasingly able to manipulate bodies and augment bodily functions, critical questions about the role of medicine's use of biotechnology arise. Is medicine's[1] mandate to restore (previously lost) bodily function and go no further? Or should the profession facilitate enhancing the biological body with synthetic devices for "better" functioning or enhanced wellness? These questions prompt us to further consider whether it is possible to clearly differentiate bodily modifications that constitute an appropriate use of biotechnology within healthcare from those that may be inappropriate, unethical, or immoral. Related is the thorny issue of societal expectations, which shape what doctors do and what healthcare provides, and so are the problems of how biological normativity is defined and determined. The ethical limits of medical practice are essential to delineate, for these parameters inform medical training, set legal constraints related to medical licensing and healthcare service provision, and ultimately shape societal expectations of doctoring. This chapter uses the topic of transhumanism as a springboard to dive into these and related issues.

Transhumanist ideology places the question of human end center stage and, as such, forces humankind to consider whether to meet this end as a passive participant lamenting its finitude, or as an active resistance-fighter using biotechnology to push back against a biologically "forced" closure to life by forging a new kind of life that transcends human experience. This proposed *trans*cending of *human* existence gives transhumanism its name. Given that medicine addresses human maladies, sets itself up

[1] My usage of the terms medicine and biomedicine is purposeful. As will be noted further in this chapter, biomedicine is a more akin to a paradigm with a body of knowledge and aligned practices, and I use this term to mean this overarching framework through which contemporary healthcare practices emerge. I use the term medicine to refer to the professional practices of clinicians within contemporary healthcare.

against premature[2] death, and deploys biotechnology to carry out its purposes, one may wonder whether contemporary healthcare is moving us toward a trans- or post-human future. Could the profession be an unwitting partner in a transhuman enterprise? Such an assertion may seem outlandish, but there may be more alignment between the practices of medicine and transhumanist goals than initially meets the eye.

Accordingly, this chapter considers the moral dimensions of medical practice as related to bodily modification. It does so by pulling at different strands of the ethical problem-space to clarify the issues at hand and then weaving together a provisional Islamic bioethical response drawn from components of an Islamic moral worldview. As part of the analysis, and in light of the larger aim of the project that motivated this chapter, namely, addressing fundamental questions about the human being in light of bioscientific and religious evidence, the chapter will also obliquely attend to the question of whether humankind is moving toward a transhuman state. This indirect addressal considers what we know about changes in human height from origin to the present day. While height is an attribute, and not an essential aspect, of the human being, investigating changes to this physical feature allow for bringing bioscientific and religious truth claims into dialog through the epistemic framework noted in Chapter 2.

[2] Premature death appears straightforward lexically but is an ambiguous concept with various measurement approaches. At the surface level, and in the way I am using the term, it is conceptualized as a death that could have been prevented should conventional, contemporary medical therapies have been accessible and applied. From an Islamic theological standpoint, no death could be classified as premature given that one's individual's lifespan (*ajal*) is predetermined. There are numerous verses of the Qur'an and Prophetic reports that substantiate God's foreknowledge of, and predetermination, of an individual's lifespan. This view does not discount the efficacy of prayer and good actions "increasing" one's lifespan, which is also referenced in numerous Prophetic reports. There are many nuances in this realm, however, one way scholars reconcile these reports is by considering the different vantage points of God and the human being; there is the fixed nature of one's overall lifespan in God's (fore)knowledge, but at the same time, the indeterminacy of such in human knowledge. The human experience of the gradual unfolding of God's ultimate decree allows for a perceived increase in lifespan. With respect to measurement, there are at least three operationalist definitions. First, there are age-based thresholds (ABTs), which classify a death as premature if it occurs under a pre-defined, absolute threshold. Second, there are age-group longevity norms (AGLNs), which quantify premature death as the gap between the time of death of an individual and the expected remaining lifetime in some chosen reference population. Third, there are lifespan disparity points (LSDPs), which classify a death as premature if its occurrence increases a population's lifespan disparity. The measurement of premature death thus involves normative claims (see Sørheim et al. 2024).

Additionally, if humans are considered solely as biological machines (a view that coheres with strong naturalistic accounts), significantly manipulating physical features could change the nature of the entity in question. Said another way, if our experiences of self are epiphenomena that derive or emerge from a biological substrate, physical changes to that substrate could alter those phenomena.[3] Thus, this chapter links to notions of human nature covered in Chapter 4. Furthermore, this chapter connects with Chapter 3's discussions on human origin and evolution as one of the scriptural texts I present discusses the height of the first human and prophet, Adam. Moreover, human height can be modified through biomedical interventions, and as such, the case study dovetails with the chapter's main focus on the moral dimensions of medicine.

Consequently, the chapter proceeds as follows. In the first part, I describe transhumanistic ideology and practices and overlay that with a discussion of biomedicine and contemporary healthcare. In this section, I introduce the "disease" of idiopathic short stature to highlight concepts of normativity and the sociocultural construction of disease. I also cover distinctions between therapy, enhancement, and alteration—categories that are used to differentiate ethical medical practice from activities aligned with transhumanist impulses. In the next section, I attend to changes in human beings by considering what Islamic scriptural and bioscientific sources tell us about human height over time. Admittedly, the example does not precisely speak to transhumanism, that is, drastic changes in human nature, but human height does appear to "naturally" change, and changes can also be induced through the application of biomedical techniques. Hence, there are sufficient merits for using this example as a case study and teaching tool (as also noted above).

Thereafter, I attend to questions of the moral boundaries of medical practice by initiating a multidisciplinary and multitier Islamic bioethical deliberation over the role of medicine. Here, I bring together different stands of Islamic bioethical writing and introduce the reader to ethical guideposts such as the theological notion of changing the creation of

[3] Coined by American philosopher David Chalmers, there is a "hard problem" to resolve regarding human consciousness. Specifically in the areas of philosophy of mind, it is unclear how physical matter, that is, the brain, generates the mind and produces phenomenal consciousness or subjective experiences. Supervenience and emergence refer to different notions, with several subtypes among them, which attempt to explain how physical phenomena generate mental phenomena.

God, which is used in Islamic ethico-legal literature to proscribe human enhancement and alteration activities. In concluding the chapter, I outline an intervention that links various Islamic moral sciences to theoretical, practical, and applied bioethical deliberation processes.

Transhumanism, Biomedicine, and Contemporary Healthcare

What Is Transhumanism?

First coined by biologist Julian Huxley (Mobayed 2017), brother of the famed British author Aldous Huxley, transhumanism has been variably described as an ideology, a social movement, and a set of processes. In reality, transhumanism is all of these things. Linked to Enlightenment philosophies that privilege human rationality in discerning the good and true (Jotterand 2010; Mobayed 2017; Hughes 2010), and ones that hold empirical science as the key to human progress, transhumanism entered popular and academic discourses in the 1980s and 1990s (Manzocco 2019) through the establishment of the World Transhumanist Association, the release of several manifesto-type books, and several high-profile cultural events and conferences (Bostrum 2014).[4]

Nonetheless, transhumanism is best conceived of as a loose socio-political and intellectual movement with various thought leaders, including Nick Bostrom, formerly an Oxford University-based philosopher; Natasha Vita-More, an American designer and artist, and James Hughes, an American bioethicist and associate provost at the University of Massachusetts in Boston. Allied with this movement are think tanks such as the Extropy Institute (n.d.) and ThinkH+ (n.d.), as well as academic

[4] For example, the World Transhumanist Association was established in 1998 and is now known as Humanity+ ; Natasha Vita-More, a strategic designer and artist, published "The Transhuman Manifesto" in 1983 and "The Transhuman Arts Statement" in 1992 and displays of her transhuman artwork occurred at venues such as the National Centre for Contemporary Arts in Russia and at the Telluride Film Festival shortly thereafter; Nick Bostrom, a philosopher, published *Transhumanist Frequently Asked Questions* in 1999 and later founded the Future of Humanity Institute at Oxford University in 2005; Max More, a philosopher, published his seminal essay, "Transhumanism: Toward a Futurist Philosophy" in 1990, and founded the Extropy Institute which held conferences dedicated to transhumanism in the 1990s; short essays such as "A Manifesto for Cyborgs" (1985) by Donna Haraway, a scholar in the fields of science, technology, and feminist studies, and "The Post-Human Movement Manifesto" (1988) by Steve Nichols, a scientist.

centers such as the Future of Humanity Institute (n.d.) and the Institute for Ethics and Emerging Technology. Furthermore, multiple technology companies, such as Elon Musk's *Neuralink* (2024) and Google's *DeepMind*, participate in projects that emerge from transhumanist ideas.

Organizing the loose movement are several aims, a set of philosophical commitments, and a group of proposed projects to accomplish the targeted aims. Concerning aims, in its boldest version, transhumanism aspires to a post-human future where humanity *trans*itions from its current species and *trans*cends into some grander form of life that is melded, infused, or wholly ensconced within technology. While not all adherents aim at such a future, they share the objective of redressing human frailty through technology. Human biology is seen as limiting one's lifespan, intellectual prowess, and experiences. Thus, technological modifications are sought to make human lives that are longer and fuller. As Bostrom notes, transhumanism is about "evaluating the opportunities for enhancing the human condition… opened up by the advancement of technology," including the "radical extension of human health-span, eradication of disease… augmentation of human intellectual, physical, and emotional capacities" and possibly "creating superintelligent machines… that could profoundly alter the human condition" (Bostrom 2005: 3).

Concerning philosophical commitments, transhumanists claim continuity with European Enlightenment thought. Specifically, transhumanists give primacy to rationalism, humanism, and scientism. These linkages are readily apparent in the writings of its promulgators. Hughes notes that transhumanism is "an ideological descendant of the Enlightenment," and its core belief is that "science can be used to transcend the limitations of the human body and brain" (Hughes 2010: 622). Bostrom writes, "Transhumanism has roots in secular humanist thinking," which values personal autonomy and human rationality (Bostrom 2005: 4). Max More, another transhumanist philosopher, adds that transhumanism imports its optimism for curtailing human finitude through science-based developments and its valuation of this life, rather than a supernatural afterlife, from secular humanism (More 1990). Consequently, organized religion is viewed with a modicum of disdain. Illustratively, Bostrom exclaims, "empirical science and critical reason—rather than revelation and religious authority—are ways of learning about the natural world and our place within it and of providing a grounding for morality" (Bostrom 2005: 4–5). Not surprisingly, over 90% of the World Transhumanist Association's members classify themselves as atheists (Jackson 2020; Hughes 2007).

Nonetheless, there has been some credible dialog between transhumanism and religion, particularly Christian voices, given shared goals of relieving human suffering and self-improvement.[5] In summation, transhumanism upholds materialist ideology and rationally derived moral norms and believes in the power of technoscience (Jotterand 2010).

These philosophical views accord with envisaging the human being as a product of haphazard evolution; a work-in-progress that took eons to emerge. Transhumanists seek to forcibly shape humanity's future by guiding the next phase of evolution in which human reason illuminates what societal and species flourishing looks like, personal visions of one's bodily future are advanced, and innovative technologies facilitate bodily change.

Three principal technology-mediated endeavors are vital to the transhumanist vision. The first involves genetic engineering through which inborn diseases are eliminated, aging-related physiological declines are curtailed, and the body and mind are enhanced[6] for optimal functioning within one's environment and/or to push beyond species-specific capacity limitations. The second set of activities dovetail with the first to encompass physical modifications that restore previously lost functioning and/or augment biological functioning. A cyborg may represent the ultimate end of such activities. Prosthetics that restore eyesight while also enabling "seeing" beyond the visible spectrum as well as those that allow the user to "hear" colors, and neural implants that allow for controlling extra body mechanical devices exemplify augmentations that mark moves toward becoming a transhuman. The third endeavor, mind-uploading, targets death directly. The aspirational goal is to copy and download one's mind—the aspect of one's being that is the repository of one's identity—onto a mechanical device and upload it into a computer or a synthetic body, thereby allowing one to 'live' on. Popularized in movies such as *CHAPPiE* (Blomkamp 2015) and *Avatar* (Cameron 2009), this idea is confronted by multiple scientific and philosophical challenges. For example, it remains scientifically unclear whether consciousness and one's

[5] An August 2005 special issue of the *Journal of Evolution and Technology* titled "Religion and Transhumanism" documents such engagements (Campbell and Walker 2005); a book section titled "Philosophy and Religion" in *The Transhumanism Handbook* edited by Newton Lee (2019) also details religious outlooks that may overlap with transhumanism.

[6] A particularly provocative argument for the necessity of moral enhancement is found in Persson and Savulescu (2012).

experience of self are emergent properties of the biological brain. Moreover, if, as some transhumanist philosophers contend, the core of human nature is a substrate-independent mind, that is, a mysterious mind that is neither reducible to properties of the biological brain nor dependent upon it, then it is unclear what mind-uploading entails since it would not involve the reproduction of physical state of a brain but something else entirely. Furthermore, it is not philosophically resolved whether the reproduction of the data state of the biological brain within a mechanical interface or another biological entity would produce a different "carbon-copy" self or the "original" person. Committed as they are to a materialist, naturalistic account of humanity, transhumanists aspire to solve this riddle through biotechnological means.[7]

While mind-uploading may be the realm of science fiction, some aspects of the genetic engineering and body modification envisaged by transhumanists are part of current biomedical research and practice. Research on genetic and genomic therapies abounds, and according to a recent review, over 2600 gene therapy trials are underway across the globe. These trials target various inherited and acquired disorders and utilize multiple platforms to effect gene editing (Belete 2021). The technical capacity to screen for genetic disorders was achieved decades ago, and our ability to induce changes in the genome continues to expand through the development of CRISPR Cas-9 and other biotechnological platforms. This exponentially increasing capability has renewed concerns over eugenic uses of genetic engineering (Paul 2005).

Cybernetic implants and function-enhancing prosthetics are also readily available today. A biohacking subculture abounds where individuals implant computer chips and sensors under their skins for functional and aesthetic purposes.[8] Heralding these sorts of interventions was Kevin Warwick, an English bioengineer, dubbed Captain Cyborg for experiments he conducted by implanting various devices into his body with the aim of becoming a cyborg (Warwick et al. 2003). These experiments spanned several years and allowed Kevin to control lights and other devices using a radio frequency transmitter implanted in his arm, as well

[7] The processes of forestalling death by taking the core essence of human nature, the mind, and preserving and transferring it into another entity is conceived as involving multiple different steps including mind copying, mind transfer, mind preservation, and whole brain emulation.

[8] For some examples, see Adams (2017).

as to manipulate and obtain sensory input from a robotic arm through a neural interface embedded in his hand. A more ambitious experiment allowed Warwick and his wife's nervous systems to interact with each other using radiotelegraphy.

While Warwick was motivated by a desire to augment the capacity of his non-diseased body, Peter Scott-Morgan, an English robotics expert, was spurred to utilize technology by the onset of amyotrophic lateral sclerosis. In successively replacing his body parts with mechanical substitutes, Scott-Morgan sought to stave off being crippled by his disease. To prevent losing his voice and infectious complications from losing his ability to swallow, Scott-Morgan had his larynx removed and utilized a computer interface to speak, and he underwent eye surgery to optimize his eyesight for using computer screens to mobilize his wheelchair. In advance of losing muscular control for alimentation and excretion, he opted for being fed via a gastric tube and for a urinary catheter and colostomy bag (Scott-Morgan 2021; Kingsley 2022; Smith 2022; Goldsberry 2022; Littlejohn 2020; Durlacher and Pelly 2020). Though Scott-Morgan ultimately succumbed to the disease, he was dubbed the Human Cyborg, and his efforts brought worldwide attention to the plight of those with motor neuron disease and the possibility of using biotechnologically electively to reduce disease morbidity. Scott-Morgan's journey exemplifies how transhumanist ideals have, arguably and with some limits, aligned with the goals of biomedicine.

Connections Between Transhumanism, Biomedicine, and Contemporary Healthcare

Although ubiquitous today, the term biomedicine was first used in early twentieth-century Europe and the United States to describe medical and scientific research related to radioactive materials. The term has expanded to represent "the umbrella theoretical framework for most health science and health technology work done in academic and government settings [globally]" (Valles 2020). More precisely, biomedicine signifies a theoretical framework together with a body of multidisciplinary knowledge that is used to identify and address human maladies.[9]

[9] One could argue that biomedicine is also applied to the care of animals (veterinary medicine) and would be applied to organisms discovered elsewhere in the universe (exobiology). In this book, our concern is with the application of biomedicine to humans.

Like transhumanism, biomedicine inherits its views on human nature and what is knowable from the European Enlightenment thought (Ali 2018). Both systems share a naturalistic account of human nature and experiences, rely on empirical approaches to glean truths, and hope in technology-infused solutions to human maladies. As such healthcare practices growing out of the biomedical paradigm are susceptible to becoming instrumentalized for transhumanist goals.[10]

Concerning biomedicine, social epidemiologist Nancy Krieger delineates three hallmark features of biomedicine's approach to human maladies. First, diseases are tied to biological, chemical, and physical phenomena. Second, biomedicine privileges knowledge garnered through laboratory and clinical experimentation. Third, biomedicine commits itself to, both philosophically and methodologically, reductionism. Thus, biomedicine holds that the human being is best understood through an examination of its subsidiary parts, and, in concert, the phenomenon of human illness is primarily explained by disorder, malformation, maladaptation, and dysfunction of these constituent parts (Krieger 2011; Valles 2020). These features are mutually reinforcing; a naturalistic account of disease requires methodological naturalism be employed in research and, in turn, lends explanatory power to reductionism. Extending the line a bit further, the biomedical worldview generates a research paradigm that delivers a body of knowledge that, in turn, furnishes the grounds for a specific form of healthcare practice.

Accordingly, the etiology of disease as well as the grounds for cure are found in the body; the body's anatomical structures, biochemical functions, and physiological processes contain "truths" about human disease and cure. Primacy is given to the "hard" sciences of biology and chemistry

[10] Related is the idea of biomedicalization of societal issues. The process of biomedicalization involves referring to a problem in terms of concepts, practices, and technologies that are endemic to biomedicine. When problems are framed in this way, biomedically oriented solutions often follow. For example, the use of illicit drugs repeatedly can be viewed in many ways, from a personal failure resulting from the loss of willpower, to a disease resulting dysfunctional neurosignaling attributable to chronic drug exposure. When biomedical explanations for a condition or behavior become dominant in society then biomedicine is also relied upon to generate solutions. Similarly, social practices that can be conceived of as having been biomedicalized are ones where the practice is justified in terms of health or illness considerations. In other words, medical rationales and meanings become primary supports for the practice.

as materialism is biomedicine's principal grounds for knowledge. Technology is used to examine and probe the human body so that it reveals its secrets, and biodata in the form of chemical markers and physiological thresholds are sought to ascertain the presence of disease, its status/progression, and its resolution. Biomedicine insists upon causal chains that specify pathogenesis in a language of structural flaws and malfunctioning bodily mechanisms. Indeed, biomedicine focuses on objectifying, classifying, and quantifying the human body because deviations from posited norms mark supposed disease.[11]

Biomedicine is intrinsically linked to healthcare delivery across the globe. Indeed, the physician's daily work of identifying, diagnosing, and treating patients is wedded to a naturalistic, reductionist, mechanistic understanding of the human being. Ideally, the clinician identifies and diagnoses human maladies by obtaining biological data and examining bodily structures. Only after interrogating the body in this way should the physician proceed to select an appropriate, that is, a biostatisticly validated, therapeutic regimen to treat the malady. Moving upwards from the patient–clinician encounter, patient-level biomedical data is aggregated into larger population and societal-level data stores, which, in turn, are used to rate healthcare quality and rank how well countries meet their citizens' healthcare needs. Aggregate data also sets up the boundaries of biological normativity and allows for the sociocultural construction of disease states. Once the disease label is applied to a condition, pharmaceutical and biotechnological companies take up the cause to develop therapies.

While biomedicine and transhumanism share philosophical and epistemological commitments, I am not claiming that contemporary healthcare wholly operates out of a transhumanist vision. Certainly, there is no cabal of transhumanists insidiously pulling societal strings to co-opt healthcare. Rather, I contend that biomedicine and transhumanism share a common intellectual heritage, and that the goals of biomedicine

[11] Diseases are, at least in part, socioculturally constructed. Societal expectations shape norms in terms of human behavior, traits, and appearance. Should an individual fall outside of these expected norms, they could be classified as diseased. A much-discussed example is the "disease" of idiopathic short stature, a topic discussed in greater detail a bit further in this chapter. Other exemplars of the sociocultural construction of disease include infertility and congenital deafness. Implicit to labeling such conditions as diseases are assumptions about the normative body, and fueling cultural expectations that these conditions be termed disease states are biotechnological remedies that can be had.

and transhumanism, namely, reducing human mortality and suffering, overlap. Simultaneously, societal hopes for biotechnological solutions fuel increased alignment between transhumanism and contemporary healthcare. Long-held caricatures of physicians as "men against death" have set up the expectation that biomedical research and healthcare treatments will continually increase human longevity (De Kruif 1932). Similarly, recent technological marvels and leaps in computing sciences and biological mimicry give hope that science and technology will rid humankind of problems related to bodily dysfunction and disorder. Small gains in mortality and modest technological breakthroughs fuel expectations of further gains, thus creating a self-reinforcing cycle of social expectations, and such expectations, in turn, shape the goals and practices of healthcare. Hence, societal aspirations of solving aging through biotechnology bring healthcare practices into closer alignment with transhumanist aspirations.

Boundary Management: Differentiating Therapy from Enhancement and Alteration

Numerous bioethicists and intellectuals have warned against healthcare embracing transhumanism (Jones 2016; Porter 2017; McNamee and Edwards 2006; Thompson 2017; Bauer 2010; Fukuyama 2004). In varied ways, these scholars seek to define the practices of medicine proper as related to restoring lost abilities and functions, facilitating equality of opportunity, and acknowledging human finitude. They also caution against desiring the impossible, for example, ending human death, and raise concerns about intellectual hubris with respect to the capacities of science and technology. Accordingly, scholars have used the concepts of therapy and enhancement to draw a border between practices that fall within the purview of medicine and those that do not. The distinction is explicated in at least three ways, each giving primacy to a different stakeholder in drawing the line.

The first approach puts the onus on policymakers and licensing bodies, assumed to gain their authority from the polity, to demarcate therapy from enhancement. This view, termed the professional domain approach, holds that any practice that a physician engages in, or in other words, any good or service that a clinician provides, is therapeutic. It follows that activities that occur outside of a professional relationship between a patient and a clinician may be considered an enhancement. For example, the prescription of testosterone within the confines of a patient–doctor relationship is therapeutic. In contrast, the taking of testosterone on the advice of an athletic trainer may be considered enhancement. This view coheres with the idea that medical practice is based on a social contract and that patients' values regarding what is done to their bodies outline the parameters for the social structuring of healthcare delivery (Juengst 2004).

The second approach entails the construction of normativity. Proponents argue that healthcare, at its core, is a means to equality of opportunity for society's members. Hence, medicine's role is to assist individuals in overcoming their inherent biological and psychological, and at times social, curtailments that frustrate achievement. Accordingly, restoring or improving the patient's abilities to an "equality of opportunity" threshold is medicine's mandate and judged to be therapeutic; moving beyond that reference to create an advantage is considered enhancement. The crux of the matter relates to defining norms. Tying normativity to equality of opportunity introduces society-specific elements to the definition of treatment, for what therapy is in one social context might be considered enhancement in another. Alternatively, one could ground normativity in biological or psychological function. This would require identifying species-specific capacities, developing tools to measure these capacities, and then mapping data onto a bell curve-type distribution to set up thresholds for what is classified as treatment and what is deemed enhancement. Social and biological/psychological data could also be combined by linking biological/psychological features to data streams that delineate access to opportunities within a given society, thus generating a society-specific "normal" distribution.

Illustratively, one could measure levels of attentiveness among five-year-olds within a given society and then determine how these levels map onto school performance. Assuming that school performance is linked to opportunities for flourishing, one could define treatment as that which is done to assist five-year-olds in obtaining the normal range of attentiveness. This example illustrates the sociocultural construction of disease and the many judgments made while doing so. Certainly, there may be biological and psychological determinants to the "abnormal" attentiveness exhibited by a five-year-old. However, the social implications of the biological/psychological nature turn it into a disease that requires treatment. The arbiters of what is categorized as treatment are biomedical and social scientists who make data-based arguments for norms of function, capability, and opportunity. To be sure, the constructed dimension of disease gives critics pause as it makes the profession of medicine susceptible to being an instrument of social engineering through the biomedicalization of social problems.

In a third approach, termed the disease-based approach, the state of the human being acted upon distinguishes treatment and enhancement. Treatments target individuals experiencing a malady, a health problem tied to a disease, disability, dysfunction, or disorder. On the other hand, enhancement activities are applied to an individual who has normal traits, functioning, and the like. This approach is intuitively based on the idea that maladies are empirically verifiable and, hence, the appropriate targets for treatment. Yet, this view inherits the same normativity challenge as the previous approach. It also places the authority to define, and responsibility to police, the boundary between therapy and enhancement squarely within the profession of medicine, and in so doing, creates the potential for abuse.

Differentiating therapy from enhancement is viewed as critical to the defense against transhumanistic ideals instrumentalizing healthcare. Yet, the dividing line is fuzzy, and it is unclear which stakeholder group presently manages the boundary. Fabrice Jotterand, a philosopher-ethicist, attempts to address ambiguities by adding a third conceptual category, alteration. Driven by the concern that therapy and enhancement do not accurately portray the possibilities of radical transformation envisaged by transhumanism, he notes that "the goal [of transhumanism] is to transcend biological boundaries through technological means to alter human capacities" (Jotterand 2008: 17). He thus slightly redefines therapy and enhancement to accommodate a third category of alteration and to make the distinctions between each category sharper. Therapy primarily entails restoring capacities lost due to disease but may include modifications that raise individuals who have inborn diminished capacities to a posited biological norm. Enhancement involves activities that augment an individual's biological capacities beyond species-typical norms. Alteration, on the other hand, ventures further by traversing biological limits such that the resulting state is no longer considered human. Within the realm of neurobiology, an example of treatment is to provide medications that reverse cognitive declines related to Alzheimer's disease. Substances that allow an individual to process and retain information beyond the typical capabilities of the human brain would fall into the category of enhancements. Creating a neural implant that transfers memories, images, and facts between one's biological brain and an external storage device would be an alteration because biological limits have been overcome, and the resulting being may no longer be part of the human species. Notably, Jotterand's schema requires biomedical scientists to fix biological norms, which, as noted above, is a value-laden endeavor.

The "Disease" of Idiopathic Short Stature

Thus far, I have provided an overview of transhumanism and biomedicine, discussed how they share philosophical and epistemological commitments, and highlighted how healthcare practices are central to the debate regarding what constitutes an appropriate use of biotechnology for bodily modifications. Some see transhumanism as a noble endeavor that assists humankind in conquering finitude, while others see this vision as a Frankenstein-ish abomination that must be rejected. The battleground is contemporary healthcare for, through its practices, worldviews can be legitimated, cultural norms instantiated, and a specific future for the human body implemented. To further illustrate the cultural power of medicine and set up the forthcoming discussion of the changing of humankind, let me introduce the "disease" of idiopathic short stature. Indeed, idiopathic short stature exemplifies the sociocultural construction of a disease based on statistically driven notions of normality. Its remedy involves interventions which, depending on one's perspective, could be either treatment or enhancement.

Idiopathic short stature is defined as height below two standard deviations of the average for one's age with no known organic cause. This definition highlights features of the biomedical approach. Notably, empiricism is reflected in labeling a disease state based on biological data; in this case, societal data concerning height and naturalism is hinted at by the term idiopathic, which stands in for the claim that the scientific method should have uncovered a biological cause for the disease. Even though the biomedical approach, properly applied through a reductionist probing of the body, yielded no cause for the short stature, the social implications of a particular biological state can be deemed by biomedicine as sufficient grounds for labeling a certain condition a disease. Indeed, psychological and social harm for individuals may ensue should that person's body habitus or behaviors be too "deviant" from others within a given society. Notwithstanding the primacy given to biomaterial considerations as grounds for disease, biomedicine acknowledges social harms related to morphology and behavior and allows for constructing diseases based on them.

In my view, four related and somewhat overlapping concepts furnish grounds for attaching the label of disease to a human malady: disease, disorder, dysfunction, and disability. Each of these ideas join together, albeit variably, biological and social considerations. First, the notion of being ill at ease, or dis-eased. On this view, an individual's subjective experience serves as the basis for labeling a condition a disease. While there may be biological analogs or causes underlying one's experience, the experience is given primacy. For example, psychological distress can be considered a disease. Second is the concept of a disorder. The disorder might be related to psychological or biological features of the human being. In common vernacular, something is not as it was expected to be. Having a sixth toe could be subsumed of the idea of disorder. Disorder is determined by biomedical scientists in relation to their view of the normative body and/or human personality. Closely related is dysfunction. While a disorder may have no functional correlate, dysfunction is rooted in the observable. The working of a body part, or the capacity of the entire organism, is reduced. Just like with the notion of disorder, biomedical science sets up the threshold and measures for abnormal functioning. Kidney failure would be classified as a dysfunction. Finally, we have the idea of disability as undergirding a diseased state. This concept is tied to the notion that healthcare creates equality of opportunity for individuals within society. Thus, disabilities must be ascertained in the context of a particular society. Where biological features of a human being engender less than equal opportunity within a society, those features support the labeling of a disability. Individuals with reduced mobility or poor eyesight can be labeled as disabled, and biomedical measures may be used to address that disability. The various grounds for specific disease states implicate the type of treatments offered by healthcare and may create the impetus for alternative enhancements or biotechnological alterations provided outside of the healthcare system.

Returning to the sociocultural construction of idiopathic short stature, are there social harms to being too short? Biomedical and sociological research provides evidence of several detriments. For example, studies reveal that individuals of short stature are less likely to advance in their education and thus have, on average, lower incomes than those of median height within a society (Judge and Cable 2004; Tyrrell et al. 2016). Individuals of short stature may suffer from ridicule and teasing, particularly at young ages, which negatively impacts their psychological growth and well-being (Bullinger et al. 2009; Keselman et al. 2000). Studies examining the behavior profile of short-statured children also find higher levels of behavioral problems than normal-statured individuals and the general population overall (Stabler et al. 1998; Steinhausen et al. 2000). Short stature among men may be associated with fewer romantic prospects (Salska et al. 2008; Shepperd and Strathman 1989; Gillis and Avis 1980); and in turn, we know that married individuals tend to have longer lives (Kaplan and Kronick 2006). These and other observed correlations, such as evidence that taller people on average have higher education, higher earnings, and possibly even a higher social position, provide the grounds for intervening upon idiopathic short stature in early childhood (Adair et al. 2013; Stulp et al. 2015; Barker et al. 2005; Strauss and Thomas 1998; Chen and Zhou 2007; Case and Paxson 2008).[12]

Truth be told, the medical community is divided as to whether idiopathic short stature should be considered a biological disease. Given that a biological cause is ruled out by definition, some argue that idiopathic short stature is simply a descriptive label. While this view might be technically accurate, functionally the condition is treated as a disease because the house of medicine offers remedies for it. Harkening back to the professional domain approach for distinguishing treatment from enhancement, idiopathic short stature is an entity that is addressed by medical treatments. And those treatments are applied because of the social implications of being "too" short.

Human growth hormone supplementation may be offered to parents of a young child should a child's projected full adult height be too short. In the United States, this corresponds to an adult height of less than 63 inches (~160 centimeters) for males and 59 inches (~150 centimeters) for females. A more controversial treatment restricted to adults and adolescents who have reached their maximum height, that is, their long bones have fused, is leg-lengthening surgery. This option is not widely available, given that many physicians and regulatory bodies consider such

[12] Some may argue that one's livelihood, marriage prospects, and the like are predetermined, that is, part of God's decree. Hence, the associations discovered by sociological analyses should not be accorded much weight. Such a perspective, in my view, short circuits the exact sort of conversation this book attempts to foster, namely, dialog over religious and scientific evidences through a cogent epistemic framework. With respect to aspects of humanity that are predetermined the reader is directed to discussions in Chapter 5.

modification to be an enhancement. One can certainly understand the concern that offering such surgery would legitimize the practice as part of healthcare based on the professional domain view of what constitutes treatment. However, such a procedure would not be considered enhancement according to the normalcy and disease-based approaches, nor according to Jotterand's schema, as it does not traverse biological limits.

Is Humankind Changing?

Addressing this question necessitates unpacking a few embedded assumptions and concepts. Recall that the question's purpose is to allow for juxtaposing religious and bioscientific knowledge regarding fundamental aspects of the human being. As such, the question was a pedagogical device for organizing didactic content and deliberative activities within the pre-medical student seminars from which this book project emerged (see Chapters 1 and 10). Within that curriculum, and in this essay, the question's framing allows for connecting discussions about different "Big Questions" such as, What is Human Nature? and What is the Origin of Humankind?

In attending to the question of change, human height stands in for the larger concept of humankind.[13] Bioscientific narratives operating out of a naturalistic and reductionist frame consider the human being to be a purely biological entity. This biological composition is the foundation for human consciousness and experiences of self, albeit explanations for how that comes about are varied. Additionally, the conception of humans as empirical agents that absorb sense data to take reasoned action appears to breakdown when tasked to explain qualia and the influence of emotion on decision-making (Jackson 2020). Nonetheless, some philosophers and scientists support biological determinism and contend that choice is mostly a fallacy; we are but biological automata (Sapolsky 2023; Plomin 2019; Pinker 2004).[14]

While the "hard" problem of consciousness continues to beguile biomedical explanation, transhumanist aspirations accentuate the

[13] For a fuller discussion on human nature, the reader is directed to Chapter 3.

[14] There are variants to biological determinism and almost no scholar holds the view that we are completely determined.

problem. Within transhumanism, the mind is the locus of one's identity and the essential aspect of human nature. The aspiration of being able to upload one's mind and, in so doing, skirt death underscores transhumanism's goal of expunging human frailty and biological limitations. Despite being beholden to a naturalistic account of humanity, the mind and brain are seen as separate within transhumanism; the substrate-independent mind (SIM) is that which is displaced into an alternative mechanical construction or a computational platform such that the new transhuman you is you (Jackson 2020). From a philosophical vantage point, this scientific fiction is not entirely satisfying, for if the new entity is you, then your (human) nature has not substantively changed, and a post-human has not yet been created. Perhaps this new state, an initially human SIM within a non-human mechanical entity, marks the transhuman, but the trouble with linking human essence with the "current" mind persists. Would we have to discard the mind to be a post-human? And the brain–mind problem persists in the background for any type of mind-uploading, and the reconstituting of such requires an understanding of its nature and how it is linked (or not) to the biological matters of the brain. Bostrom notes that the post-human is marked by wholly different experiences, values, and, indeed, nature, and as such, its being may be beyond our present ability to fathom (Bostrom 2005). Thus, clarity may emerge only when we are further along in the process.

Within religion, particularly the Abrahamic faiths, the concept of the soul explains human nature. Humans are composed of biological material and something otherworldly, a soul. This dualistic nature explains what biology, heretofore, cannot. In the Islamic tradition, scholars debate the soul's nature, with the two most-dominant positions, at least in Sunnī theology, being that it is: (i) a subtle body (*jism laṭīf*); or (ii) an immaterial substance (*jawhar mujarrad*). The former position holds that a physical locus of the soul is found within the body, traditionally thought to be in the heart. In contrast, the latter view contends that the soul is immaterial and nonphysical and, while (temporarily) connected to the body, it is not part of it (Brown 2013; Qazi et al. 2018). The first view demonstrates that classical Islamic theologians, just like secular scientists today, sought to explain the human being by recourse to material and physical causes; importantly, however, they were not beholden to naturalism given they were operating within a broader Islamic ontological schema. Delving much further into these camps is beyond our scope here. The point I want to underscore, however, is that human beings remain a mystery to both religious scholars and bioscientists, and there is plurality within each disciplinary camp.

Concerning bodily change, human height reflects the biological components of the human being and results from interactions between the internal biological components and interactions between one's biology and the surrounding environment. Said another way, it is partially inherited and partially subject to external factors and conditions. While height may not be essential to human nature, it does "emerge" from the biological body, and documenting changes in this observable aspect of the human could be a helpful analog for changes to other parts of the body, for example, the brain, which may be more closely linked to what marks a human being as human.

As noted previously, discussing bodily change connects our discussions with controversies over neo-Darwinian evolution. This theory claims that the human species is an interval product of an eons-old process of gradual random mutations and natural selection. If this is the case, then humankind is constantly changing, and each change by itself may be inconsequential or unobservable. Yet, the theory holds that over long periods, the successive changes lead to observable transformations such that *homo sapiens* will transform into a newer species of a different nature. Those differences will be marked differences of kind, not simply of degree, and maybe morphological. As noted above, transhumanism builds out from this view to champion humans taking control of the purportedly haphazard and slow process by purposively molding and speeding humankind's transition into a post-human.

Bioscientific Evidence Regarding Human Height

At least over the past century and a half, there is no doubt that average human height has increased. While there is variance according to sex, race, ethnicity, and geography, the overall trend of increasing height is undeniable. Using the framework presented in Chapter 2, the proposition that human height has increased is an empirical (*ʿādī*) one with data streams that evidence it to be a probable (*ẓannī*), in this case a highly probable (*ẓann rājiḥ*)[15] truth claim based on inferential (*naẓarī*) judgments. As will be seen shortly, related scientific concepts support a rational (*ʿāqlī*)

[15] A somewhat analogous term used in Islamic law is *ghalabat al-ẓann*, which refers to preponderant probability of truth.

proposition that human height has an upper bound, that is, humans cannot have been giants. This second proposition can be classified as possible (*jā'iz*) but not necessary (*wājib*).

There is multiple evidence for this phenomenon: archaeological, socio-historical, and empirical. If one visits houses built during nineteenth-century Europe, they will find the ceilings lower and the doorways smaller, evidencing that the inhabitants were likely shorter. The same can be gleaned by visiting long-standing graveyards across the globe, where burial plots from previous eras are smaller and shorter than those of today. Detailed studies affirm these casual observances.

Surveys of European skeletons over the past two millennia reveal that adult male height remained around 170 centimeters (5 feet 6 inches) until the 1800s, after which steadily increasing height among the remains of individuals is found (Steckel 2001; Koepke and Baten 2005). This empirical evidence is probabilistic, given that measurement error is involved in radiocarbon dating skeletons. There is also sample bias because the measured skeletons belonged to soldiers, criminals, slaves, and servants, and thus, the data is not representative of the general population. Yet, corroborating evidence comes from surveys conducted by multiple governmental and non-governmental agencies that the University of Tübingen has collated. These data show a nearly universal upward trend in average male height across the globe from 1810 through 1980. For example, the average male height in the USA, the United Kingdom, China, and Yemen was 171.1, 165.6, 166.2, and 162.3 centimeters in 1850, respectively. These heights increased to 179, 176.8, 171.5, and 167.6 centimeters by 1980 (Baten and Blum 2014). Research conducted by a global network of health scientists, the non-communicable disease risk factor collaboration (NCD-RisC), lends further support. This group analyzed data from nearly 1500 population-based studies sourced in 200 countries and encompassing over 18.6 million persons from 1896 to 1996 to examine country-specific variances and thus make finer claims. They concur that while there have been significant increases in European men's heights in the nineteenth and twentieth centuries, the same could be said for Japan, and that the largest height gains from 1896 to 1996 occurred in South Korean women and Iranian men, who became 20.2 and 16.5 centimeters (13.3–19.7) taller, respectively (NCD Risk Factor Collaboration 2016). However, height gain among individuals living in sub-Saharan

Africa and South Asia has been more modest in the range of 5 centimeters. Overall, human beings are, on average, 5% taller than their ancestors a century ago (NCD Risk Factor Collaboration 2016).

Notably, the available archaeological evidence does not suggest wide variations concerning human height before the common era. For example, research suggests that in the Mesolithic period in the Eastern Mediterranean region, males averaged 172.5 and females 158.7 centimeters, and heights were similar in the Bronze Age as well (Angel 1984).

Multiple factors contribute to the observed increase in human height over the past century and a half. Environmental factors, specifically greater access to and more nutritious food, are held to be the largest factor. Proper caloric intake during adolescence, when most of one's growth occurs, is critically important. The evidentiary basis for this assertion comes from examining skeletal remains which show relatively modest variance in the heights of men from foraging societies in the previous millennium and those from pre-industrial societies in the late 1800s and early 1900s (Steckel and Prince 2001). Even today, human height is positively correlated with living standards, lending further support to the strong influence of economic factors. That said, height gains have largely plateaued in developed nations, suggesting a possible biological upper limit to height (Schönbeck et al. 2013). While genetic factors influence height, and individuals with abnormal growth hormone secretion suffer from the breakdown of long bones and joints, there is no precise estimate for optimal human height. Some consider the recently observed plateau attributable to poorer-quality food intake, inadequate access to preventive health resources, and more sedentary lifestyles (Freedman et al. 2003; Berkey et al. 2009).

While there is no consensus on optimal human height, there are upper limits mandated by physiology and physics. For example, Galileo's square-cube "law" mathematically relates an object's height, width, length, and volume when scaled up or down. Doubling an individual's height while maintaining proportionality of body parts entails doubling one's length and width and increasing one's surface area by four and volume by eight. This scaling rule effectively limits human height because weight is proportional to volume and strength to area, and the taller one gets, the relatively weaker one would be, making it challenging to maintain erectness or be mobile (Tall.Life, n.d.; Samaras 2007; Zaman 2022). Indeed, the tallest known humans, for example, Robert Ludlow, required mechanical support to walk as their bones and muscles could not adequately support locomotion.

Similarly, human physiology curtails one's height. Recall that the heart works against gravity to pump blood throughout the body. Adequate circulation is needed for organismal functioning. Taller humans would require bigger hearts to pump blood over longer distances effectively, yet the increased size would alter body habitus as the heart would need to be proportionally larger. Similarly, height would be limited by the speed of neurotransmission from the brain to one's nerves. Neurotransmission velocity is relatively fixed, based on the physical composition of our nerve cells and the chemicals involved in transmission; if one's brain were further from one's limbs due to excessive height, the transmission would be slower, and one's response time would be slower. These examples illustrate that an altered physiology would be needed to sustain a human being beyond a certain height. While there is no precise estimate of the upper bound of height based on current human physiology and Earth-based physical laws, a limit can be reached.[16]

In summary, bioscientific data and concepts evidence that human height was relatively stable from at least the beginning of the common era until the 1800s and that it has trended upward significantly in the past century and a half. These claims carry the epistemic weight of dominant probability to near certainty. Before the common era, the available empirical evidence suggests humans of somewhat shorter heights, although data is scant. Hence, this feature of humankind, based on presumptive evidence of rational nature, has changed and is changing.

Scriptural Views on Human Height

Before turning to the scriptural texts, the reader will benefit from a brief overview of terms related to the Islamic scriptures and their epistemic evaluation. In the Islamic tradition, the Qurʾan and Prophetic Sunnah are considered to be two parts of a single revelatory event. The former, termed the recited revelation, *waḥy matlū*, and the latter, termed the revelation that is not recited, *waḥy ghayr matlū*, represent the Islamic scripture and mutually reinforce each other. The Sunnah, as the statements, actions,

[16] Yet, one could argue that prehistoric humans had different physiology. Such an argument would require us to precisely define what we mean by human.

and tacit approvals of the Prophet Muhammad, is captured in *ḥadīth* collections and is generally accorded an explanatory role with respect to the Qur'an.

The Qur'an and *ḥadīth* are both classified as a type of knowledge based on testimony, that is, statements that are heard and then transmitted. Said differently, revelatory sources of knowledge are received based on the report of others having heard the statements and these chains of transmissions eventually end with the personage of the Prophet Muhammad conveying the statements. Reports are a valid source of knowledge within scientific and religious epistemic frameworks as requiring one to directly observe natural phenomena in order to have confidence in it or to make inferences from it would render scientific investigation unable to build upon the knowledge of others. Similarly, requiring immediate sensory perception, for example, hearing or seeing, of the Sunnah in order to have confidence in it would render revelatory knowledge inaccessible to those beyond the time of the Prophet. Concerning revelation, over the ages, Islamic scholars developed highly sophisticated systems of verification, recording, and classifying these transmitted statements.[17] Simultaneously, nuanced hermeneutical approaches were crafted to derive meaning from scripture. Hence, Islamic scholars fixed their epistemic gazes on both the chains through which revelatory knowledge was transmitted and the methods by which knowledge was produced from such transmissions.[18] Nomenclature-wise, each *ḥadīth* has two parts: the *isnād*, or chain of transmission; and the *matn*, the statement, recorded action or tacit approval.

[17] For an overview of *ḥadīth* sciences in the English language the reader is directed to the following texts: *Ḥadīth Literature: Its Origin, Development, and Special Features* by M. Zubayr Ṣiddīqī (1993); *Studies in Ḥadīth Methodology and Literature* by M.M. Aʿẓamī (1978); and *The Authority of Sunnah* by Justice Muḥammad Taqī ʿUsmānī (1991).

[18] In Islamic parlance, scholars assess the epistemic value of the *thubūt* (transmission) by interrogating the accuracy and probity of each narrator along the chain in consideration of the exact words transmitted forth within each chain, the connections between narrators, as well as the repute of the narrator. With respect to the *dalāla* (semantic import of the source-text), they examine the words, the meanings they could convey, as well as how the words are used in other scriptural text and similar considerations. In Chapter 6, the authors use the English terms establishment and indication in reference to *thubūt* and *dalāla*, respectively.

Concerning chains of narration, each verse of the Qurʾan has been reported in the same way by multiple narrators across successive generations, and thus, it is classified as a *tawātur*, a mass-transmitted report.[19] Epistemically, testimonial reports of this type are accorded the highest evidentiary status based on the idea that it is impossible for such a large number of people across time and space to have colluded together to contrive a false statement. With the chain of transmission being ironclad, Qurʾanic texts that are univocal in meaning are deemed to convey certain knowledge (*yaqīn* or *qaṭʿī*).

Ḥadīth, on the other hand, are variegated, with some reaching the status of *tawātur*, but most are classed as *āḥād*, which lexically means a single report, but within *ḥadīth* sciences indicates that the number of chains of narration is less than that needed to be *tawātur*. These reports are grounds for probabilistic knowledge, ranging from highly probable to improbable. As referenced in Chapter 2, although conventional ranges are offered, Islamic scholars avoid attaching precise mathematical probabilities to these categories. A doubtful proposition represents equipoise, whereby the intellect does not assent to the proposition nor to its opposite, as both have an equal possibility of being true. This midpoint is termed *shakk*. When evidence pushes one to a greater possibility of the claim being true, we are in the zone of probabilistic/presumptive knowledge, *ẓann*, and if the intellect is pushed closer to believing the contradictory proposition is likely to be true, then we are in the zone of improbable or illusory claims, *wahm*.

A common division classifies *ḥadīth* as *ṣaḥīḥ*, *ḥasan*, and *ḍaʿīf*. *Ṣaḥīḥ*, rigorously authentic, reports have unbroken chains of narrators, that is, the narrators are historically contiguous, and all are judged to have upright character and sound memory. Some *ḥadīth* scholars add other qualifications to classify a *ḥadīth* as rigorously authentic, and among the group there is variance regarding the ways in which memory and character are judged to consider a narrator as reliable. *Ḥasan*, good/fair, reports have unbroken chains but fail to meet the standards of *ṣaḥīḥ* due to questions regarding one or more of the narrators, significant inconsistencies in wording across chains, or features that clash with other *ḥadīth*. *Ḍaʿīf*, weak, *ḥadīth* have further faults. Again, *ḥadīth* scholars apply various

[19] I acknowledge that there are variant readings of certain verses of the Qurʾan and contentions around various abrogated verses. Delving into these matters is extraneous to the aims of this chapter.

criteria to place *ḥadīth* within these categories, and some scholars add subcategories under these three. While it is beyond this chapter's scope to detail the sciences of *ḥadīth* classification, authentication, and analysis, it is important for the reader to appreciate that *ḥadīth* analyses involve specific methodologies and are performed within communities of practice that have established norms of assessment and interpretation.

The Height of Prophet Adam and Subsequent Humans

A direct reference to human height is found in Prophetic reports concerning the first human, Prophet Adam. Within the Sunnī canon, several reports are linked to the Companion Abū Hurayra, in which the Prophet Muhammad discusses individuals entering paradise. A version recorded in Ṣaḥīḥ al-Bukhārī reads:

> The first group of people who will enter Paradise, will be glittering like the full moon and those who will follow them, will glitter like the most brilliant star in the sky. They will not urinate, relieve nature, spit, or have any nasal secretions. Their combs will be of gold, and their sweat will smell like musk. The aloes-wood will be used in their centers. Their wives will be *houris*. All of them will look alike and will resemble their father Adam (in stature), sixty cubits tall. (Ṣaḥīḥ al-Bukhārī no. 3327)

The term cubit refers to an ancient unit of measurement equal to a forearm's length (the Arabic text is precise in the reference to cubits) and was between 43 and 56 centimeters. An individual 60 cubits in stature would be approximately 90 feet tall. The versions recorded in Ṣaḥīḥ Muslim (*ḥadīth* 2834c) and Sunan Ibn Majah (*ḥadīth* 4333) are nearly identical. Another report contained with Ṣaḥīḥ al-Bukhārī mentions Prophet Adam's height and reads:

> Allah created Adam, making him 60 cubits tall. When He created him, He said to him, "Go and greet that group of angels, and listen to their reply, for it will be your greeting (salutation) and the greeting (salutations of your offspring)." So, Adam said (to the angels), *As-Salamu Alaikum* (i.e., Peace be upon you). The angels said, "*As-salamu alaika wa rahmatu-l-lahi*" (i.e., Peace and Allah's Mercy be upon you). Thus, the angels added to Adam's salutation the expression, "*wa rahmatu-l-lahi*," Any person who will enter Paradise will resemble Adam (in appearance and figure). People

have been decreasing in stature since Adam's creation (*Falam yazali al-khalqu yanquṣu ḥatta al-āna*) (Ṣaḥīḥ al-Bukhārī no. 3326)

This statement corroborates two pieces of content in the other reports: first, the height of the Prophet Adam is 60 cubits in both; and second, the inhabitants of paradise will be of that height. This report has an added truth claim of particular relevance to our discussion: "People have been decreasing in stature since Adam's creation." Of note, this report is also transmitted by Abū Hurayra.

For our purposes, the propositions of import are: (i) Prophet Adam was 60 cubits tall; and (ii) Humans have been decreasing in stature ever since. Concerning epistemology, these claims are based on *ḥadīth* that are graded *ṣaḥīḥ*, and thus, the statements are highly probable to have been voiced by the Prophet Muhammad, given the strength of the transmission chains. With respect to the meaning of the statements, the verbiage is relatively straightforward as they embed the truth claims explicitly. Hence, based on these scriptural accounts and their evidential grade, one could claim that the first human being, Adam, was nearly 90 feet tall, with dominant probability.[20] With respect to the claim of decreasing human height since then, we have another probabilistic truth claim. This picture gets a bit more nuanced once we probe deeper into chains and possible meanings of the source text.

The Height of Prior Nations

Turning to the Qur'anic insights regarding human height, human height is alluded to when discussing the might and power of the communities of prior prophets. Notably, the Qur'an mentions two successive nations, *ʿAād* and *Thamūd*, who were the descendants of the people of Prophet Nūḥ and were assigned Prophet Hūd[21] and Prophet Ṣāliḥ, respectively. As evidence of their skill and strength, the Qur'an notes that these communities built gigantic castles and mountain dwellings (see Q.26:128–129;

[20] To be sure it is a probabilistic claim but given that multiple chains corroborate each other in reference to Prophet Adam's height, the weight of the evidence delivers near-certainty.

[21] HūD has sometimes been identified with Eber, an ancestor of the Ishmaelites and the Israelites, who is mentioned in the Old Testament (see Noegel and Wheeler 2010: 140).

Q.54:20; Q.69:7), the remains of which are found in the Arabian Peninsula. In the Qur'anic narrative, these accomplishments did not prevent them from being destroyed by strong winds and storms after they rejected their prophets' monotheistic message. Further evidence of the existence of the *Thamūd* is found in building inscriptions, preserved tablets, letters, and historical chronicles from Assyrian and Roman times.

During one of the Prophet Muhammad's journeys, he and his Companions passed by the dwellings of *Thamūd*, whereupon he counseled his followers not to tarry, to adopt a demeanor of sadness lest they endure the calamity that befell that community, and not to partake of a water well in the area. These admonitions are captured in multiple *ḥadīth* within the canonical *Ṣaḥīḥ* collections of Muslim and Bukhārī (Riyāḍ as-Sāliḥīn no. 955; Ṣaḥīḥ Muslim no. 2980b; related to not drinking etc. Ṣaḥīḥ Muslim no. 2981a), and the chains of narration appear to converge and originate with the reputable Companion ʿAbdullāh bin ʿUmar. In light of our epistemological framework, Qur'anic verses reflect the highest level of evidence of reported knowledge and, as such, the existence of these nations is a certainty. The additional Prophetic reports are rigorously authentic and provide probabilistic evidence of the ruins of these nations being encountered by the Prophet and his Companions during an excursion and thus support the claim that these nations existed.[22]

While the tribes' existence is certain, the Qur'anic narrative is but an allusion, and the Prophetic report does not mention the height of these peoples. As such, at best, a probabilistic claim that prior nations were taller could be supported. Given the lack of explicit evidence from scripture, doubt would be more appropriate.

As I did not discuss sociological and historical evidence related to *Thamūd* in the previous section on the bioscientific evidence regarding human height, a few comments here may aid our analysis. Islamic scholars

[22] One could ask whether the Prophet Muhammad was certain that the ruins belonged to *Thamūd*. Recall that the Qur'an and the Sunnah are considered part of a singular revelation. As such, the Prophet's statements and actions are deemed Divinely inspired and as such classified as certain knowledge; indeed the Qur'anic verse Q.53:3 affirms Divine inspiration of Prophetic statements. Moreover, the Prophet was known to tell his Companions when he was offering his own, that is, not Divinely inspired, opinion. In this case, he made no such mention. One may further inquire whether the conveyance of knowledge by the Prophet was precise in word and meaning. That sort of analysis undergirds the hermeneutical side of *ḥadīth* studies. The *ḥadīth* in question are judged to be univocal.

such as Ibn Khaldūn and Ibn Ḥajar al-ʿAsqalānī have commented on the fact that the ruins of *Thamūd* are not particularly gargantuan, though appear to have larger doorways and chambers than typical (Zaman 2022). One could hazard that the tallness was due to the need to accommodate giant beings. Yet, such a judgment would be inferential and possible, rather than self-evident and necessary. Many other prehistoric nations, such as the Mayans or ancient Egyptians, built large pyramids, yet their habitus was not exceedingly large. Alternatively, another plausible claim could be that the largeness of the dwellings demonstrates civilizational might and skill. The dwellings' size could have been motivated by a desire to strike awe and wonder, just as certain castles, governmental buildings, and cultural buildings do today. This interpretation is somewhat supported by the Qurʾanic account of these nations being prideful and powerful. Furthermore, the dwellings we observe today are few in number. Hence, it is unclear if these were typical dwellings, or rather, they were the grandest among those to have withstood the effects of time and were deliberately left standing by God as a warning and reminder. Based on the empirical evidence, a claim that the people of *Thamūd* were taller than today's humans is improbable or, at best, doubtful.

The Bioscientific and Religious Evidence Juxtaposed

Generally speaking, one can take many different approaches to the dialog between religion and science. As highlighted in this book's introduction, one could treat the two domains as nonoverlapping magisteria, *a la* Stephen Jay Gould (1997), and thus not pay much attention to perceived areas of conflict or consonance. Somewhat related, one could suspend judgment altogether, a hermeneutical strategy termed *tawaqquf* in Islamic moral theology. The pause does not mean that one leaves off critically assessing the evidence; rather, such an evaluation precedes the suspension of judgment and logically follows from the analysis. One withholds rendering judgment as to which side has the stronger claim, that is, is likely to be true, because they judge the evidence to be insufficiently persuasive for either side.

On the other hand, one could find themselves confronting conflicting claims and thus must privilege one side or the other, or work to find a harmonious or integrative solution where both claims are fine-tuned so as to be true. For the purposes of this chapter, and the teaching exercises that spurred this book project, our preconceived notion was that there was harmony to be had. Said another way, we held the view that in actuality, scientific and scriptural claims, or in different parlance, rational, and transmitted proofs, that are conclusively certain cannot be contradictory. Islam upholds that knowledge has one ultimate source, God, and the natural world, as well as revelation, are indicants pointing toward Him.[23] Accordingly, there can be no true conflict between scriptural and scientific evidence; rather, limited human understandings of each generate perceived conflict. Relatedly, much of the public perceptions of conflict between religion and science stems from unfounded metaphysical assumptions that are smuggled along with scientific claims, that is, the reality can only be known through empirical investigation or that all that exists is the observable universe, or from faulty reasoning utilized to understand scientific and scriptural evidence. Nonetheless, an apparent conflict may present itself and, in these situations, a robust epistemic framework through which to evaluate and resolve tensions among truth claims is particularly valuable.

Inspired by Chapter 2, in our instructional pedagogy, the first step in evaluating truth claims is to specify the claim based on which type of evidence is being used to furnish the claim and then to grade the evidence according to its epistemic value. Such a grading may resolve the perceived tension, as it could be that one side asserts a claim based on certain knowledge while the other is grounded in probabilistic knowledge. Therefore, priority must be given to the side that has more certain evidence. Said more plainly, prioritization entails assessing the relative certainty attached to each truth claim, and, based on this evaluation, privileging the truth claim of one or the other domain.[24] In between these two poles of suspension of judgment and prioritization is the possibilities

[23] As noted in the introduction, classical scholars across the theological spectrum such as Imam al-Ghazālī, Ibn Rushd, and Ibn Taymiyya, all propounded this view in treatises. Omar Qureshi references this view in Chapter 2 as well. This has been popularized with the term *tawḥīdic* worldview in the contemporary era (see, for example, Bakar 2008).

[24] Within various Islamic legal sciences, prioritization is termed *tarjīḥ*, as in privileging one legal opinion within a school over another due to the relative strength of the evidence related to that opinion. Some legal scholars utilize this term also when prioritizing evidence from across disciplines, though such usage is less conventional. For example, Mufti Muntasir Zaman does so in his book *The Height of Prophet Adam: At the Crossroads of Science and Scripture* (Zaman 2022); and Mufti Kamaluddin Ahmed does so in his essay “Science in the Framework of Islamic Legal Epistemology: An Exploratory Account in Islam and Biomedicine” (Ahmed 2022).

of harmonization or integration.[25] In my view, harmonization follows a similar line of analysis where the evidence from each domain is interrogated, and an attempt is made to hold both claims as true insofar as possible. Such an approach may involve multiple analytic, interpretive, and hermeneutical methods including metaphorical interpretation, *taʾwīl*, of revelatory texts. Critically, metaphorical interpretation is only available as a method in specific circumstances, such as when the scriptural evidence is probabilistic in its semantic import, *ẓannī al-dalāla*, and the scientific evidence is more certain. Relatedly, the choice of approach and method may depend on one's preconceptions about religion and science. By that, I mean one could approach a religion and science dialog as a critical realist or an anti-realist with respect to science aiming to scrutinize scientific evidence, or as a religious apologetic seeking to defend religious truth claims. As American philosopher Thomas Nagel asserted, there are no views from nowhere (Nagel 1986).

Given that preconceptions influence approaches to the dialog, one must adopt intellectual humility when engaging in religion and science discourse. Our team's general approach[26] tends to seek harmony where possible and prioritization is used only when necessary. In what follows, this approach is taken to evaluate claims about human height.

Lining up the bioscientific and scriptural narratives results in a few areas of tension. One concerns Prophet Adam's height, as what we know of human physiology and biophysics would appear to preclude a human being of such stature. Islamic scholars have adopted various strategies to deal with this perceived problem. Some scholars such as Anwar Shāh Kashmīrī (d. 1933), a master theologian of the Indian subcontinent, and ʿAbd al-Raḥmān al-Muʿallimī (d. 1966), a Yemini *ḥadīth* expert, held the statement to indicate the Prophet Adam's height at the time of his creation in the heavens (Zaman 2022). These scholars note that the inference of his height being the same on Earth is not necessary. As other *ḥadīth* speak about trees being of immense size in Paradise, Kashmīrī suggests that human height in Paradise was, is, and will be larger

[25] Harmonization is related to the ideas of *jamʿ*, *taʿarud*, and *taṭbīq* within the Islamic ethico-legal sciences. In a general sense, the core idea is to develop a hermeneutical approach that resolves the tension.

[26] As illustrated in the various chapters of this book.

(Zaman 2022).[27] In this way, they harmonize the evidence. Another way to harmonize, although less of an integrative model, would be to consider his height, just like his overall creation, to be miraculous and not subject to the laws of physics and physiology.

Other scholars suspend judgment, and some significantly interrogate both the chains and source texts to uncover defects that reduce the epistemic certainty regarding Prophet Adam's height being 60 cubits. This latter approach leads them to prioritize bioscientific evidence over the scriptural account. A contemporary Islamic jurist and *ḥadīth* expert, Muntasir Zaman, comprehensively details the varied approaches classical and contemporary scholars have taken to this issue (Zaman 2022). Since we are less focused on Prophet Adam's exact height on Earth than on the idea that human height has been changing, the reader is directed to that work for greater details on potential resolutions.

The second area of perceived tension, directly relevant to the question of whether human height has/is changing, is the claim that human height has decreased since the creation of the Prophet Adam. Here, we find a probabilistic revelation-based truth claim appearing to counter archaeological and empirical evidence, as well as rational proofs related to scaling laws, that is, Brobdingnagians and Lilliputians[28] would be physiologically impossible. While each piece of evidence is probabilistic and inferential, when combined, the bioscientific evidence moves us toward dominant probability or near certainty that human height has been relatively stable over the past two millennia and has had a marginal uptick in the modern era. Both sides of the equation carry near-equal epistemic weight. If anything, the scriptural account may be graded slightly lower, given that the report is solitary and linked to a single narrator from the Prophet Muhammad. With respect to approaching the conflict, suspending judgment does not quite fit the case given that the proofs are neither ambiguous in their claims nor doubtful in their certainty. Deciding between a privileging or harmonizing approach entails delving deeper into the chains of narration to re-evaluate the status of the clause

[27] Recall that one of the *ḥadīth* mentioning Prophet Adam's height explicitly refers to human beings entering paradise with enhanced statures.

[28] Brobdingnagians and Lilliputians are mythical humans, the former gargantuan in size and the latter miniature, that are part of Jonathan Swift's satirical tale *Gulliver's Travels* (1726).

and considering hermeneutical and lexical strategies that may prove that a different meaning is intended by the clause.

Zaman's exhaustive analysis of the narrators and wording of various reports yields several important findings (Zaman 2022). First, while multiple different *ḥadīth* comment on Prophet Adam's height, there is but one *ḥadīth* (reproduced above) with several chains of transmission that contains the clause, "People have been decreasing in stature since Adam's creation." As noted above, the report containing human height information originates with the Companion Abū Hurayra. Interestingly, multiple narrators report the same *ḥadīth* from Abū Hurayra without the controversial clause. There are varied gradings of these other versions, as only the chain of narration that proceeds with Hamman bin Munabbih from Abū Hurayra that contains the added clause makes its way into the canonical Ṣaḥīḥ al-Bukhārī and Ṣaḥīḥ Muslim compendia. Nonetheless, the existence of other versions reduces confidence that the controversial clause came from the Prophet Muhammad. There are several possibilities for the omission in other versions: it could be that that other narrators did not transmit the questionable clause because they were concerned whether they heard it correctly from Abū Hurayra; or the truncation originated with *ḥadīth* scholars who gathered *ḥadīth* in their compendia and found evidence that the clause did not originate with the Prophet Muhammad. It is also possible that Abū Hurayra recounted the *ḥadīth* on multiple occasions and, at times, he shared the controversial clause and, at others, he did not. All of these are possible explanations for the omitted clause in the other versions. Zaman notes that legal theorists consider the more preponderant version of a *ḥadīth* to have greater epistemic weight, especially if the individual narrators are more trustworthy and significant personalities (Zaman 2022). Hence, there is precedent to lower the probability of the truth claim contained in the clause. A more significant detractor is an aspect of Abū Hurayra's life. Abū Hurayra, though deemed universally trustworthy and reliable by *ḥadīth* scholars, is known to have narrated stories of Jewish, Christian, or Zoroastrian origin from a former Jewish convert to Islam, Kaʿb al-Aḥbār. Some scholars consider the clause in question to originate from such a report. Given these detractions, while the *ḥadīth* remains rigorously authentic,[29] the truth claim regarding human height decreasing is now less certain. Importantly, however, the preceding analysis does not suggest the *ḥadīth* to be unreliable or improbable and, as such, it cannot be dismissed off-hand.

These added details provide grounds for privileging the bioscientific data that human height has not decreased significantly over the scriptural account. To be sure, both sides are making probabilistic claims, but the certainty of the scriptural claim has lessened. Zaman adopts this approach and believes the bioscientific evidence is epistemically weightier (Zaman 2022). Concerning harmonization, one could pursue a reinterpretation of the clause, considering it metaphorical. For example, one could posit that height refers to honor and that the Prophet Adam was most honored since he was created by God's hands. This reinterpretation may run into difficulties given the honored status of other prophets and the scriptural statements that suggest the Prophet Muhammad was the pinnacle of creation. As a non-specialist, I am not seeking to advance a particular reading; I am simply illustrating the type of hermeneutical exercise that must be undertaken.

Another tactic of harmonization could involve timing considerations; the Prophet Muhammad's statement is dated to the 600s and could refer to the history of humankind up until that time. Thus, human height could have decreased until the era of the Prophet and thereafter stabilized. Notably, his statement does not preclude an upward trend in height in the future. Previously, I noted that based on the empirical evidence, a claim that the people of *Thamūd* were taller than today's humans is improbable or, at best, doubtful. Yet, scientific evidence regarding human height before the common era is exceedingly scant, and little is known about humans of lore. Indeed, a contemporary Indian scholar of *ḥadīth*, Zakariyya Kandhlawi (d. 1982) suggested that decreases in human height were more substantial during the times of previous prophets. Based on an inverse analogy of how humans grow most in their early years, he reasoned that the decrease in height occurred most prominently between the times of Prophets Adam and Nūḥ and that by the time of *Thamūd* and the Prophet Muhammad it plateaued (Zaman 2022). This picture of

[29] There could be a cause to judge the report as unreliable if we were to find evidence of the insertion of non-Islamic elements and commentary, that is, stories of Jewish, Christian, or Zoroastrian origin, in other *ḥadīth* by individuals along the chain of this report; see Abu Hurayra (2022).

humanity would allow for considering both the Prophetic statement and the available bioscientific evidence to be true. Regarding the evidence of recent gradual increases in human height, the *ḥadīth* in question paints a general picture and cannot be seen as contradicting a trend observed only in the past century and a half. It could be that human height declines once again in the coming centuries, rendering the upward trend just a blip on a more significant downward-sloping trend over millennia. Additionally, it could be that the present human height has stabilized, as evidence shows a plateauing of height curves in developed nations. The Prophetic report does not preclude such stabilization since it refers to a general trend and is silent on whether human height will continue to decrease over time. Taken together, there is ample opportunity to harmonize the "competing" truth claims into a more comprehensive picture regarding human height across time.

In summary, the forgoing bioscientific accounts provide knowledge with the status of dominant probability that human height was relatively stable from the time of Jesus until about the nineteenth century. Since then, human stature has modestly increased. The Islamic scriptural narrative relates that the Prophet Adam was 60 cubits tall, and this claim is buttressed by scriptural texts of the highest grade, leading this claim to be nearly certainly true. The square-cube and scaling laws, however, rule out humans being so tall. Scholars have harmonized these evidences by considering the clause about Prophet Adam's height to be referring to his height in heaven and not on earth. With respect to human height since that time, scriptural texts claim that humankind has been decreasing in stature based on a revelatory source with the probability of being true. However, there are shortcomings in the scriptural evidence as the idea of decreasing stature is found in a single version of a *ḥadīth* from a single narrative chain; other versions do not contain the clause. Moreover, archaeological data reveals no giant human remains, and positing the largesse of prior nations based on the large size of their ruins is dubious. Thus, some scholars prioritize the bioscientific evidence of no substantial decrease in human height over the scriptural claim.

Alternatively, a harmonization strategy is also possible by considering the statement of decreasing height to be time-bound and referring to the period between the Prophet Adam and the Prophet Muhammad. Critically, bioscientific evidence revealing that human height has increased in recent centuries does not counter the Prophetic statement when interpreted in this way. Hence, both the prioritization and harmonization approaches to the perceived conflict regarding scriptural and bioscientific evidence about human height remain available.

Let us now return to the question posed at the start of this section: Is humankind changing? The reader needs no evidence to answer affirmatively. Our living standards, cultural mores, knowledge, and capacities constantly change. However, the question was not referring to these aspects; rather, it concerned the change in the natural kind. As an approximation, the case of human height also suggests the affirmative insofar as our physical nature is concerned. Both scientific and scriptural evidences indicate changes in human height over known/recorded time.

As to whether human height has optimized; if one gives credence to the evolutionary account, it is quite possible that human height may change through the processes of natural selection as we adapt to altered and different environmental conditions. Islamic scriptures, on the other hand, provide no definitive clues as to the future of human height in this world. Given the unknown future, and the current detractions of short stature, should the work of medicine involve altering height?

What Should Medicine's Role Be in Inducing Bodily Change?

As discussed previously, contemporary bioethics discourse uses the concepts of therapy and enhancement to separate appropriate uses of medicine from inappropriate ones. More clearly, the concept of therapy governs the ethical uses of biotechnology within contemporary healthcare. To counter the impulses of transhumanism, in addition to enhancement, another category, alteration, has been added to the equation. Through biotechnology, alteration aims at augmenting human capacities beyond species-specific norms, thereby modifying human nature. It is also aligned with transhumanism. Enhancement activities straddle the fence by improving capacities which may abut biological upper limits. Some bioethicists argue that enhancement is a moral obligation to create flourishing lives for ourselves and our progeny (Powell and Buchanan 2011; Murphy 2012, 2014; Persson and Savulescu 2012; Buchanan 2009). In contrast, others consider these activities to represent a slippery slope that fuels societal expectations of clinicians helping humankind push past biological limits.

In this section, I will undertake an Islamic bioethical assessment of the matter. Islamic bioethics is a discourse that uses the Islamic tradition to address moral questions and ethical issues arising from biomedical sciences and allied health practice. The discourse commonly occurs between a discursive pairing: an individual (or institution) from the biomedical arena seeking Islamic guidance and an expert respondent (or group) representing the Islamic tradition (Padela 2021a). Islamic jurists and Muslim physicians are the typical dialogical partners. Nowadays, however, the discourse is becoming multidisciplinary due to the ethical problem-space being multidimensional in involving considerations beyond the clinical encounter, such as policymaking and technical challenges, and because appropriately mapping out the problem-space requires interrogating social mores, cultural values, and legal precedents. Hence, multiple experts are brought to the table of dialog to evaluate, as comprehensively as possible, the problem-space and to generate holistic solutions. Thus, we increasingly see juridical councils admitting policymakers and social scientists alongside physicians to exercise collective *ijtihād* (ethico-legal deliberation). It bears mention that multidisciplinarity is not only required to address the context, but also to draw upon multiple fonts of moral guidance within the Islamic tradition. Indeed, in addressing an issue of global import, one might convene representatives from different theological sects, Islamic legal schools, and juristic communities of practice. Just as crucial to a comprehensive Islamic bioethical evaluation is the inclusion of Islamic scholars with expertise in moral sciences beyond law so that moral assessments of the act(s), the end goal(s), and the agent(s) can be effected. Islamic bioethics deliberation must not focus only on a specific action related to a particular moral agent within a localized context; its outputs must provide an overarching vision for biomedicine and humanity.

Before proceeding further, let me offer some provisos. Islamic bioethics is a nascent field with a small but growing body of literature and experts. Comprehensive Islamic bioethics expositions on any number of issues remain wanting. Tending to this current state will require considerable institutional and intellectual resources. What follows is a modest intervention that sets up a fuller analysis concerning the moral boundaries of medical practice in light of the "threat" of transhumanism. Accordingly, I stitch together three pieces of Islamic bioethics literature: (i) key expositions regarding the medical profession from classical Muslim physicians[30]; (ii) select extant juridical opinions addressing the licitness of body modification; and (iii) academic research writings of Muslim thinkers who directly engage with transhumanism. To be sure, my sample is eclectic as these are pieces that target divergent audiences. Nonetheless, weaving aspects of these writings together facilitates a provisional moral assessment.

The Medical Calling

Abū ʿAlī al-Ḥussein ibn ʿAbdullah Ibn Sīnā (d. 1037), the pre-eminent Muslim philosopher and physician and famed author of the Canon of Medicine, defines medicine as "a science by which we learn about the conditions of the human body. Its purpose is to preserve health when well and restore health when ill" (Aligabi 2020: 310). His less famed, yet learned, contemporary, Abū al-Faraj ʿAlī ibn al-Ḥusayn ibn Hindū (d. 1032), in his *The Key to Medicine and A Guide for Students* (*Miftāḥ al-ṭibb wa-minhāj al-ṭullāb*), concurs that medicine includes a scientific component but notes that "medicine is a profession that deals with human bodies and brings them good health" (Hindū 2010: 15). Anchoring the definition to a professional practice (*ṣināʿa*) and not a science (*ʿilm*) is critical, for the care of bodies entails skill alongside knowledge. Explicating the phrase "brings them good health," he notes that it refers to preserving health when present and restoring it if absent, which matches Ibn Sīnā's view of the goals of medicine. Another shared feature of the two definitions is their focus on eliciting information from the human body to ascertain health status. Corroborating the idea of health inhering in the body, the Prophet Muhammad, in a rigorously authenticated tradition, relates a scene from the Day of Judgment when God interrogates humankind about the bounties He provided them saying, "Did We not give health to your body?" (Jāmiʿ at-Tirmidhī no. 3358).

Since medieval Muslim physicians operated out of a humoral view of disease, they considered restoring balance an essential function of doctoring (Pormann and Savage-Smith 2007). They also readily acknowledged relationships between the physical body and psychological health.

[30] The writings of these physicians may perhaps be better termed as Islamicate writings; literature emerging from authors inhabiting the lands and epochs associated with the Islamic civilization; see Marshall Hodgson (d. 1968), *The Venture of Islam: Conscience and History in a World Civilization* (Hodgson 1974).

Indeed, pre-dating both Ibn Sīnā and Ibn Hindū, Abū Zayd al-Balkhī (d. 934), a Muslim polymath physician and scientist, explicated how psychological diseases manifest in the body in his tome, *Sustenance for Body and Soul* (*Maṣāliḥ al-abdān wa-al-anfus*). These passages indicate that the medical profession attends to the human body and its practices entail preventive and restorative interventions. The bodily focus aligns with the biomedical framework. However, that Muslims did not give exclusivity to bodily interventions as psychological and spiritual therapies can also found in several treatment manuals (Rāzī 1950; al-Balkhī 2013). Additionally, the focus on health restoration coheres with therapeutic applications of pharmaceuticals and technologies as medicine's prime directive, rather than using available means to enhance humankind. As these definitions set up the contours of medical practice, they also speak to the "rightness" of intentions that physicians should have as they live out their vocation.

Juridical Views on Body Modification

Moving from classical writings to more contemporary ones, Islamic jurists address the contours of modern medical practice through the edifice of Islamic law, *fiqh*. Using the medium of *fatāwā* (pl., sing. *fatwā*), jurists respond to queries from patients and practitioners as to whether a specific modification or intervention is morally permissible. Arguably, the craft of *fatwā* is an ad hoc enterprise and a partial window into Islamic morality. As I have written elsewhere, *fatāwā* are necessarily context dependent, may utilize strained legal reasoning, and are provisional responses. Additionally, *fatāwā* research is methodologically flawed due to publication bias and the impossibility of comprehensively reviewing *fatāwā* in their various mediums and languages (Padela 2013, 2020; Padela et al. 2011). While these features give pause to developing Islamic bioethical theory based on *fatāwā*, the responsa literature provides valuable insights into scriptural source texts, legal precedents, and other juridical considerations that shape Islamic morality.

Bearing these caveats in mind, several common themes are found when perusing *fatāwā* related to genetic engineering, cosmetic surgery, and intervening upon human height. Citing a Qur'anic verse in which Satan promises to command humankind to "change the creation of God (*taghyīr bi-khalq Allāh*) (Q.4:119)," jurists draw a red line by prohibiting body modifications that significantly alter the human body and/or impact human nature. The line, nonetheless, is somewhat fuzzy given the casuistic nature of the discourse. In the realm of genetic engineering, germline modifications are viewed with suspicion, given the distaste for eugenics (Lala 2020). Explicating the conditions under which genetic engineering may be permitted in 1998, the Islamic Fiqh Academy of the Muslim World League (IFA-MWL) stated that genetic engineering is licit when aimed at preventing or curing disease, or at reducing its harms (Jurisprudence 2019; Al-Bar and Chamsi-Pasha 2015). Moreover, the council noted that genetic therapy must operate under the principle of dire necessity (*ḍarūra*) and hence must be the most effective or the only treatment for a disease that has significant morbidity and/or mortality (Al-Bar and Chamsi-Pasha 2015). Therapeutic uses are defined as treatments that "restore the original (sound) condition" (Shabana 2022: 11) of the human being because ailments whether inborn or acquired are exceptions from the norm, namely, that the Divine created human in the best of shapes (Q.95:4). While therapeutic purposes warrant stipulated permissibility, a majority of Islamic jurists prohibit the use of genetic intervention for enhancement (Shabana 2022). In addition to tampering with God's creation, they view germline genetic intervention to violate human dignity and the desire for enhancement as tantamount to challenging Divine decree for human traits and characteristics are part of one's provision (Shabana 2022).[31]

In response to a questioner who appears to desire a religious ruling on transhumanism when asking whether it was permissible to alter genes to "improve humans" so that they can "think much faster" and "have extra muscle, bigger and stronger bones" and possibly become "new humans," a prominent online juridical academy deferred stating that "leave it until it happens" (Islamweb 2008). This "wait and see" approach is emblematic of Islamic juridical responses to biotechnology given their spatial and intellectual dislocation from the centers of biomedical and biotechnological advancements. In contrast, another prominent scholar asserts that enhancement is licit. He supports this view by noting that there is no

[31] Shabana's piece also refers to minority views that utilize scriptural texts to argue for the permissibility of enhancement (Shabana 2022). For example, they contend that the verse regarding changing of God's creation refers not to physical features or capacities of the human being but rather to changing the monotheistic faith tenets. They also bring into the conversation *ḥadīth* that note God loves beauty and as such beautification purposes should be licit. The paper demonstrates how scriptural texts contain a multiplicity of meanings and how scholars develop arguments based on them.

explicit scriptural text that denounces bodily augmentation[32] and that augmenting one's capacities, such as intelligence and strength, would benefit the individual and the entire Muslim community; he advances his argument on the basis of *maṣlaḥa*.

Further insights are obtained from deliberations over cosmetic surgery. In this instance, a rigorously authenticated Prophetic report details what constitutes the changing of God's creation. The Prophet is reported to curse women who obtain tattoos and have their teeth separated for the sake of beauty, noting that these are examples of those who change the creation of God (Sunan an-Nasāʾī nos. 5099; 5109; 5107; 5253; 5108; 5254). Similar statements are recorded in the *Ṣaḥīḥ* collection of al-Bukhārī (Ṣaḥīḥ al-Bukhārī no. 5943). Based on these texts, jurists consider beautification inadmissible as the sole reason for cosmetic surgery. Illustratively, Shaykh ʿAttiyah Saqr (d. 2006), the former head of Al-Azhar's Fatwā Committee, cites both the verse mentioned above and the Prophetic tradition stating:

> surgery that is normally used to change the creation of Allah and attempt to look more beautiful such as breast enlargement or reduction, facelifts, spacing between teeth (just for beautification), reduce the size of lips and nose, this kind of surgery is not permitted in Islam because they alter the creation of Allah. (About Islam 2018)

Modern Egyptian jurists largely echo this opinion in prohibiting surgery that enhances one's features and permitting surgeries that restore disfigured bodily form (Rispler-Chaim 1993). The same view is voiced by contemporary British jurist, Muḥammad ibn Adam al-Kawtharī, who cites legal precedents of the Ḥanafī school of law to consider surgery for beautification purposes unlawful (About Islam 2023). The carveout for Islamically permitted cosmetic surgery involves the restoration of a defect, for example, surgery to repair facial features after injury, or removal of an extra body part, for example, a sixth digit, should there be psychological harm due to it.

Several *fatāwā* respond to an individual desiring to increase their height through surgery. On the widely utilized online *fatwā* forum, Islam Question and Answer, the juridical committee notes that the option of using growth hormone for short stature is licit, while:

[32] This point is highly debatable given the scriptural texts that will be quoted shortly.

> with regard to surgery, it is done by breaking the bones and carries a risk of fibrosis of the muscles or infection. The basic principle is that it is *haraam* (impermissible) for a person to bring harm upon himself, unless it is for the purpose of medical treatment. (Islam Q&A 2011a)

The opinion references an extant ruling issued by Dr. Ṣāliḥ ibn Muḥammad al-Fawzān, a member of the Permanent Committee for Islamic Research and Issuing Fatwā and senior scholar in Saudi Arabia, that using biotechnology to break healthy bones contravenes the sanctity of the human body. Violation of this sanctity is only permitted for necessary medical treatment; the desire to be taller is judged to be insufficient justification (Islam Q&A 2011a). Interestingly, another *fatwā* on the site authorizes the use of growth hormone for short stature, noting that it "comes under the heading of medical treatment" (Islam Q&A 2011b). Hence, the group differentiates between surgical and hormonal treatments for the malady. Perhaps hormonal treatments are viewed as less invasive and thus less violating of human dignity, and deploying hormonal treatments subtly frames the problem as a physiological deficiency which coheres with the notion of medicine being restorative. Relatedly, in discussing genetic modifications, the Egyptian jurist Muḥammad Raʾfat ʿUthman, appears open to the possibility of using the technology in order to treat children who have a genetic disposition for short stature (Shabana 2022). These verdicts suggest a potential expansion of the notion of treatment as an activity that is not simply restorative in nature, rather, at least in children, it can also assist with achieving the threshold of normality within a society.

Generally speaking, as can be gleaned from these rulings, Islamic jurists consider medicine's role to be restorative, are cautious about authorizing interventions that disrupt the body, and prohibit bodily modifications that are purely cosmetic. The theological assertion that some bodily modifications represent changing the creation of God is an overarching concern. Commonly, beautification purposes are viewed as aligning with this condemnable notion, as are bodily modifications that significantly alter the human form, capacities, or nature. Given the medium, jurists do not fully explicate notions of the normative body, nor expound on conceptions of human nature within their *fatwā*, thus leaving these discussions to other scholars and venues.

The judgments heavily rely on treatment and enhancement distinctions. Responses to concerns over short stature reveal an acceptance of hormonal and genetic treatments, while surgical options are considered to violate dignity. In meting out licitness, the jurists emphasize the concept of treatment, implicitly giving policing authority to the profession in determining a medical need. Yet, such a disposition could lead to abuse and does not clarify the ethics of the matter; the mere declaration of a medical necessity by a clinician or the societal labeling of a condition as a disease could place a certain bodily modification within the licit category. It appears that Islamic jurists consider there to be a clear dividing line, whether grounded in biology, sociology, or professional ethics, between therapy and enhancement. Yet, given our preceding discussions related to the sociocultural construction of disease states, contentions over what constitutes therapy and what enhancement, and the psychological grounds for disease states, such a view is highly problematic. Indeed, in researching juridical councils' views on genetic intervention, Mutaz al-Khatib, Associate Professor of Methodology and Ethics in the College of Islamic Studies at Hamad Bin Khalifa University, decries the lack of specificity and nuance regarding what constitutes alteration of human nature, as well as the concepts of treatment and enhancement. He suggests that Islamic theological discourses must be added upon juridical exercises in order to map out the (Islamic) ethical limits of genetic intervention (Al-Khatib 2019).

Before proceeding, a further comment on the Qurʾanic verse Q.4:119 may help clarify the fuzzy nature of the proverbial red line. Exegetes hold various opinions about the phrase in question. As noted above, some consider the phrase to refer to turning people away from monotheism, as this is the main thrust of Satanic work. Another camp interprets the verse as prohibiting aesthetic enhancements and tattooing, as supported by the Prophetic narration, and witnessed in the juridical deliberations, glossed above. Others interpret the verse as a warning that the changes inspired by Satan would involve significant changes in human form, attributes, or disposition. This reading bolsters Islamic moral arguments against transhumanism (Lala 2020).

Transhumanism and Muslim Thought

There are but a few scattered academic pieces at the intersection of Islam and transhumanism. While the juridical writings address clinical practices that could be utilized for transhumanist goals, these reflections chiefly analyze the underlying motivations for the transhumanism project. While these intellectuals take different stances concerning the ultimate compatibility of transhumanism with Islamic morality, they all voice, for the lack of a better term, theological concerns with transhumanism.

The first sticking point is the rejection of an afterlife. Once this potential metaphysical reality is discarded, transhumanism sets its focus on maximizing one's contentment in the present reality and forestalling biological death. Islam, like other monotheistic religions, gives primacy to an eternal afterlife and considers our present reality to be a temporally bounded place of toil for afterlife merit-making. These radically different perspectives generate dissonance between Islam and transhumanism. Illustratively, Hamid Mavani, an Associate Professor of Islamic Studies at Bayan Islamic Graduate School, writes that embracing the idea of immortality "would challenge Islam's deeply anchored and accepted eschatological doctrines related to death, resurrection, and the afterlife," where both "body and soul will be resurrected" and granted eternal bliss or damnation (Mavani 2014: 78). Anke Bouzenita, a Professor of Islamic Sciences at Sultan Qaboos University, says the transhumanist vision of human nature as materialistic is the root cause of such denial and sets in motion the pursuit of an eternal, this-worldly, life (Bouzenita 2018). Such aspirations speak to transhumanism's spiritual bankruptcy notes IlhamAllah Ferrero, an Italian Muslim intellectual. Rather than elevating humanity through spiritual exercises that work against the desires of the (lower/bestial) self, transhumanism advances a vision of human progress that aims at the prolongation of (this worldly) life and an increase in the functional abilities of humans. She argues that such goals are truncated conceptions of flourishing that reject metaphysical realities and ignore the spiritual dimension of the human being. In her view, religions, Islam included, aim to "rescue people from the condemnation of being entangled in the world ad infinitum" (Ferrero 2021: 216). Transhumanism is thus directly opposed to Islam as it aspires to leave individuals interminably trapped in a disembodied virtual world or in a reformed unhuman body, with no "higher" purpose to be had. Since the belief in an afterlife orients a Muslim's present behaviors, these thinkers offer that the transhumanistic project offers no orientation for self-improvement and, as such, is not aligned with an Islamic moral vision. Rather, the denial of an afterlife sets up the transhumanist quest for immortality along with activities that seek to radically refashion, and possibly disconnect oneself from, the body. In sum, these writers note that dissonance between the Islamic tradition and transhumanism as an ideology stems from the denial of an afterlife, and this rejection must be accounted for by Islamic jurists when they opine on the bodily transformations that transhumanists may want to undertake.

Building up from the missing focus on quasi-transcendence through the moral cultivation of the self, a second area of theological concern focuses on developing vices. Hinted above in Ferrero's views, Muslim thinkers note that the Islamic moral tradition aims to cultivate intellectual and civic virtues such as humility, personal responsibility, and stewardship. Adopting a transhumanist worldview, in their view, entails self-aggrandizement, hubris, and hedonism. Tamim Mobayed, a doctoral candidate at the University of Oxford, associates transhumanist impulses with characters like Pharaoh, who sought supremacy in the earth and self-deification. Citing a Qurʾanic verse (Q.55:9), he further contends that transhumanism seeks to "transgress the [Divinely set] balance" related to humankind and nature (Mobayed 2017). The Qurʾan exclaims that every soul shall taste death (Q.3:185, Q.21:35, and Q.29:57) hence, the project of seeking immortality, even if it is technological immortality[33] and not true human immortality, is hubris filled and morally suspect. Multiple authors underscore that Satan enticed the Prophet Adam to partake of the forbidden tree with a promise of immortality (Q.7:20). Hence, the promise of worldly immortality is illusory and contrary to God's plan (Bouzenita 2018). Genetic manipulation for augmentation purposes carries the same moral disapproval since God created humankind in the "best of molds" (Q.95:4). Moreover, transhumanist activities like genetic manipulation are based on a belief in non-directed, haphazard evolution. Such a view suggests that the Creator was/is not fully aware of His creation or was faulty in His design. Both notions are contrary to Islamic creedal views on God (Bouzenita 2018). These views can lead to the cultivation of moral vices among transhumanist ideologues, which color the activities they pursue. Muslim writers thus dismiss the project as one that may result in a (trans)human being that would be both a moral and a physical abomination.

[33] Ronald Cole-Turner, an emeritus Professor of Theology and Ethics at Pittsburgh Theological Seminary, differentiates between human immortality and technological immortality (Cole-Turner 2011). He considers human immortality to be about the continuity of the spiritual self, while technological immortality focuses on preserving the biological or cognitive self through artificial means. Relatedly, he notes that human immortality deals with questions of ultimate morality and spiritual fulfillment, while technological immortality focuses on extending life and societal impacts of doing so.

Somewhat related to the transhumanists' rejection of the purpose of life and moral cultivation are the concerns of Serdar Kurnaz, a Professor of Islamic Law at Humboldt University of Berlin. He notes that Islamic rituals such as prayer and fasting are predicated on bodily experiences and that the spiritual elevation and merit-making of worship are, at least in part, tied to the physical exertion related to these activities. Accordingly, transcending the human body not only poses significant questions for Islamic law related to worship, it calls into question the ultimate purposes of such activities (Kurnaz 2024). He remains hopeful that the inherent flexibility within Islamic legal constructs would allow for rendering appropriate rulings for virtual or disembodied worship activities but highlights that the *telos* of the human body and of worship must be carefully weighed.

That said, some writers attempt to reframe and salvage aspects of transhumanism to align with Islamic morality. Regarding activities, radical life extension instead of immortality, is seen as consonant. Sara Hejazi, an Italian journalist, and former scholar at the University of Torino, argues that as long as such life extension benefits the Muslim community, it should be pursued. She cites the long lives of prophets as a precedent for such a goal (Hejazi 2019). Hamid Mavani holds a similar view (Mavani 2014). Dr. Yunus delves into the arguments for and against such a view in the next chapter, Chapter 8, in this volume. In an extended critique of Muslim thinkers who view transhumanism as antithetical to the tradition, Ali Ghandour, Professor of Islamic Theology at the University of Hamburg, argues that the quest for immortality and enhanced bodies is not contrary to the Islamic tradition. Rather, these ideas animate the quest for paradise and will be actualized there (Ghandour 2024). As such, he contends that there is no reason to negate such pursuits using the means we have available. He bolsters his argument by noting the long lives attributed to the Prophets Nūḥ and Khidr.[34] Ghandour further contends that the body-related miracles attributed to the Prophets Abraham and Moses, the inability of a great fire burning the body of the former and the radiant hand of the latter, could be seen as symbolic representations that cohere with the possibility of the human body being radically transformed. Given that the prophets are ideals, he contends that Muslim should not view bodily enhancement negatively either. In

[34] There is disagreement among Islamic scholars as to whether Khidr was a prophet and as to whether he is still alive.

his view, Muslim thinkers should not adopt essentialist and static views of human nature; rather, there is the possibility of alignment between transhumanist ideology and Islamic theology if human nature were viewed as malleable.

Another reframe envisages transhumanism as becoming more morally and intellectually perfect. This enhanced self-cultivation and character formation is intrinsically aligned with religious goals. Insofar as technology facilitates this goal, it should be welcomed. Multiple authors, including Mobayed (2017), hint at this view, but it is aptly dealt with in a monograph by Roy Jackson, a senior lecturer in the philosophy of religion at the University of Gloucestershire. Indeed, he ends his monograph by drawing up a provisional list of affirmations for a future Muslim Transhumanist Association, which includes the affirmation, "We believe that the intentional use of technology will empower us to transcend our current state and move towards perfectibility, guided by the example of the Prophet Muhammad as the Perfect Human" (Jackson 2020: 176).

Tying the Threads Together to Address the Threat of Transhumanism: An Islamic Bioethics Intervention

What is an Islamic perspective on the moral parameters of medical practice? The disparate and multidisciplinary inputs above come together in the following way. Classical Muslim theorists see physicians' practices as restoring patients' health, fixing that which has been lost due to injury, and preventing illness. Islamic jurists adopt the same normative view in regulating the extent to which Muslim physicians (and patients) should modify the human body. Physicians must treat diseases that can be objectively assessed; therapy entails correcting a physical or physiological defect, whether acquired or inborn. Moving beyond these thresholds to augment capacities or to enhance physical features is judged to contravene commandments that prohibit changing creation and violating human dignity. Muslim intellectuals address transhumanism directly by demonstrating that its philosophical commitments to materialism and naturalism clash with the morally edifying mission of Islamic tradition. Bereft of the notion of an afterlife and of a self that needs to be cultivated, the transhumanist vision appears to have no higher ideals as human nature is rendered as a soulless mind that strives to continue experiencing this-worldly existence without a *telos*. Consequently, a medical practice that adopts human immortality as its ultimate goal and utilizes biotechnology toward this goal would be morally bankrupt.

This picture becomes clear and cohesive when the different strands of Islamic bioethics discourse are connected. While the jurists comment on the purposes of medicine, they are predominantly focused on specific medical activities and may not see these as connected to a larger biomedical paradigm with distinctive epistemic and philosophical commitments. When opining on surgery and genetic therapy, they may scarcely be aware that these are two potential tools through which transhumanist ideals can be actualized in medical practice. Muslim thinkers, on the other hand, comment on moral imperatives through the lens of theological tenets. They do not consider medical practices directly, nor the relationship between theology and law. Medieval manuals on medicine are obviously disconnected from the contextual realities of biomedicine and contemporary healthcare. Indeed, neither the jurists nor the thinkers excerpted above comment on the sociocultural construction of disease states, nor on the contemporary bioethical debates over treatment, enhancement, and alteration.

In my view, a multidisciplinary Islamic bioethical evaluation of contemporary doctoring and the goals of medicine is required. Transhumanism is a harbinger of various ideologies and worldviews that will shape the future of medicine and healthcare. Jurists, clinicians, bioethicists, policymakers, social scientists, and other relevant experts must come around the dialog table to discuss the epistemic, philosophical, and practical aspects of an "Islamic" medical practice in the present and future contexts. I have outlined a process model for such multidisciplinary deliberative exercises and illustrated how such evaluations must be multitiered in that they consider the multidirectional relationships between state and local policies and the ethics of bedside practices elsewhere (Padela 2021b; Padela et al. 2021). I have also commented on the necessity of bringing together agent, act, and outcome-based moral frameworks from within the Islamic tradition when conducting bioethical analyses (Padela 2024). While the reader is directed to these pieces for a better sense of the "how" with respect to a comprehensive Islamic bioethical assessment, I offer an added remark on the "what" in concluding this chapter.

Specifically, a fuller Islamic bioethical evaluation must involve at least three types of analysis: *theoretical/conceptual*, *practical*, and *applied*.[35] Take the case of idiopathic short stature, assessing that entity from an Islamic lens requires *theoretical* and *conceptual* work. Within the Islamic tradition, one must draw upon theology to specify notions of disease, health, and the normative body. Only after these concepts have been fleshed out can one consider whether idiopathic short stature is a human malady that could be categorized as a disease within an Islamic worldview.

Moving to the *practical* register entails considering the medical profession's role, namely, its embeddedness within society and its instrumentalization for societal goals. Drawing upon Islamic theology and philosophy, one would detail the normative doctor and outline their remit vis-à-vis patient health and societal goals. Theological concepts such as changing the nature of creation and human nature would link the theoretical with the practical as they would flesh out the contours of medical practice with nuance and specificity. Similarly, the conceptual and practical distinctions between therapy, enhancement, and alteration must be demarcated, as a fuller Islamic bioethical evaluation must link up with an overall metaphysics and philosophy for medical practice. A practical analysis of the goals of medical practice must also be undertaken in light of views on the future of humanity, specifically per eschatological doctrines.

An *applied* bioethics evaluation can be undertaken only after identifying the practical dimensions of Islamic medical practice. Applied Islamic bioethics would involve using casuistry to address whether a specific intervention for a specific patient within a specific context is licit according to Islamic law. This level mirrors the *fatwā*-making enterprise but has the added advantage of the prior conceptual and practical work. Instead of just looking to prior legal precedent, jurists could connect their judgments to Islamic views on health, disease, and doctoring; in other words, their judgments would flow out of a larger moral vision for healthcare. In my view, Islamic bioethical deliberation that is multidisciplinary, multitier, and multilevel will be more robust in attending to the ethical problem spaces of modern biomedicine and contemporary healthcare. And once the discourse adopts this multiplexity, the field of Islamic bioethics will be fully born.

[35] In other publications, I have used the terms theoretical, practical, and applied to refer to genres within Islamic bioethics literature. Herein, I use the terms to differentiate between tiers of ethical/moral assessment.

References

About Islam. 2018. “What’s Islam’s Stance on Plastic Surgery?” December 26. https://aboutislam.net/counseling/ask-the-scholar/health-science/whats-islams-stance-plastic-surgery/.

About Islam. 2023. “Is Laser Eye Surgery Haram?” December 23. https://aboutislam.net/counseling/ask-the-scholar/health-science/corrective-laser-eye-surgery-permissible/.

Abu Hurayra. 2022. “Review of Muntasir Zaman’s ‘The Height of Prophet Adam’: Towards a Universal Framework for Resolving Conflicts between Science and Scripture.” *Science, Philosophy, and Islamic Theism*, July 17. https://synoptically-islam.blogspot.com/2022/07/review-of-muntasir-zamans-height-of.html.

Adair, Linda S., Caroline H.D. Fall, Clive Osmond, Aryeh D. Stein, Reynaldo Martorell, Manuel Ramirez-Zea, Harshpal Singh Sachdev, Darren L. Dahly, Isabelita Bas, Shane A. Norris, Lisa Micklesfield, Pedro Hallal, and Cesar G. Victora, for the COHORTS Group. 2013. Associations of Linear Growth and Relative Weight Gain During Early Life with Adult Health and Human Capital in Countries of Low and Middle Income: Findings from Five Birth Cohort Studies.” *Lancet* 382 (9891): 525–534. https://doi.org/10.1016/S0140-6736(13)60103-8.

Adams, Dallon. 2017. “8 Bold Biohacks that Blur the Line between Human and Machine.” *Digital Trends*. May 20. https://www.digitaltrends.com/cool-tech/coolest-biohacking-implants/.

Ahmed, Kamaluddin. 2022. “Science in the Framework of Islamic Legal Epistemology: An Exploratory Account.” In *Islam and Biomedicine*, edited by Afifi al-Akiti and Aasim I. Padela, 255–270. Cham: Springer International Publishing.

Ali, Muna. 2018. “The “Bio” in Biomedicine: Evolution, Assumptions, and Ethical Implications.” In *Islamic Perspectives on the Principles of Biomedical Ethics: Muslim Religious Scholars and Biomedical Scientists in Face-To-Face Dialogue with Western Bioethicists*, edited by Mohammed Ghaly, 41–67. Qatar: World Scientific/Imperial College Press.

Aligabi, Zahra. 2020. “Reflections on Avicenna’s Impact on Medicine: His Reach beyond the Middle East.” *Journal of Community Hospital Internal Medicine Perspectives* 10 (4): 310–312. https://doi.org/10.1080/20009666.2020.1774301.

Angel, J. Lawrence. 1984. “Health as a Crucial Factor in the Changes from Hunting to Developed Farming in the Eastern Mediterranean.” *Paleopathology at the Origins of Agriculture*, 51–74. New York: Academic Press.

Aʿẓamī, Muḥammad Muṣṭafā. 1978. *Studies in Hadīth Methodology and Literature*. Oak Brook, IL: American Trust Publications.

Bakar, Osman. 2008. *Tawhid and Science Islamic Perspectives on Religion and Science*. Shah Alam, Malaysia: Arah Pendidikan Sdn Bhd.

Al-Balkhī, Abū Zayd. 2013. *Sustenance of the Soul*. Translated by Malik Badri. London: International Institute of Islamic Thought.

Al-Bar, Mohammed Ali, and Hassan Chamsi-Pasha. 2015. *Contemporary Bioethics: Islamic Perspective*. Cham: Springer.

Barker, D.J.P., J.G. Eriksson, T. Forsén, and C. Osmond. 2005. " Infant Growth and Income 50 Years Later." *Archives of Disease in Childhood* 90 (3): 272–273. https://doi.org/10.1136/adc.2003.033464.

Baten, Joerg, and Matthias Blum. 2014. "Why Are You Tall While Others Are Short? Agricultural Production and Other Proximate Determinants of Global Heights." *European Review of Economic History* 18 (2): 144–165. https://doi.org/10.1093/ereh/heu003.

Bauer, Keith A. 2010. "Transhumanism and Its Critics: Five Arguments against a Posthuman Future." *International Journal of Technoethics* (IJT) 1 (3): 1–10. https://doi.org/10.4018/jte.2010070101.

Belete, Tafere Mulaw. 2021. "The Current Status of Gene Therapy for the Treatment of Cancer." *Biologics: Targets and Therapy* 15: 67–77. https://doi.org/10.2147/BTT.S302095.

Berkey, Catherine S., Graham A. Colditz, Helaine R.H. Rockett, A. Lindsay Frazier, and Walter C. Willett. 2009. "Dairy Consumption and Female Height Growth: Prospective Cohort Study." *Cancer Epidemiology Biomarkers & Prevention* 18 (6): 1881–1887. https://doi.org/10.1158/1055-9965.EPI-08-1163.

Blomkamp, Neill. (dir.). 2015. *CHAPPiE*. USA.

Bostrom, Nick. 2005. "Transhumanist Values." *Journal of Philosophical Research* 30 (Supplement): 3–14. https://doi.org/10.5840/jpr_2005_26.

Bostrom, Nick. 2014. "Introduction—The Transhumanist FAQ: A General Introduction." In *Transhumanism and the Body: The World Religions Speak*, edited by Calvin Mercer and Derek F. Maher, 1–17. New York: Palgrave Macmillan.

Bouzenita, Anke Iman. 2018. "'The Most Dangerous Idea?' Islamic Deliberations on Transhumanism." *darulfunun ilahiyat* 29 (2): 201–228. https://doi.org/10.26650/di.2018.29.2.0031.

Brown, Jihad H. 2013. "The Problem of Reductionism in Philosophy of Mind and its Implications for Theism and the Principle of Soul: Framing the Issue for Further Islamic Inquiry." *Tabah Paper Series* 7: 1–30. Abu Dhabi: Tabah Foundation.

Buchanan, Allen. 2009. "Human Nature and Enhancement." *Bioethics* 23 (3): 141–150. https://doi.org/10.1111/j.1467-8519.2008.00633.x.

Al-Bukhārī, Abū ʿAbd Allāh. n.d. "The Creation of Adam and his Offspring." Ṣaḥīḥ al-Bukhārī no. 3327. https://sunnah.com/bukhari:3327.

Al-Bukhārī, Abū ʿAbd Allāh. n.d. "Prophets." *Ṣaḥīḥ al-Bukhārī* no. 3326. https://sunnah.com/bukhari:3326.

Bullinger, Monika, Maria Kołtowska-Häggström, David Sandberg, John Chaplin, Hartmut Wollmann, Meinolf Noeker, and Anna Levke Brütt. 2009. "Health-Related Quality of Life of Children and Adolescents with Growth Hormone Deficiency or Idiopathic Short Stature—Part 2: Available Results and Future Directions." *Hormone Research in Paediatrics* 72 (2): 74–81. https://doi.org/10.1159/000232159.

Cameron, James. (dir.). 2009. *Avatar.* USA/UK.

Campbell, Heidi, and Mark Walker. 2005. "Religion and Transhumanism: Introducing a Conversation." *Journal of Evolution and Technology* 14 (2). https://www.jetpress.org/volume14/specialissueintro.html.

Case, Anne, and Christina Paxson. 2008. "Stature and Status: Height, Ability, and Labor Market Outcomes." *Journal of Political Economy* 116 (3): 499–532. https://doi.org/10.1086/589524.

Chen, Yuyu, and Li-An Zhou. 2007. "The Long-Term Health and Economic Consequences of the 1959–1961 Famine in China." *Journal of Health Economics* 26 (4): 659–681. https://doi.org/10.1016/j.jhealeco.2006.12.006.

Cole-Turner, Ronald. 2011. "Transhumanism and Christianity." *Transhumanism and Transcendence: Christian Hope in an Age of Technological Enhancement*, edited by Ronald Cole-Turner, 193–203. Washington, DC: Georgetown University Press.

De Kruif, Paul. 1932. *Men Against Death.* New York: Harcourt.

Durlacher, Chris, and Matt Pelly (dirs.). 2020. "Peter: The Human Cyborg." Channel 4. documentary. https://www.channel4.com/programmes/peter-the-human-cyborg. Accessed May 30, 2021.

Extropy Institute. n.d. "Prologue: What is the Purpose of the Principles of Extropy?" https://www.extropy.org/About.htm.

Ferrero, IlhamAllah Chiara. 2021. "Islam and Transhumanism." In *Enhancement Fit for Humanity: Perspectives on Emerging Technologies*, edited by Michael Baggot, Alberto García Gómez, Alberto Carrara, and Joseph Tham, 215–222. Abingdon: Routledge.

Freedman, David S., Laura Kettel Khan, Mary K. Serdula, William H. Dietz, Sathanur R. Srinivasan, and Gerald S. Berenson. 2003. "The Relation of Menarcheal Age to Obesity in Childhood and Adulthood: The Bogalusa Heart Study." *BMC Pediatrics* 3: 1–9. https://doi.org/10.1186/1471-2431-3-3.

Fukuyama, Francis. 2004. "Transhumanism." *Foreign Policy* (144): 42–43.

Future of Humanity Institute. n.d. University of Oxford. Closed on April 16, 2024. https://www.fhi.ox.ac.uk/.

Ghandour, Ali. 2024. "Beyond Muslim Concerns About Transhumanism: Challenging Muslim Skepticism and Advocating a Pro-Transhumanism Perspective." *Journal of Ethics and Emerging Technologies* 34 (2): 1–19. https://doi.org/10.55613/jeet.v34i2.157.

Gillis, John Stuart, and Walter E. Avis. 1980. "The Male-Taller Norm in Mate Selection." *Personality and Social Psychology Bulletin* 6 (3): 396–401. https://doi.org/10.1177/014616728063010.

Goldsberry, Jenny. 2022. "'Human Cyborg' Peter Scott-Morgan Dead at 64." *Washington Examiner*, June 15. https://www.washingtonexaminer.com/news/2878337/human-cyborg-peter-scott-morgan-dead-at-64/. Accessed June 15, 2022.

Google. n.d. *DeepMind*. https://deepmind.google/.

Gould, Stephen Jay. 1997. "Nonoverlapping Magisteria." *Natural History* 106 (March): 16–22.

Haraway, Donna. 1985. "A Manifesto for Cyborgs: Science, Technology, and Socialist Feminism in the 1980s." *Socialist Review* 15 (2): 65–107. https://monoskop.org/images/4/4c/Haraway_Donna_1985_A_Manifesto_for_Cyborgs_Science_Technology_and_Socialist_Feminism_in_the_1980s.pdf.

Hejazi, Sara. 2019. "'Humankind: The Best of Molds'—Islam Confronting Transhumanism." *Sophia* 58 (4): 677–688. https://doi.org/10.1007/s11841-019-00755-7.

Hodgson, Marshall G.S. 1974. *The Venture of Islam: Conscience and History in a World Civilization*. London: University of Chicago Press.

Hughes, James J. 2007. "The Compatibility of Religious and Transhumanist Views of Metaphysics, Suffering, Virtue and Transcendence in an Enhanced Future." *Global Spiral* 8 (2).

Hughes, James J. 2010. "Contradictions from the Enlightenment Roots of Transhumanism." *The Journal of Medicine and Philosophy: A Forum for Bioethics and Philosophy of Medicine* 35 (6): 622–640. https://doi.org/10.1093/jmp/jhq049.

Ibn Hindū, Abū al-Faraj ʿAlī ibn al-Ḥusayn. 2010. *Key to Medicine & a Guide for Students*. Translated by Aida Tibi. Reading: Garnet Publishing.

Ibn Majah "Zuhd." Sunan Ibn Majah no. 4333. https://sunnah.com/ibnmajah:4333.

Islam Q&A. 2011a. "Ruling on Cosmetic Surgery to Make One Taller." *Islam Question and Answer*, August 6. https://islamqa.info/en/answers/161799/ruling-on-cosmetic-surgery-to-make-one-taller.

Islam Q&A. 2011b. "Using Medicine to Increase Growth Hormone and Treat Short Stature." *Islam Question and Answer*, November 22. https://islamqa.info/en/answers/149617/using-medicine-to-increase-growth-hormone-and-treat-short-stature.

Islamweb. 2008. "Conditions for Using Genetic Engineering." April 20. https://www.islamweb.net/en/fatwa/107188/conditions-for-using-genetic-engineering.

Jackson, Roy. 2020. *Muslim and Supermuslim: The Quest for the Perfect Being and Beyond*. Cham: Springer Nature.

Jones, Richard A.L. 2016. *Against Transhumanism: The Delusion of Technological Transcendence*. http://www.softmachines.org/wordpress/?p=1772. Accessed June 14, 2024.

Jotterand, Fabrice. 2008. "Beyond Therapy and Enhancement: The Alteration of Human Nature." *NanoEthics* 2: 15–23. https://doi.org/10.1007/s11569-008-0025-z.

Jotterand, Fabrice. 2010. "At the Roots of Transhumanism: From the Enlightenment to a Post-Human Future." *Journal of Medicine and Philosophy* 35 (6): 617–621. https://doi.org/10.1093/jmp/jhq050.

Judge, Timothy A., and Daniel M. Cable. 2004. "The Effect of Physical Height on Workplace Success and Income: Preliminary Test of a Theoretical Model." *Journal of Applied Psychology* 89 (3): 428–441. https://doi.org/10.1037/0021-9010.89.3.428.

Juengst, Eric T. 2004. "Enhancement Uses of Medical Technology." In *Encyclopedia of Bioethics*, 3rd ed., edited by Stephen G. Post, 753–757. New York: Macmillan.

Jurisprudence, Saudi Society for Studies in Medical. 2019. *Handbook of Islamic Medical Ethics and Jurisprudence*. Saudi Arabia: Imam Muhammad Ibn Saud Islamic University.

Kaplan, Robert M., and Richard G. Kronick. 2006. "Marital Status and Longevity in the United States Population." *Journal of Epidemiology & Community Health* 60 (9):760–765. https://doi.org/10.1136/jech.2005.037606

Keselman, Ana, Alicia Martinez, Liliana Pantano, Cesar Bergadá, and Juan Jorge Heinrich. 2000. "Psychosocial Outcome in Growth Hormone Deficient Patients Diagnosed during Childhood." *Journal of Pediatric Endocrinology and Metabolism* 13 (4): 409–416. https://doi.org/10.1515/JPEM.2000.13.4.409.

Al-Khatib, Mutaz. 2019. "The Ethical Limits of Genetic Intervention: Genethics in Philosophical and Fiqhi Discourses." In *Islamic Ethics and the Genome Question*, edited by Mohammed Ghaly, 169–200. Leiden: Brill.

Kingsley, Thomas. 2022. "Dr Peter Scott-Morgan Dies: British Scientist who became 'World's First Cyborg' Passes Away." *The Independent*, June 15. https://www.independent.co.uk/news/uk/home-news/dr-peter-scottmorgan-first-cyborg-dies-b2101629.html. Accessed June 18, 2022.

Koepke, Nikola, and Joerg Baten. 2005. "The Biological Standard of Living in Europe during the Last Two Millennia." *European Review of Economic History* 9 (1): 61–95. https://doi.org/10.1017/S1361491604001388.

Krieger, Nancy. 2011. *Epidemiology and the People's Health: Theory and Context.* New York: Oxford University Press.

Kurnaz, Serdar. 2024. "Transhumanism as a Challenge for Islamic Law." *Journal of Ethics and Emerging Technologies* 34 (2): 1–15. https://doi.org/10.55613/jeet.v34i2.158.

Lala, Ismail. 2020. "Germ-Inating Solutions or Gene-Rating Problems: An Islamic Perspective on Human Germline Gene Editing." *Journal of Religion and Health* 59 (4): 1855–1869. https://doi.org/10.1007/s10943-019-00770-5.

Lee, Newton. (ed.) 2019. *The Transhumanism Handbook.* Cham: Springer.

Littlejohn, Georgina. 2020. "Peter Scott-Morgan: The Extraordinary True Story Told in the Documentary *Peter: The Human Cyborg* on Channel 4 Tonight." *Inews.co.uk*, August 26. https://inews.co.uk/culture/television/peter-scott-morgan-true-story-told-documentary-peter-human-cyborg-channel-4-tonight-610781.

McNamee, Mike J., and Steven D. Edwards. 2006. "Transhumanism, Medical Technology and Slippery Slopes." *Journal of Medical Ethics* 32 (9): 513–518. https://doi.org/10.1136/jme.2005.013789.

Manzocco, Roberto. 2019. "Timeline of Transhumanism." In *Transhumanism: Engineering the Human Condition*. Chichester: Praxis Publishing. https://timelines.issarice.com/wiki/Timeline_of_transhumanism#cite_note-Manzocco-19.

Mavani, Hamid. 2014. "Islam—God's Deputy: Islam and Transhumanism." In *Transhumanism and the Body: The World Religions Speak*, edited by Calvin Mercer and Derek F. Maher, 67–83. New York: Palgrave Macmillan.

Mobayed, Tamim. 2017. "Immortality on Earth? Transhumanism through Islamic Lenses." December 11. https://yaqeeninstitute.org/read/paper/v1/immortality-on-earth-transhumanism-through-islamic-lenses. Accessed June 12, 2024.

More, Max. 1990. "Transhumanism: Towards a Futurist Philosophy." *Extropy* 6 (6): 11.

Murphy, Timothy F. 2012. "The Ethics of Impossible and Possible Changes to Human Nature." *Bioethics* 26 (4): 191–197. https://doi.org/10.1111/j.1467-8519.2010.01851.x.

Murphy, Timothy F. 2014. "Genetic Modifications for Personal Enhancement: A Defence." *Journal of Medical Ethics* 40 (4): 242–245. https://doi.org/10.1136/medethics-2012-101026.

Musk, Elon. 2024. "Prime Study Progress Update." *Neuralink*, April 12. https://neuralink.com/blog/prime-study-progress-update/.

Muslim ibn al-Hajjāj al-Naysābur. n.d. "The Book of Paradise, its Description, its Bounties and its Inhabitants." No. 2834c. https://sunnah.com/muslim:2834c.
Muslim ibn al-Hajjāj al-Naysābur. n.d. "The Book of *Zuhd* and Softening of Hearts." No. 2981a. https://sunnah.com/muslim:2981a.
Muslim ibn al-Hajjāj al-Naysābur. n.d. "The Book of *Zuhd* and Softening of Hearts." No. 2980b. https://sunnah.com/muslim:2980b.
Nagel, Thomas. 1986. *The View from Nowhere*. New York: Oxford University Press.
Al-Nawawī. Yahya ibn Sharaf. n.d. "The Book of Visiting the Sick." No. 955. https://sunnah.com/riyadussalihin:955.
NCD Risk Factor Collaboration (NCD-RisC). 2016. "A Century of Trends in Adult Human Height." *eLife* 5: e13410. https://doi.org/10.7554/eLife.13410.
Nichols, Steve. 1988. "The Post-Human Movement Manifesto." *Games Monthly*. http://dpya.org/en/index.php/1988_-_The_Post-Human_Movement_Manifesto_-_Steve_Nichols.
Noegel, Scott B., and Brannon M. Wheeler. 2010. "Hud." In *The A to Z of Prophets in Islam and Judaism*. Lanham, MD: Scarecrow Press.
Padela, Aasim I. 2013. "Islamic Verdicts in Health Policy Discourse: Porcine-Based Vaccines as a Case Study." *Zygon: Journal of Religion and Science* 48 (3): 655–670. https://doi.org/10.1111/zygo.12036.
Padela, Aasim I. 2020. "Islamic Bioethical Positions on Organ Donation and Transplantation: Stressing Rigor and Caution in Fatwa Reviews." *Transplant Direct* 6 (8): e586. https://doi.org/10.1097/TXD.0000000000001023.
Padela, Aasim I. 2021a. "An Introduction to Islamic Bioethics Its Producers and Consumers." In *Medicine and Shariah: A Dialogue in Islamic Bioethics*, edited by Aasim I. Padela, 1–38. Notre Dame, IN: University of Notre Dame Press.
Padela, Aasim I. 2021b. "Jurists, Physicians, and Other Experts in Dialogue: A Multidisciplinary Vision for Islamic Bioethical Deliberation." In *Medicine and Shariah: A Dialogue in Islamic Bioethics*, edited by Aasim I. Padela, 227–234. Notre Dame, IN: University of Notre Dame Press.
Padela, Aasim I. 2024. *Maqasid al Shariah and Biomedicine: Bridging Moral, Ethical, and Policy Discourses*. Herndon, VA: International Institute for Islamic Thought.
Padela, Aasim I., Mansur Ali, and Asim Yusuf. 2021. "Aligning Medical and Muslim Morality: An Islamic Bioethical Approach to Applying and Rationing Life Sustaining Ventilators in the COVID-19 Pandemic Era." *Journal of Islamic Ethics* 7 (1–2): 129–164. https://doi.org/10.1163/24685542-12340061.
Padela, Aasim I., Hasan Shanawani, Ahsan Arozullah, and M. Kholwadia. 2011. "Using Fatawa within Islamic and Muslim Bioethical Discourse: The Role of

Doctrinal and Theological Considerations–A Case Study of Surrogate Motherhood." *Islam and Bioethics*, edited by Berna Arda and Vardit Rispler-Chaim, 151–167. Ankara: Ankara University.

Paul, Diane B. 2005. "Genetic Engineering and Eugenics: The Uses of History." In *Is Human Nature Obsolete? Genetics, Bioengineering, and the Future of the Human Condition*, edited by Harold W. Baillie and Timothy K. Casey, 123–151. Cambridge, MA: The MIT Press. https://doi.org/10.7551/mitpress/3977.003.0008.

Persson, Ingmar, and Julian Savulescu. 2012. *Unfit for the Future: The Need for Moral Enhancement*, 1st ed. Oxford: Oxford University Press.

Pinker, Steven. 2004. *The Blank Slate: The Modern Denial of Human Nature*. New York: Viking.

Plomin, Robert. 2019. *Blueprint: How DNA Makes Us Who We Are*. Cambridge, MA: MIT Press.

Pormann, Peter E., and Emilie Savage-Smith. 2007. *Medieval Islamic Medicine*. Washington, DC: Georgetown University Press.

Porter, Allen. 2017. "Bioethics and Transhumanism." *Journal of Medicine and Philosophy* 42 (3): 237–260. https://doi.org/10.1093/jmp/jhx001.

Powell, Russell, and Allen Buchanan. 2011. "Breaking Evolution's Chains: The Promise of Enhancement by Design." In *Enhancing Human Capacities*, edited by Julian Savulescu, Ruud ter Meulen, and Guy Kahane, 49–67. Chichester: Blackwell Publishing.

Qazi, Faisal, Don Fette, Syed S. Jafri, and Aasim I. Padela. 2018. "Framing the Mind–Body Problem in Contemporary Neuroscientific and Sunni Islamic Theological Discourse." *The New Bioethics* 24 (2): 158–175. https://doi.org/10.1080/20502877.2018.1438835.

Rāzī, Abū Bakr Muḥammad ibn Zakarīyā. 1950. *The Spiritual Physick of Rhazes*, 1st ed. Translated by Arthur J. Arberry. London: J. Murray.

Rispler-Chaim, Vardit. 1993. *Islamic Medical Ethics in the Twentieth Century*. Leiden: Brill.

Salska, Irmina, David A. Frederick, Boguslaw Pawlowski, Andrew H. Reilly, Kelsey T. Laird, and Nancy A. Rudd. 2008. "Conditional Mate Preferences: Factors Influencing Preferences for Height." *Personality and Individual Differences* 44 (1): 203–215. https://doi.org/10.1016/j.paid.2007.08.008.

Samaras, Thomas T. 2007. "Human Scaling and the Body Mass Index." In *Human Body Size and the Laws of Scaling: Physiological, Performance, Growth, Longevity and Ecological Ramifications*, edited by Thomas T. Samaras, Andrzej Bartke, and Christopher David Rollo, 17–31. New York: Nova Science Publishers.

Sapolsky, Robert M. 2023. *Determined: Life Without Free Will*. London: Vintage.

Schönbeck, Yvonne, Henk Talma, Paula van Dommelen, Boudewijn Bakker, Simone E. Buitendijk, Remy A. HiraSing, and Stef van Buuren. 2013. "The

World's Tallest Nation has Stopped Growing Taller: The Height of Dutch Children from 1955 to 2009." *Pediatric Research* 73 (3): 371–377. https://doi.org/10.1038/pr.2012.189.

Scott-Morgan, Peter. 2021. *Peter 2.0: The Human Cyborg*. London: Michael Joseph.

Shabana, Ayman. 2022. "Between Treatment and Enhancement: Islamic Discourses on the Boundaries of Human Genetic Modification." *Journal of Religious Ethics* 50 (3): 386–411. https://doi.org/10.1111/jore.12404.

Shepperd, James A., and Alan J. Strathman. 1989. "Attractiveness and Height: The Role of Stature in Dating Preference, Frequency of Dating, and Perceptions of Attractiveness." *Personality and Social Psychology Bulletin* 15 (4): 617–627. https://doi.org/10.1177/0146167289154014.

Ṣiddīqī, Muḥammad Zubayr. 1993. *Ḥadīth Literature: Its Origin, Development, and Special Features*. Cambridge: The Islamic Texts Society.

Smith, Colleen. 2022. "Dr Peter Scott-Morgan Dies: Sadness at Death of World's First 'Fully Robotic Person.'" *DevonLive.com*, June 15. https://www.devonlive.com/news/devon-news/dr-peter-scott-morgan-dies-7212028. Accessed June 15, 2022.

Sørheim, Preben, Mathias Barra, Ole Frithjof Norheim, Espen Gamlund, and Carl Tollef Solberg. 2024. "Premature Death as a Normative Concept." *Health Care Analysis* 32: 88–105. https://doi.org/10.1007/s10728-023-00471-x.

Stabler, Brian, Patricia T. Siegel, Richard R. Clopper, Catherine E. Stoppani, Peter G. Compton, and Louis E. Underwood. 1998. "Behavior Change after Growth Hormone Treatment of Children with Short Stature." *The Journal of Pediatrics* 133 (3): 366–373. https://doi.org/10.1016/S0022-3476(98)70271-9.

Steckel, Richard H. 2001. "Health and Nutrition in the Preindustrial Era: Insights from a Millennium of Average Heights in Northern Europe." National Bureau of Economic Research, Cambridge, MA. https://doi.org/10.3386/w8542.

Steckel, Richard H., and Joseph M. Prince. 2001. "Tallest in the World: Native Americans of the Great Plains in the Nineteenth Century." *American Economic Review* 91 (1): 287–294. https://doi.org/10.1257/aer.91.1.287

Steinhausen, Hans-Christoph, Helmuth G. Dörr, and Roland Kannenberg. 2000. "The Behavior Profile of Children and Adolescents with Short Stature." *Journal of Developmental & Behavioral Pediatrics* 21 (6): 423–428. https://doi.org/10.1097/00004703-200012000-00004.

Strauss, John, and Duncan Thomas. 1998. "Health, Nutrition, and Economic Development." *Journal of Economic Literature* 36 (2): 766–817.

Stulp, Gert, Louise Barrett, Felix C. Tropf, and Melinda Mills. 2015. "Does Natural Selection Favour Taller Stature among the Tallest People on Earth?"

Proceedings of the Royal Society: Biological Sciences 282 (1806): 20150211. https://doi.org/10.1098/rspb.2015.0211.

Tall.Life. n.d. "The Evolution of Height." https://tall.life/the-evolution-of-height/.

Taqī ʿUsmānī, Justice Muḥammad. 1991. *The Authority of Sunnah*. New Delhi: Kitab Bhavan.

ThinkH+. n.d. "About Us." https://thplus.org/about/.

Thompson, Joel. 2017. "Transhumanism: How Far Is Too Far?" *The New Bioethics* 23 (2):165–182. https://doi.org/10.1080/20502877.2017.1345092.

Tyrrell, Jessica, Samuel E. Jones, Robin Beaumont, Christina M. Astley, Rebecca Lovell, Hanieh Yaghootkar, Marcus Tuke, Katherine S. Ruth, Rachel M. Freathy, and Joel N. Hirschhorn. 2016. "Height, Body Mass Index, and Socioeconomic Status: Mendelian Randomisation Study in UK Biobank." *BMJ* 352: i582. https://doi.org/10.1136/bmj.i582.

Valles, Sean. 2020. "Philosophy of Biomedicine." In *The Stanford Encyclopedia of Philosophy*, edited by Edward N. Zalta. https://plato.stanford.edu/archives/sum2020/entries/biomedicine/.

Vita-More, Natasha. 1983. "The Transhuman Manifesto." *Humanity+*. https://www.humanityplus.org/the-transhumanist-manifesto.

Warwick, Kevin, Mark Gasson, Benjamin Hutt, Iain Goodhew, Peter Kyberd, Brian Andrews, Peter Teddy, and Amjad Shad. 2003. "The Application of Implant Technology for Cybernetic Systems." *Archives of Neurology* 60 (10): 1369–1373. https://doi.org/10.1001/archneur.60.10.1369.

Zaman, Muntasir. 2022. *The Height of Prophet Adam: At the Crossroads of Science and Scripture*. Oldham: Beacon Books.

CHAPTER 8

Radical Life Extension: Intersecting Bioscientific Possibilities and Islamic Sensibilities

Raudah M. Yunus

Introduction

As biomedical technologies and scientific discoveries push the boundaries of science fiction into potential reality, humans begin to question the unthinkable: What if we don't age? Is death a necessity? Can we live much longer, or indefinitely? Could this form of life enhancement become humanity's future? Certainly, such inquiries—especially those pertaining to the desire for immortality—are not new. The Sumerian *Epic of Gilgamesh* and Homer's *Odyssey*, for instance, eloquently narrate how death—or the lack thereof—remains a central theme in human thought and imagination (Masci 2013). What is new, however, is the means, that is technology, through which humans are hoping to achieve extreme longevity, and perhaps, infiniteness (Bradshaw 2009). Radical Life Extension (RLE) has been defined in different ways and, in this chapter, it refers to the pursuit of profound life extension—not immortality—as championed by Western transhumanist philosophy, and not other forms of RLE as understood in other traditions such as Buddhism

R. M. Yunus (✉)
Medical College of Wisconsin, Milwaukee, WI, USA
e-mail: raudah.yunus@gmail.com
Universiti Teknologi MARA, Selangor, Malaysia

© The Author(s), under exclusive license to Springer Nature Switzerland AG 2026

A. Padela (ed.), *Bridging Bioscience and Islam*, Islam & Science,
https://doi.org/10.1007/978-3-032-09996-9_8

and Hinduism. Australian philosopher and ethicist Margaret Somerville (1942–) describes RLE as "life extension beyond the postulated human species maximum lifespan, which is believed to be about 120 years and where the extension is brought about by human intervention" (Somerville 2009). Sociologist Bryan Turner (1945–) on the other hand, defines RLE as "the expansion of a person's lifespan far beyond what is currently humanly possible by eradicating aging and disease" (Turner 2009). Some scholars distinguish between moderate and radical life extension. For instance, Campbell places moderate life extension in the realm of 100–200 years and RLE in the territory of 400 years or more (Campbell 2020), while American sociologist Shahin Davoudpour and philosopher John Davis view moderate extension as a life expectancy of 120 years, and RLE as the complete cessation of aging in which humans can live up to 1000 years (Davoudpour and Davis 2022).

The current, modern pursuit of RLE is a gradual phenomenon that results from the accumulation of decades of advancement in public health (mainly), medicine, and biotechnology—all of which have contributed to the decline in global mortality and the increase in human lifespan. Public health and technology-driven measures such as food security, water purification, sanitation, vaccination, antibiotics, assistive devices, and cancer chemotherapy have substantially impacted the milestones in human longevity (Vijg and De Grey 2014). In addition, recent progress in the understanding of aging processes at the molecular levels and the ideological beliefs about the nature and trajectory of human evolution have led many scientists and philosophers, especially the transhumanists, to adopt a highly "optimistic" view toward the human lifespan and challenge the inevitability of traditionally accepted notions of aging and death. This new attempt to redefine human destiny is based on a set of assumptions that underpin the transhumanist and life extension movement. First, humans are not necessarily designed for aging or death; they are, in fact, malleable (Schuler 2021). Therefore, becoming old and dying need not be part of human destiny. Second, since humankind is not defined by old age and death, aging and mortality are not "natural," but rather are seen as diseases to be overcome and transcended (Schuler 2021). Remedying the symptoms of, and diseases resulting from, old age, therefore amounts to rectifying human deficiencies (Lombard 2023). Third, our current understanding of health, medicine, and therapy are narrow and distorted; health should be viewed as not simply restoring what is "normal," rather, it should also aim at maximizing human potentialities through bodily enhancements,[1] as freeing humans from physical vulnerabilities reflects a more accurate meaning of health and well-being (Lombard 2023). Fourth, pain, suffering, and bodily demise are considered "bad," while longevity enables humans to find joy and self-fulfillment through meaningful endeavors or loving relationships that would not be disrupted by death (Lorrimar 2018).

In the RLE discourse, emphasis is often laid on the bioscientific means through which it can be realized. Largely missing from existing debates are the motivations underlying the pursuit of RLE (Campbell 2020). In her thesis, Campbell uses the Thanatophobic and Romantic Motivational (TRM) spectrum as a framework to explain why people would strive for RLE or use its technologies should they become available. The TRM framework postulates two key motivators for RLE: Thanatophobia (fear of death) and Romanticism (love of life) (Campbell 2020). While thanatophobia refers to death anxiety that may result from beliefs about the state of being dead, fear of the unknown, images of loved ones' bereavement, and the notion of non-existence (Blomstrom et al. 2022), romanticism denotes one's passion for living associated with enjoyment, material pleasures, meaningful relationships, and a sense of importance in existence (Campbell 2020). Indeed, while these two concepts are portrayed as separate entities or two entities at the opposing end of a spectrum, they are also inextricably linked, because one's fear of non-existence could be tied to his/her love for material pleasures that would be curtailed by death.

The Bioscience Behind Longevity and Life Extension

There are still mysteries as to why and how exactly humans become old but scientists have begun to understand better the mechanisms that lead to aging, a process often described as a gradual decline in structures and functions of the body (Chang et al. 2022). In the pursuit to "conquer" and ultimately "defeat" this biological phenomenon, transhumanists propose two streams of life prolongation. The first is "mind uploading," defined as the process of transferring one's consciousness[2]

[1] For the definition of transhumanism, enhancement, and therapy, see Chapter 7.

[2] For details on the human mind, consciousness, and the mind–body problem, refer to Chapter 4, "Scientific Advances in Understanding of the Human Nature: An Islamic Occasionalist Perspective" by Nazif Muhtaroğlu and Macksood Aftab.

into a non-biological entity, spanning across the internet or another body (Doorani 2023). Here, the aim is to keep our minds indefinitely operative after death, thus liberating people from their bodily confinement (Gaitán 2019). The second is the extension of embodied human lives through measures that aim to attenuate or stop aging. This mainly comprises advanced biomedical interventions such as caloric restriction (CR), genetic manipulation and stem cell therapy, and administration of pharmaceuticals (Verdaguer et al. 2012; Fontana, Partridge, and Longo 2010). In this chapter, I will focus on the latter (extension of embodied human lives) and not the former (mind uploading). This is because mind uploading is still very much highly speculative (Warby 2021), and its goal is to achieve (digital) immortality, which is beyond the remit of RLE.

The second stream of longevity emphasizes the preservation of human corporeality as long as possible, by slowing down, or arresting, or reversing the aging process. To date, anti-aging strategies can be largely condensed into four groups: caloric restriction (CR), pharmacological intervention, gene therapy, and stem cell therapy (Magalhães 2012; Tacutu et al. 2012). Defined as "reduction of dietary intake below energy requirements while maintaining optimal nutrition" (Flanagan et al. 2020), CR exerts its effects by lowering metabolic rate and oxidative damage, improving markers of age-related diseases, altering neuroendocrine activities and promoting stem cell self-renewal and tissue regeneration (Green et al. 2022; Heilbronn et al. 2006). Even though CR experimentations have been largely conducted in animal models, studies involving human subjects have corroborated the role of CR in enhancing longevity. For instance, research among Japanese adults in the Okinawa Island establishes associations between lower caloric consumption and a healthier and longer lifespan. The island is said to have five times higher number of centenarians than in any other part of the world, as its residents consume 17 percent fewer calories than the average Japanese, and 40 percent less than the average American adult (Suzuki et al. 2001; Most et al. 2017). The biggest CR trial on humans known as the Comprehensive Assessment of Long-term Effects of Reducing Intake of Energy (CALERIE) study demonstrated that moderate CR improved aging biomarkers (Dorling et al. 2021). In real-life settings, CR is achieved through prescription of reduced caloric intake by about 20–40 percent below the usual free dietary consumption, which may come in the form of intermittent fasting, time-restricted feeding or fasting-mimicking diets (Balasubramanian et al. 2017; Shetty et al. 2018).

Pharmacological interventions largely comprise drugs that act by mimicking the effects of CR. Known as "CR mimetics," they target the metabolic pathways affected by CR but without requiring actual calorie reduction (Ingram et al. 2006). The lack of requirement for prolonged, mandatory dietary restrictions—or in other words, lifestyle changes—makes the pharmacological approach a much more attractive and convenient option for clinicians and patients. Anti-aging pharmacological interventions can also be classified into four groups according to their target or level of interventions: (1) senolytics; (2) rapalogs; (3) metabolic; and 4) anti-diabetic drugs (Petr et al. 2024). The first group, senolytics, are a class of drugs that clear senescent cells (SC) (Kirkland and Tchkonia 2020). SCs are cells that have stopped multiplying but do not die off. They accumulate in the body with aging, and release chemicals that trigger inflammation, interrupting healthy cells (Childs et al. 2017). The second group, rapalogs, refer to a class of allosteric inhibitors that intervene with the mammalian target of rapamycin (mTOR) pathway (Dittmer et al. 2012). The latter is a cellular signaling pathway involved in various physiological functions in the body and plays a vital role in regulating lifespan and aging (Papadopoli et al. 2019). The third group, metabolic drugs, comprise drugs that largely mimic the effects of CR. The last one, anti-diabetic drugs, are medicines used to control blood sugar levels in diabetic patients, among which is metformin (Petr et al. 2024).

Gene therapy is a technique that alters a person's gene. In anti-aging research, it is considered more promising than pharmacological interventions because drug efficacies often fade with progressive conditions and increasing doses can cause adverse reactions. Gene therapy focuses instead on the primary markers of aging by directly addressing aging-related genes (Yu et al. 2023). There are three main categories of gene therapy according to the methods and level of manipulation: gene replacement, gene editing, and epigenetic modification (Yu et al. 2023). Gene replacement refers to the substitution of a faulty gene with a healthy or engineered/corrected one (Pandey and Balekar 2018). Meanwhile, gene editing involves techniques that alter specific DNA sequences by inserting a healthy/modified sequence and replacing the pathological one (Scanlon 2020; Yu et al. 2023). Genome editing tools include transcription activator-like effector nuclease (TALEN), zinc-finger nuclease (ZFN), and the clustered regularly interspaced short palindromic repeats (CRISPR) technology, the latter being the most efficient and widely used (Shamshirgaran et al. 2022). Epigenetic modification operates differently; it modifies the expression of genetic information without affecting the nucleic acid sequences (Solís-Paredes et al. 2014). This is achieved through inhibiting or activating the expression of certain genes and adjusting imbalances in gene expression (Li et al. 2020; Li et al. 2021). An example of gene therapy application for longevity is injecting the telomerase reverse transcriptase (TERT) gene into animal models to improve their aging biomarkers and increase their median lifespan (Bernardes de Jesus et al. 2012).

On the other hand, stem cell (SC)[3] therapy is a modality of regenerative medicine that utilizes the unique properties of viable human stem cells, such as self-renewal and differentiation, to replenish and replace damaged cells and tissues with new and functional cells (Biehl and Russell 2009). Sources of stem cells include embryonic tissues, fetal/perinatal tissues, adult tissues, and differentiated somatic cells. The latter are adult cells that have been genetically reprogrammed to act like embryonic stem cells and are called the induced pluripotent stem cells (iPSCs) (Bacakova et al. 2018). SC therapies can be self-to-self (autologous), in which patients' own cells are used, or they can be derived from a healthy donor (allogeneic). Potential applications of SC therapy have been demonstrated in aging-related diseases such as Parkinson's, Huntington's, and stroke (Nguyen et al. 2019). In Parkinson's disease research, prior experimentation using animal models shows that transplanting stem cells into the brain allows them to re-establish neural capacity through differentiation and release of neurotrophic factors (Lindvall et al. 2012), and that transplanted intracerebral stem cells can reprogram host cells to display astroglial characteristics, promoting the emergence of dopaminergic neurons (Niclis et al. 2017). Additionally, adipose-derived mesenchymal stem cells (ADMSC) have been found to effectively repair damaged skin cells and stimulate cell regeneration, improving aging-related defects such as wrinkles, skin thickness, and ultraviolet-induced injury (Mazini et al. 2020). Needless to say, SC therapy can be used in tandem with gene therapy; this combination is said to have greater effects on tissue regeneration and healing (Rafi 2011). Table 8.1 summarizes the four main strategies behind current anti-aging endeavors.

[3] Stem cells are cells that have the capacity to self-renew and differentiate into more specialized types of cells and tissues. Sources of stem cells include embryonic stem cells, adults stem cells, induced pluripotent stem cells (which are adult stem cells that have been altered to act like embryonic stem cells), and perinatal stem cells.

Table 8.1 The four main strategies behind the science of anti-aging

	Caloric restriction (CR)	*Pharmacological intervention*	*Gene therapy*	*Stem cell therapy*
Definition	Reduction of dietary intake below energy requirements while maintaining optimal nutrition.	Drugs that act by mimicking the effects of CR, known as "CR mimetics."	A technique that alters a person's gene to slow down or reverse aging, by directly addressing aging-related genes.	A modality of regenerative medicine that utilizes the unique properties of human stem cells.
Potential mechanism(s)	Lowers metabolic rate and oxidative damage, improves markers of age-related diseases, promotes stem cell self-renewal, and tissue regeneration.	Operate by targeting the metabolic pathways affected by CR but without requiring actual calorie reduction.	Gene replacement (substitution of a faulty gene with a healthy one), gene editing (alteration of DNA sequences), and epigenetic modification (modification of the expression of genetic information).	Self-renewal and differentiation to replenish and replace damaged cells and tissues with new and functional ones.
Evidence/ Examples	• Okinawa Island study • CALERIE study • Intermittent fasting • time-restricted feeding	• Senolytics • Rapalogs • Metabolic drugs • Anti-diabetic drugs	• Transcription activator-like effector nuclease (TALEN) • Clustered regularly interspaced short palindromic repeats (CRISPR) • Injection of telomerase reverse transcriptase (TERT)	• Transplantation of intracerebral stem cells in Parkinson's disease experiment. • Use of adipose-derived mesenchymal stem cells (ADMSC) to improving aging-related skin defects.

Limitations in the Pro-longevity Research Evidence

Research evidence that supports the possibility of life extension is burgeoning in number and sophistication. Nevertheless, despite the rising optimism about life extension stemming from extant and ongoing studies, pro-longevity evidence is constrained by several drawbacks and therefore needs to be viewed with critical eyes. First, while several experiments have indicated a positive impact of biomedical interventions on lifespan, these studies predominantly used animal models rather than human subjects. For instance, a meta-analysis by Liang et al. that looked into survival records from longevity investigations reported that CR and CR mimetics were indeed more impactful and superior to other types of anti-aging interventions (Liang et al. 2018). However, all the studies included in this meta-analysis used animal models, such as Drosophila (a type of fruit fly) and Caenorhabditis elegans (a type of nematode) (Liang et al. 2018). In a similar vein, two systematic reviews by Cho et al. and Wang et al. which corroborate the anti-aging effects of herbal medicines only included non-human subjects such as mice, Caenorhabditis elegans, silkworm, and yeast (Cho et al. 2023; Wang et al. 2021). Likewise, the pro-longevity effects of anti-diabetic drugs (metformin and acarbose) and rapalogs (rapamycin), and the anti-aging effects of SC therapy (e.g., mesenchymal SC transplantation) on cognitive function have been mainly demonstrated in animal models (Garay 2021; Shetty et al. 2018).

Second is the question of safety, adverse effects, and the efficacy of the interventions used in anti-aging research. Given the relative newness of this field, it is crucial to underscore that most investigations either were started only recently or are currently ongoing. For the studies that have been completed, their treatment modalities are still largely in the clinical trial phase(s),[4] with very few having been approved by the Food

[4] Clinical trials are research that involve people and test new ways to prevent, detect, diagnose, or treat diseases. There are usually four phases: (1) in Phase I, researchers test a drug or treatment in a small group of people (20–80) for the first time—the aim is to study drug safety and identify side effects; (2) in Phase II, the new drug or treatment is given to a larger group of people (100–300) to determine its effectiveness and to further study its safety; (3) in Phase III, the drug is tested among large groups of people (1,000–3,000) to confirm its effectiveness, monitor side effects, and compare it with standard treatments; 4) in Phase IV, once a drug is approved by the FDA and made available to the public, researchers track its safety in the general population, seeking more information about the treatment benefits and optimal use (see source: NIH 2022).

and Drug Administration (FDA). A recent review of stem cell therapies to slow or reverse aging that gathered and appraised clinical trials from databases in the United States, Europe, China, Japan, and the World Health Organization (WHO) showed that the vast bulk of studies are Phase I/II clinical trials (Garay 2021). This denotes that while extant results may appear encouraging, the safety (and efficacy) of SC therapies in these studies are yet to be ascertained (NIH 2022). For example, a key finding in this review was the potential efficacy of intravenous Lomecel-B, an allogeneic bone marrow stem cell preparation, to treat aging frailty. Despite its modest effect, Lomecel-B was said to be the only SC therapy in Phase II trial, with no other SC preparation in late clinical development (e.g., Phase III) (Garay 2021; Golpanian et al. 2017). Another example is the potential negative side effects of prolonged CR such as hypotension, infertility, osteoporosis, and depression (Dirks and Leeuwenburgh 2006).

Additional safety-efficacy issues have been attributed to the lack of standard protocol, absence of consensus on efficacy outcomes and diagnostic tools, inadequate dosing information, and scarcity of pharmacokinetic studies of intravenous MSC in human subjects (Garay 2021; Ghazizadeh 2018). Similarly, gene therapy has been associated with the genotoxic potential of viral vectors that can upregulate proto-oncogenes, predisposing subjects to cancer (David and Doherty 2017). Serious side effects have been documented in gene therapy in the past, for instance, the incident of Jesse Gelsinger's death in 1999 due to a systemic inflammatory response caused by adenoviral gene transfer (Raper et al. 2003), and the tragic incidence of leukemia among children treated for severe combined immunodeficiency (Hacein-Bey-Abina et al. 2010). While these two anecdotes are not directly related to gene therapies used for anti-aging, they exemplify side effects and safety issues which might remain unknown until new techniques are tested on humans. Moreover, some experts hold that even if gene and cellular therapies can delay or prevent the onset of age-related diseases, they do not necessarily affect or slow the overall process of aging (Rattan 1998; Kassem 2006).

Considering the above-mentioned limitations, I contend that research on longevity and life extension is at a rudimentary stage, with most studies or clinical trials being initiated in recent years (Fukuda et al. 2023; Garay 2021). In addition, substantiation of evidence on efficacy is limited by factors such as inadequate human subjects, safety concerns and adverse effects, and scarcity of late-stage clinical trials. Thus, as it stands, whether human beings can achieve RLE or extreme longevity in the near future is a question to which science can only provide a partial answer for now. Until all the described hurdles have been overcome, RLE and the impact of life-extension technologies remain highly uncertain and debatable.

The Human Life Journey from an Islamic Perspective

In order to fully capture the Islamic perspective on aging, longevity, and death, one needs to know how Islam views the human life course. In his book *The Lives of Man*, the well-known Islamic scholar Imam Abdullah ibn ʿAlawi al-Haddad (1044–1142 AH) described the five stages of human existence (al-Haddad 1991). The first stage is life before conception. This began with the creation of Adam, after which all his progenitors (in their essential form, as souls) were brought forward before God to take the covenant that they recognized (His) Unity and Lordship (Q.7:172). The second stage is the life in this world, which involves the soul being merged with a body to create an ensouled being and begins with human conception and birth and ends with death (defined as a separation of the human soul from the human body in such a way that the body is no longer capable of enacting the soul's commands).[5] It is in this life that humans are tested and given the opportunity to perform virtuous deeds on the basis of which they will be held accountable for their actions. This stage is further divided into several substages including childhood, youth, maturity, old age, and others. The third stage is the intermediary life (*Barzakh*); a phase that begins upon death and continues until the Day of Resurrection. The fourth stage is life of the Day of Judgment, where all of humankind will stand before God to give an account of their lives in this world. In the last stage—also known as life of the eternal abode—every human being either enters heaven or hell (al-Haddad 1991; Sulaiman 2020). Figure 8.1 illustrates these five stages of human existence and positions RLE in relation to them.

[5] The specifics of how the soul and the body are connected are subject to continued theological and empirical deliberation. A bioethical topic that offered the occasion to renew these pre-modern debates was the advent of "brain death." While Muhtaroğlu and Aftab touch upon these questions as related to the soul–mind–body problem in Chapter 4, other authors have written on Islamic views on brain death more directly, including the editor of this book, Aasim I. Padela.

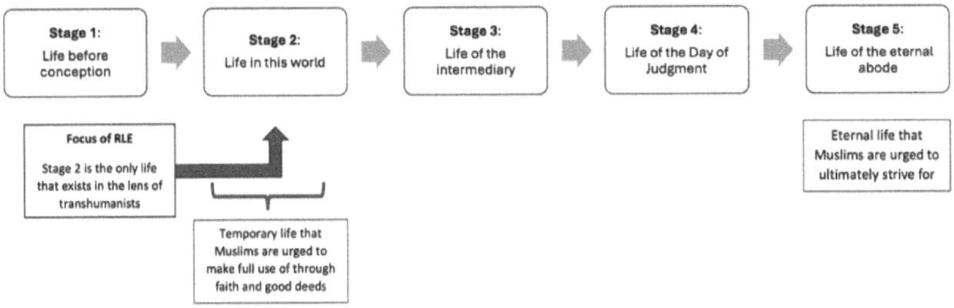

Fig. 8.1 The five stages of human existence from the Islamic perspective

From Fig. 8.1, it is clear that the second life—life of this world—is but a "brief portion" of the full spectrum of human existence. The endeavor to extend life as espoused by transhumanists or RLE advocates is fully concentrated in this stage; as transhumanists generally refute the unseen or the metaphysical world, it only makes sense that they focus on maximizing longevity through which all life opportunities can be realized and physical pleasures can be enjoyed. From the Islamic perspective, there are two ways of understanding this phenomenon. First, given its brevity and the ultimate human destiny being the hereafter, this "second life" is viewed as trivial or less important. Therefore, Muslims tend to question the enthusiasm and desire to "live so long," when this worldly life is described as illusion and deception (Q.57:20). Second, and in contrast to the former view, this "brief portion" is the only stage of human existence that allows humans to gather virtuous deeds and prove their worth in the sight of God. In other words, the relatively short life of this world is the sole opportunity for merit-making; this phase is the main determinant of one's destiny in the afterlife. Thus, some Muslims may contend that living longer is praiseworthy, for it gives man more time and chance to repent, do good, and rectify their affairs. As such, there is nothing wrong with efforts to slow down the aging process, if that can make humans live longer and healthier in piety and good faith.

Islamic teachings—sourced from the Qur'an and *ḥadīth*—view aging as a natural phenomenon that is part of the human life cycle. Aging reminds humans of, and is a path toward, death and the next life. Scriptural sources attribute meaning to the aging process at different levels. The first level pertains to the physical or biological nature of aging; becoming old is depicted as a natural process, characterized by a gradual decline in physical and mental strength (Q.19:4; Q.30:54). The second level is social or relational, where notions of aging are tied to relations between the older persons and surrounding people, and responsibilities of community members for their care and welfare. For instance, the Qur'an emphasizes a child's duty toward his/her old parents, highlighting that caring and honoring them comes second after worshiping God (Q.17:23). Similarly, several *ḥadīth* expound the Divine command of treating older people with respect and humility, underscoring collective obligations to take care of elderly members in society. The last (third) level involves the spiritual domain, in which those who attain old age are portrayed as having a special rank before God, thus cherishing and extending support to them is a means to attain God's pleasure. An example is a *ḥadīth* that illustrates how God's mercy, forgiveness, and leniency increases for His servants as they age,[6] and another two *ḥadīth* that highlight kind treatment and respect for the elderly as a means to obtain Divine reward and being accepted into the community of believers.[7]

Juxtaposing Scriptural with Bioscientific Evidence on Longevity

The Qur'an does not specify the human lifespan, but rather stresses in multiple verses that this is a matter of the unseen and ultimately resides within God's province (Q.21:35) (Musa 2009). The Qur'an also clearly

[6] This *ḥadīth* says:

"No one reaches forty years in Islam except that Allah protects him from madness, leprosy, and leucoderma. When he is fifty, Allah makes easy his reckoning. When he reaches sixty, Allah grants him repentance. And when he reaches seventy, Allah loves him and the inhabitants of the heavens love him. And when he reaches eighty, Allah accepts his good deeds and overlooks his sins. When he reaches ninety, Allah forgives his past and future sins, he is named 'the prisoner of Allah on earth' and is granted intercession (for forgiveness) for his family."

(Musnad Ahmad no. 5626 and 13279)

[7] Anas bin Malik reported: Messenger of Allah (pbuh) said, "*If a young man honours an older person on account of his age, Allah appoints someone to show reverence to him in his old age.*" (Riyad as-Salihin no. 359).

Narrated by Abdullah ibn Amr ibn al-'As: The Prophet (pbuh) said: "Those who do not show mercy to our young ones and do not realise the right of our elders are not from us" (Sunan Abi Dawud no. 4943).

mentions that God blesses some humans with long lives, and decrees 'early' deaths for others according to His power and wisdom (Q.7:34; Q.10:49; Q.16:61; Q.35:45). Accordingly, it is difficult to pinpoint specific verses that explicitly spell out the prospect of extreme longevity or inspire humankind to pursue it in the general sense. It is perhaps not far-fetched to think of the overall Qur'anic outlook on longevity or the possibility of radically prolonged lifespan as one of "indifference." This apparent lack of importance attached to extreme longevity—as opposed to the enthusiasm for anti-aging research seen in modern science today—should not be viewed in a pejorative sense. It is not because Islam does not value life, rather, this reflects the Qur'an's attempt to redirect the human focus from quantity (number of years lived) to quality (good action in lifetime) (Q.101:6–9).

Nonetheless, the Qur'an does not negate the possibility of humans having the capacity to survive extraordinarily long years. Several accounts in the Qur'an, in fact, confirmed its occurrence. The story of Prophet Nūḥ is the most frequently cited example of extreme longevity. The Qur'an (Q.29:14) narrates the struggle of this noble man, with an explicit mention of "*a thousand years bar fifty*" as the duration in which he lived with his people. It states: "*And, indeed, [in times long past] We sent forth Noah unto his people, and he dwelt among them a thousand years bar fifty; and then the floods overwhelmed them while they were still lost in evildoing.*" Scholars of the Qur'an have various opinions on Prophet Nūḥ's actual lifespan; some stated that he lived for 950 years, while others suggested that he might have lived between 1020 and 1700 years (IslamQ&A 2013). The second account is the story of "The Men of the Cave" (*Ashāb al-kahf*) described in Chapter 18 of the Qur'an about seven men who fled into a cave to escape persecution while maintaining their faith in God. While this event can be regarded largely as a miracle, the verses clearly indicate that the men survived for an extended duration—approximately three centuries according to some scholars (Fudge 2007).

Several *ḥadīth* are often quoted in discussions on aging and the human lifespan. This chapter does not contain an exhaustive list of *ḥadīth* on death and old age/aging, rather I focus on two well-known Prophetic sayings. In the first one, Abū Hurayra reported that the Prophet Muhammad described the lifespan of his *ummah* to be between

60 and 70 years,[8] with a small number exceeding that range. In the second narration, the Prophet is reported to have said that God does not create a disease without providing a remedy, except for one which is old age.[9] Al-Tirmidhī categorized the former *ḥadīth* as *ḥasan* and the latter as *ṣaḥīḥ*. At first sight, these two Prophetic traditions seem to suggest that the human lifespan is somewhat static or fixed, thus nullifying the possibility of "treating" or defying the aging process. One might then wonder: Does this mean life extension beyond its current threshold is impossible? Does this not contradict extant scientific evidence of anti-aging biomedical interventions that seem promising? What about the Qur'anic accounts?

Here I argue that the two Qur'anic narrations confirm the occurrences of extreme longevity, thus proving that living extraordinarily long lives is not impossible. However, the narratives are not straightforward indicators that human beings are naturally endowed with the capacity for "extreme" longevity because both stories have a miraculous element.[10] Prophet Nūḥ's lifespan of 950 years is said to be special Divine gift upon him (see Sedick 2023), and this is nothing strange for the Islamic tradition recognizes that God's messengers and prophets are sometimes blessed with miracles or extraordinary abilities that contradict the laws of nature to prove their truthfulness.[11] Similarly, the story of *Ashāb al-kahf* is an obvious miracle and Divine intervention that was bestowed on the seven youths for their faith in, and loyalty to, God.[12] Considering the

[8] Abū Hurayra narrated that the Messenger of Allah (pbuh) said: "The lifespan for my Ummah is from sixty years to seventy [years]" (Jami' at-Tirmidhī no. 2331). In another narration, it was reported slightly differently: The lifespan for my Ummah is from 60 years to 70 years and very few surpass this (Sunan Ibn Majah no. 4236).

[9] Usamah bin Sharik said: Some Bedouins asked: "O Messenger of Allah (pbuh) shall we treat (our ill)?" He said: "Yes, O worshipers of Allah! Use remedies. For indeed Allah did not make a disease but He made a cure for it"—or—"a remedy. Except for one disease." They said: "O Messenger of Allah (pbuh)! What is it?" He said: "Old age." (Jami' at-Tirmidhī no. 2038).

[10] For more explanation on correlation, causation, and laws of nature from the Ash'arī metaphysical framework, see Chapter 4.

[11] For the explanation on the different types of miracles in Islam (*mu'jizah* and *karaamah*), see IslamQ&A (2012).

[12] For more details on the exegesis and explanation of this story, see Imam Ghazali Institute (n.d.)

miraculous status of these two anecdotes, I will limit myself to the two *ḥadīth* in my evidence appraisal exercise in this section and the next.

The first *ḥadīth* that specifies a range "between 60 and 70" can be deemed problematic as published data show that the global average life expectancy has surpassed 70 since 2011, currently stands at 73.3, and is expected to hit 77.0 by 2050 (Macrotrends n.d.). In other words, its trend has been predominantly upward for the last several decades and will continue to do so. However, I maintain that these statistics cannot negate the *ḥadīth* content for the following reasons. First, the terms *umr* and *ummah* used in the narration may be conceptualized differently than today's modern terminologies. For example, in its simplest form, a layperson would understand *umr* as the duration in which an individual survives, but since the *ḥadīth* refers to *ummah*—which is a collective term—the interpretation of *umr* now becomes slightly more technical and complicated. Determining the *umr* of a person is not the same as determining the *umr* of a population. The latter needs a more complex calculation, but first, the definition has to be clear and comparable to that of modern terms used in the relevant field. Does *umr* refer to the human lifespan or to average life expectancy? Or does it denote the modal age at death? Note that these demographic terminologies have a different meaning and may reflect a distinct concept. Additionally, they are calculated in different ways, and for specific purposes. While people tend to use lifespan and life expectancy interchangeably, these are not similar. Lifespan is the maximum length of time a species can be expected to survive (regardless of the number of members who achieved that age), while life expectancy refers to the average number of years a person can expect to live (Parker and Achaval 2017). The latter estimates the mean age of population members when they die. On the other hand, modal age at death means the age at which most deaths occur in a population (Canudas-Romo 2010).

What follows is my reasoning of why the *ḥadīth* content cannot be refuted based on the conceptualization of these demographic terminologies. Importantly, this list of longevity measures is not meant to be exhaustive; I select three commonly used modern metrics to demonstrate how they are not comparable to the term *umr*, and thus cannot be used to affirm or negate the Prophetic saying:

1. **Umr as Lifespan**
The term lifespan in modern demography means the maximum age that can be achieved by any member of a given species (Parker and Achaval 2017). Currently, the human lifespan stands at 122 (Robine et al. 2019). The age mentioned in the *ḥadīth* is a range and not a single value. Moreover, a maximum lifespan can be established even by one or a few members reaching an extreme age, while the *ḥadīth* clearly refers to a collective phenomenon.
2. **Umr as Average Life Expectancy**
Typically, the term average life expectancy refers to the average life expectancy at birth (LEB). As mentioned earlier, the global LEB has passed 70 since 2011, while the *ḥadīth* clearly gives an upper range of 70. A closer examination shows that comparing LEB to *umr* is problematic because LEB is a mean value[13] and it is highly influenced by infant and child mortality rates (Ortiz-Ospina 2017). Thus, LEB is less helpful if the goal is to understand longevity of older adults (Parker and Achaval 2017). In fact, compared to the other two longevity measures—median and modal ages at death—LEB is the only measure that "changes regardless of the age at which the reduction in mortality occurs" (Canudas-Romo 2010). Said another way, an LEB value does not accurately reflect longevity in old age given its calculation that depends on other factors, mainly the mortality rates in infancy.
3. **Umr as Modal Age at Death**
Given the drawback of LEB that employs a mean value, some demographers maintain that the modal age at death is a more useful measure to reflect longevity of older adults. While this might be true, the main obstacle is the lack of data on this longevity measure in many countries. Today, most countries maintain records and trends of LEB but not of modal age at death. In fact, nations that systematically document their modal age at death are few, mainly comprising high-income Western countries (Canudas-Romo 2010). Therefore, there is insufficient evidence to prove or disprove the *ḥadīth* based on the modal age of death.

In a similar vein, the word *ummah* is not uniformly defined. There are various opinions as to what *ummah* means, and to whom it refers. Even though the *ḥadīth* text is explicit about the *ummah* being attributed to Prophet Muhammad, the next question would be: Does this refer only to Muslims who believe in his message? Or does this include the global

[13] A major disadvantage of mean is that it is sensitive to extreme values/outliers, and it is not an appropriate measure of central tendency if the data distribution is skewed.

community from his time to this day? Unpacking the term *ummah* is germane if we were to make sense of the *ḥadīth* content in the light of modern knowledge. But without delving into the details of the linguistic, theological, or political debates on how *ummah* can be defined, I contend here that the term is most generally understood as being specific to the Muslim *ummah* or the community of believers (Al-Ahsan 1986). However, other scholars have underscored the inclusivity of this term,[14] which can also be extended to non-Muslims (Mohammed and Jureidini 2022). What I want to highlight is how the term *ummah*—regardless of how it is defined—cannot be used to affirm or negate the content of the second *ḥadīth*. Let's see the two scenarios below:

1. ***Ummah* as the Community of Believers (Muslims)**

   If we take this definition, the existing life expectancy statistics may not be applicable at all because most published data are global or country-based and do not differentiate people based on religious beliefs. There is almost no way to know the actual lifespan (here I am not referring to the maximum lifespan as defined in modern demography) of Muslims exclusively. Perhaps the closest is to look at the vital statistics of Muslim-majority nations; again, this is still not accurate because it is not entirely specific to Muslims. Interestingly, the Pew Research Center found that the LEB in Muslim-majority countries was approximately 68 between 2010 and 2015, and subsequently rose to 71 between 2020 and 2025 (Pew Research Center 2011), but it also reported wide variations between low-income and high-income Muslim nations (Razzak et al. 2011). As such, I maintain that the range between "60 and 70" in the *ḥadīth* text cannot be proven false because there is no evidence from contemporary demographic data to show this is true for Muslim populations.
2. ***Ummah* as the Global Community**

   If we regard *ummah* as the global community (not just Muslims), the next hurdle is to determine to what extent existing data/statistics reflect those in the past. The *ummah* of Prophet Muhammad is not restricted to contemporary modern society, rather it includes those who lived during his time and beyond his death. In short, the *ummah* began with

[14] Some scholars who claim that *ummah* includes non-Muslims often (historically) refer to the People of the Book (Christians and Jews), especially those who lived in peace with Muslims as a community. I acknowledge that details on how this term applies, or does not apply, to the modern nation-states is a more complex issue and unfortunately is beyond the scope of this chapter.

> the advent of Islam in 610 CE and will continue beyond our generation today.[15] Conversely, systematic civil registration and collection of vital statistics are a relatively modern affair. For instance, the first country reported to implement a national civil registry was France in 1539 (Scholarly Community Encyclopedia 2022). In Northern Europe, the parish register began in 1608 in Sweden, later followed by Finland and Denmark. But the organization of, and progress in, modern vital statistics only started in the nineteenth century (United Nations 1991). The USA developed its first standard for registration of vital events, including births and deaths, in 1900 (Statistics 2009). Thus, one cannot discredit the *ḥadīth* content because the lack of organized and systematic data collection before the modern civil registry limits our knowledge of how long the *ummah* in the past had lived.

Now I turn to the second *ḥadīth* which states that God provides a cure for every disease except for old age. If we accept the literal meaning of this narration, it would appear as if old age or the aging process—ignoring the difference in definition of these two terms for now—is not a condition that can be remedied. Hence, how does one make sense of the increasing scientific evidence today that demonstrates the *possibility* of aging being slowed down or delayed? True that the evidence is not conclusive, but does this *ḥadīth* mean that we should not strive toward prolonging lives or invest in anti-aging studies and technologies because the scripture clearly says there is "no cure to old age"?

Expounding on this *ḥadīth*, some scholars maintain that the word "*al haram*"—which is literally understood as old age, or the feebleness and frailty resulting from aging—was used metaphorically to refer to death.[16] In his famous book *Fatḥ al-Bārī*, Ibn Ḥajar al-ʿAsqalānī (d.1449) wrote that *al-haram* in this *ḥadīth* was symbolic of death and was used to signify death (*shabīhan bil maut*) (see Islamweb 2017). Likewise, al-Zarqānī (d.1710) in his book *Sharḥ al-Zarqānī ʿala Muwaṭṭaʾ al-Imām Mālik* likened the term *al-haram* to death; he highlighted another instance in

[15] Muslims believe that Prophet Muhammad is the final messenger of God and there is no prophet after him. This means his *ummah* continues until the end of time.

[16] This explanation was given in a *ḥadīth* website maintained by the Aldorar Alsaniyyah Foundation in Saudi Arabia, under the supervision of Shaykh Alawi bin Abdul Qadir Al-Saqqaf (see Hadith Encyclopedia n.d.).

which the Prophet narrated something similar but used a different term, "*al-Sām*," which was then defined as death (al-Zarqānī 1990; see also Islamweb 2017). Commenting on the association between *al-haram* (old age) and disease, al-Khaṭṭābī (d. 998 CE) claimed that the term *al-haram* implied weakness and decline as a result of old age; it does not signify disease in the typical sense; rather, *al-haram* was likened to a disease, because with aging comes a gradual and irreversible deterioration and damage to bodily functions—this condition is similar to a progressive disease that leads to death (see Islamweb 2017).

Is there any contradiction between this *ḥadīth* and scientific findings from the longevity and life extension research? My answer is no, and I provide the following reasons: First, if the term *al-haram* is symbolic of death, as opined by many classical Islamic scholars described earlier, the *ḥadīth* is precise; death is a certainty in Islam (Q.3:185), a universal phenomenon, and an indisputable scientific fact. To this day, there is no "cure" to death, despite all the sophisticated biomedical technologies we have acquired. Second, if *al-haram* denotes a state of irreversible weakness and frailty due to old age (as indicated by al-Khaṭṭābī), the *ḥadīth* is still accurate. This is because anti-aging measures today are predominantly focused on prevention; their aim is to preserve health and delay age-related diseases and symptoms instead of treating illnesses after they start (Tenchov et al. 2024). In other words, anti-aging therapies are meant for those in relatively good health to prolong their vitality and youthfulness. They are less applicable to older adults who are already inflicted with age-related diseases and symptoms of decline—a state similar to *al-haram*.

Bringing It Altogether—Is Radical Life Extension Possible Based on Empirical and Revelatory Evidence?

Herein I propose that the way to navigate this subject matter is by first specifying where and at which level(s) the convergence/divergence might be. Discord can occur at the level of empirical possibility, or epistemological or philosophical commitments. If we assume the conflict to reside in the realm of empiricism, the corollary question would be: Is RLE possible based on empirical and revelatory evidence? To answer this, I will now revisit the epistemological framework delineated in Chapter 2. As Omar Qureshi puts it:

> There are four possible ways rational and transmitted proofs can conflict: (1) both proofs are conclusively certain (*qaṭʿī*); (2) both are presumptive (*ẓannī*); (3) the transmitted proof is conclusively certain, while the rational proof is presumptively known; and (4) the rational proof is conclusively certain, while the transmitted proof is presumptively known.

I regard the two *ḥadīth* as transmitted proof (revelatory evidence), and the scientific findings that corroborate the effectiveness of anti-aging/ pro-longevity measures as rational proof. By rational proof, I mean empirical evidence. But before bringing together the propositions from these two knowledge sources, I will first attempt to appraise the two *ḥadīth* based on their mode of transmission (*thubūt*) and their mode of signification (*dalāla*) in order to determine their epistemic strength:[17]

1. **Ḥadīth on the *umr* of Prophet Muhammad's *Ummah***
   I categorize this *ḥadīth* as transmitted proof that has been conveyed with presumptive authenticity and whose meaning is presumptively authoritative (*ẓannī al-thubūt wa ẓannī al-dalāla*). This is because the *ḥadīth* has been rated as *ḥasan* and does not reach the level of *mutawātir*. Additionally, its meaning—especially of the terms *umr* and *ummah*—can be classified as non-definitive. In other words, those terms do not hold a single, plain meaning.
2. **Ḥadsīth on the Availability of Cure for Every Disease Except *al-haram***
   I categorize this *ḥadīth* as transmitted proof that has been conveyed with presumptive authenticity and whose meaning is conclusively certain (*ẓannī al-thubūt wa qaṭʿī al- dalāla*). This is because the *ḥadīth* has been rated as *ṣaḥīḥ* but does not reach the level of *mutawātir*. Nonetheless, its meaning can be considered definitive as the term *al-haram* was described almost uniformly by all the cited scholars, with little or no disagreement.

Harkening back to the epistemological framework that elucidates four possible ways rational (in this case, empirical) and transmitted proofs can be in conflict, and the framework proposed by Ashraf ʿAlī al-Tahānwī in his book *Al-Intibāhāt al-Mufīda fī Ḥall al-Ishtibāhāt al-Jadīda*, I propose that this apparent conflict resides in the second scenario, where both types of proofs are presumptive (*ẓannī*). According to Omar

[17] For details on grading the level of epistemic strength of scriptural/transmitted proofs, refer to Chapter 2.

Qureshi, al-Tahānwī states that when this type of scenario is encountered, "the transmitted proof is given precedence over rational [empirical] proofs" (for more explanation, see Chapter 2).

However, for this specific case example (RLE), I argue that there is no need to prioritize one proof over the other and I offer the following explanation: First, it is clear from my prior elaboration that the two *ḥadīth* do not fundamentally contradict contemporary scientific findings on life extension and longevity. While the Qur'anic accounts confirm the occurrence of extreme longevity, the *ḥadīth* neither support nor deny the possibility of humans living very long lives. At the same time, the *ḥadīth* of the *umr* of the *ummah* cannot be refuted by modern demographic data, and it also does not suggest the futility of pro-longevity or anti-aging endeavors. In fact, the "few that surpass" could refer to modern society or the future generations. Second, the *ḥadīth* on the lack of cure for "old age" does not conflict with scientific findings, as the scholars largely indicate that "old age" (*al-haram*) in the text was symbolic of death, or a state of decline that leads to death. As such, there is no dispute about the certainty of mortality. Therefore, given that the best epistemic strength attributed to both proofs are presumptive (*ẓannī*), and that fundamental conflicts do not exist (a conflict seems apparent at first, but is easily resolved), I conclude that radically extending the human life is a possibility in the future. And since there are no essential contradictions between transmitted and empirical evidence in this example, it is not necessary to bring up or explore the issue of grading presumptive evidence along a spectrum of epistemic strength, as opposed to treating it as a single and homogenous entity.

Other Potential Conflicts and Where They Reside

The discussion thus far assumes that the "conflict" occurs at the level of empirical possibility. But it is clear now that both the scientific and scriptural proofs do not fully reject or confirm that RLE can/will happen in the future. Combining the texts from Qur'anic narrations and the two *ḥadīth* shows that humans have indeed (in the past) lived extremely long lives, and it is not something impossible in the future. This likelihood is further strengthened by the ongoing research on anti-aging and longevity whose evidence is still inconclusive, but is rapidly growing.

So, is there a conflict at all? I would hazard the affirmative, but the conflict is at a different level. Decoding the details of these conflicts is beyond this chapter's scope, but I will describe them here briefly. First, the RLE movement is part of the transhumanist ideology. In Chapter 7, Aasim Padela has delineated the transhumanist philosophical underpinnings, goals, motivations, and its vision of humankind. Islamic scholars and jurists have voiced their concerns about this ideology and its potential consequences, centering their argument on the transhumanists' rejection of God and the afterlife. As such, transhumanists' predominant focus is on maximizing individual pleasure in the present reality and delaying/avoiding death.[18] This is in contrast with an Islamic moral vision that emphasizes the hereafter and considers our present reality as a temporary life for merit-making. This fundamental dissonance affects how transhumanism and Islam view life, longevity, and death. While both value human life and support health preservation and life extension, they have very disparate reasons and underlying motivations for doing so. Additionally, given Islam's greater emphasis on one's spiritual dimension, connection to the Divine, and the afterlife, Muslim jurists and scholars may not show an equal amount of enthusiasm for extreme longevity as demonstrated by RLE advocates.

Second, even though analysis of revelatory evidence implies that RLE is a possibility, the question that follows is: Does Islamic law permit radically extending the human life, if there is no clearly justified medical necessity? As elucidated in Chapter 7, Muslim jurists deploy their judgment mainly on the basis of differentiation between treatment and enhancement, with the former being permissible and the latter prohibited. RLE is a form of life enhancement as the aim is to push the human lifespan beyond its current/normal threshold. One can thus easily comprehend why RLE might not be allowed by Muslim jurists: it does not fulfill the criteria of a regular treatment (e.g., not restorative), it does not address any medical necessity, and it does not target a specific disease. Nevertheless, some legitimate questions could perhaps challenge this ruling in the future. For example, what if the definition of treatment changes, and many biomedical interventions which are now deemed enhancing are later regarded as therapeutic? What if diseases are conceptualized differently in the future? Given the consistent upward trend in

[18] For a more detailed explanation of, and criticism against, the transhumanist ideology, see Chapter 7.

human life expectancy, could it be that what is considered "a normal life span" now becomes a treatable abnormality in the future? Who gets to decide what is normal?

The list of questions could go on, but my goal here is to highlight a potential conflict between RLE (biomedical) practices and Islamic jurisprudence and bioethics, thus calling for a more nuanced bioethical evaluation[19] to resolve this tension.

Conclusion

The discussion on RLE in this chapter does not include immortality because death is a certainty in Islam (Q.29:57), a fact, and a universal truth (Koestenbaum 1972). Islamic theology views aging as a natural life process that is often, though not necessarily, accompanied by physical and mental decline. Aging is regarded as an opportunity and means to achieve higher spiritual ranks before God, not a flaw to be "corrected" or "transcended." Islamic teachings as derived from the scriptures neither praise nor condemn life extension, but tie longevity to good action and piety.[20] An appraisal of scientific evidence that supports the prospect of life-extension and of revelatory evidence reveals no fundamental conflicts, with the best epistemic strength assigned to both proofs as *ẓannī*. Therefore, RLE is a possibility for the human race in the future. At the level of philosophical/theological commitment, conflicts emerge between the RLE (transhumanist) ideology and the Islamic doctrine, and between RLE biomedical practices with Islamic law/jurisprudence.

So finally, *is humankind changing*? If we are, *should we be?* My (limited) reading and analysis thus far supports the notion that we are, in fact, changing. What is unclear is to what extent—I speculate that the changes are rather limited and restricted within certain boundaries. Some physical (and even mental) features of humankind may undergo transformations in the future but they are not necessarily or solely induced by biomedical

[19] On the proposed Islamic bioethical evaluation involving three types of analysis, see Chapter 7.

[20] Abū Hurayra reported that the Prophet, peace be upon him, said: "Shall I tell you who is the best among you?" The Companions said, "Yes, O Allah's Messenger." Thereupon he said, "The best among you are those who live long and are best in deeds." (Narrated by Ahmad and others with a sound chain; Fiqh-us-Sunnah n.d.).

interventions; rather, they can be due to a combination of factors such as public health measures and environmental changes. Nonetheless, these changes are unlikely to transform humans into a different species with extraordinary capabilities, as depicted in science fiction or imagined in the evolutionary account. As for the *should we* question, I prefer to suspend my judgment at this point; this is because more discourse and clarity is needed to unpack the whole range of potential consequences of artificially induced changes to the body. On top of that, there are other implications of RLE on ethics and social justice that I have not discussed in this chapter, which include equity of access, resource distribution, intergenerational justice, and environmental degradation. Indeed, RLE cannot be analyzed nor understood in isolation, without considering the broader questions of justice, fairness, and universal morality.

References

Al-Ahsan, Abdullah. 1986. "The Quranic Concept of Ummah." *Institute of Muslim Minority Affairs. Journal* 7 (2): 606–616. https://doi.org/10.1080/13602008608716004.

Bacakova, Lucie, Jana Zarubova, Martina Travnickova, Jana Musilkova, Julia Pajorova, Petr Slepicka, Nikola Slepickova Kasalkova, Vaclav Svorcik, Zdenka Kolska, and Hooman Motarjemi. 2018. "Stem Cells: Their Source, Potency and Use in Regenerative Therapies with Focus on Adipose-Derived Stem Cells—A Review." *Biotechnology Advances* 36 (4): 1111–1126. https://doi.org/10.1016/j.biotechadv.2018.03.011.

Balasubramanian, Priya, Porsha R. Howell, and Rozalyn M. Anderson. 2017. "Aging and Caloric Restriction Research: A Biological Perspective with Translational Potential." *EBioMedicine* 21: 37–44. https://doi.org/10.1016/j.ebiom.2017.06.015.

Bernardes de Jesus, Bruno, Elsa Vera, Kerstin Schneeberger, Agueda M. Tejera, Eduard Ayuso, Fatima Bosch, and Maria A. Blasco. 2012. "Telomerase Gene Therapy in Adult and Old Mice Delays Aging and Increases Longevity without Increasing Cancer." *EMBO Molecular Medicine* 4 (8): 691–704. https://doi.org/10.1002/emmm.201200245.

Biehl, Jesse K., and Brenda Russell. 2009. "Introduction to Stem Cell Therapy." *Journal of Cardiovascular Nursing* 24 (2): 98–103. https://doi.org/10.1097/JCN.0b013e318197a6a5.

Blomstrom, Mackenzie, Andrew Burns, Daniel Larriviere, and Jennifer Kim Penberthy. 2022. "Addressing Fear of Death and Dying: Traditional and Innovative Interventions." *Mortality* 27 (1): 18–37. https://doi.org/10.1080/13576275.2020.1810649.

Bradshaw, Debra Kim. 2009. "Radical Life Extension, a Consideration of Some Arguments: Disastrous Enterprise, or Boon to Humanity?" MA thesis, University of Manitoba, Canada.

Campbell, Lucinda April. 2020. "Motivations for Pursuing Radical Life Extension." MA thesis, Victoria University of Wellington, New Zealand.

Canudas-Romo, Vladimir. 2010. "Three Measures of Longevity: Time Trends and Record Values." *Demography* 47 (2): 299–312. https://doi.org/10.1353/dem.0.0098.

Chang, Le, Weiwen Fan, Xinghua Pan, and Xiangqing Zhu. 2022. "Stem Cells to Reverse Aging." *Chinese Medical Journal* 135 (8): 901–910. https://doi.org/10.1097/CM9.0000000000001984.

Childs, Bennett G., Martina Gluscevic, Darren J. Baker, Remi-Martin Laberge, Dan Marquess, Jamie Dananberg, and Jan M. Van Deursen. 2017. "Senescent Cells: An Emerging Target for Diseases of Ageing." *Nature Reviews Drug Discovery* 16 (10): 718–735. https://doi.org/10.1038/nrd.2017.116.

Cho, Seung-Yeon, Han-Gyul Lee, Seungwon Kwon, Seong-Uk Park, Woo-Sang Jung, Sang-Kwan Moon, Jung-Mi Park, and Chang-Nam Ko. 2023. "A Systematic Review of In Vivo Studies of the Efficacy of Herbal Medicines for Anti-Aging in the Last Five Years." *Pharmaceuticals* 16 (3): 448. https://doi.org/10.3390/ph16030448.

David, Rhiannon M., and Ann T. Doherty. 2017. "Viral Vectors: The Road to Reducing Genotoxicity." *Toxicological Sciences* 155 (2): 315–325. https://doi.org/10.1093/toxsci/kfw220.

Davoudpour, Shahin, and John K Davis. 2022. "Life Extension and Overpopulation: Demography, Morals, and the Malthusian Objection." *HEC Forum*. https://doi.org/10.1007/s10730-022-09504-9.

Dirks, Amie J., and Christiaan Leeuwenburgh. 2006. "Caloric Restriction in Humans: Potential Pitfalls and Health Concerns." *Mechanisms of Ageing and Development* 127 (1): 1–7. https://doi.org/10.1016/j.mad.2005.09.001.

Dittmer, Dirk P., Aadra P. Bhatt, and Blossom Damania. 2012. "Rapalogs in Viral Cancers." *Expert Opinion on Investigational Drugs* 21 (2): 135–138. https://doi.org/10.1517/13543784.2012.642369.

Doorani, Sana. 2023. "Mind-Uploading: A Psychoanalytic Analysis of Transcendence and Avatar and the Appeal of Artificial Immortality." Barrett Honors College thesis, Arizona State University.

Dorling, James L., Stephan Van Vliet, Kim M. Huffman, William E. Kraus, Manjushri Bhapkar, Carl F. Pieper, Tiffany Stewart, Sai Krupa Das, Susan B. Racette, and Susan B. Roberts. 2021. "Effects of Caloric Restriction on Human Physiological, Psychological, and Behavioral Outcomes: Highlights from CALERIE Phase 2." *Nutrition Reviews* 79 (1): 98–113. https://doi.org/10.1093/nutrit/nuaa085.

Fiqh-us-Sunnah. n.d. Volume 4. https://www.iium.edu.my/deed/lawbase/fiqh_us_sunnah/vol4/fsn_vol4a.html.

Flanagan, Emily W., Jasper Most, Jacob T. Mey, and Leanne M. Redman. 2020. "Calorie Restriction and Aging in Humans." *Annual Review of Nutrition* 40: 105–133. https://doi.org/10.1146/annurev-nutr-122319-034601.

Fontana, Luigi, Linda Partridge, and Valter D. Longo. 2010. "Extending Healthy Life Span—From Yeast to Humans." *Science* 328 (5976): 321–326. https://doi.org/10.1126/science.1172539.

Fudge, Bruce. 2007. "The Men of the Cave: Tafsīr, Tragedy and Tawfīq al-Ḥakīm." *Arabica* 54: (1): 67–93. https://doi.org/10.1163/157005807779575007.

Fukuda, Haruhisa, Chieko Ishiguro, Rei Ono, and Kosuke Kiyohara. 2023. "The Longevity Improvement & Fair Evidence (LIFE) Study: Overview of the Study Design and Baseline Participant Profile." *Journal of Epidemiology* 33 (8): 428–437. https://doi.org/10.2188/jea.JE20210513.

Gaitán, Leandro. 2019. "Heaven on Earth: The Mind Uploading Project as Secular Eschatology." *Theology and Science* 17 (3): 403–416. https://doi.org/10.1080/14746700.2019.1632554.

Garay, Ricardo P. 2021. "Investigational Drugs and Nutrients for Human Longevity. Recent Clinical Trials Registered in ClinicalTrials.gov and clinicaltrialsregister.eu." *Expert Opinion on Investigational Drugs* 30 (7): 749–758. https://doi.org/10.1080/13543784.2021.1939306.

Ghazizadeh, Mohammad. 2018. "Safety and Efficacy of Human Adipose-Derived Mesenchymal Stem Cell Conditioned Media." 国際抗老化再生医療学会雜誌 [*The Journal of World Academy of Anti-Aging & Regenerative Medicine (WAARM) Society*] 国際抗老化再生医療学会 編 1: 21–24.

Golpanian, Samuel, Darcy L. DiFede, Aisha Khan, Ivonne Hernandez Schulman, Ana Marie Landin, Bryon A. Tompkins, Alan W. Heldman, Roberto Miki, Bradley J. Goldstein, and Muzammil Mushtaq. 2017. "Allogeneic Human Mesenchymal Stem Cell Infusions for Aging Frailty." *Journals of Gerontology Series A: Biomedical Sciences and Medical Sciences* 72 (11): 1505–1512. https://doi.org/10.1093/gerona/glx056.

Green, Cara L., Dudley W. Lamming, and Luigi Fontana. 2022. "Molecular Mechanisms of Dietary Restriction Promoting Health and Longevity." *Nature Reviews Molecular Cell Biology* 23 (1): 56–73. https://doi.org/10.1038/s41580-021-00411-4.

Hacein-Bey-Abina, Salima, Julia Hauer, Annick Lim, Capucine Picard, Gary P. Wang, Charles C. Berry, Chantal Martinache, Frédéric Rieux-Laucat, Sylvain Latour, and Bernd H. Belohradsky. 2010. "Efficacy of Gene Therapy for X-Linked Severe Combined Immunodeficiency." *New England Journal of Medicine* 363 (4): 355–364. https://doi.org/10.1056/NEJMoa1000164.

Al-Haddad, Imam 'Abdallah Ibn Alawi. 1991. *The Lives of Man: A Guide to the Human States: Before Life, in the World, and after Death*. Louisville, KY: Fons Vitae.

Hadith Encyclopedia. n.d. https://dorar.net/hadith/sharh/67840.

Heilbronn, Leonie K., Lilian de Jonge, Madlyn I. Frisard, James P. DeLany, D. Enette Larson-Meyer, Jennifer Rood, Tuong Nguyen, Corby K. Martin, Julia Volaufova, and Marlene M. Most. 2006. "Effect of 6-Month Calorie Restriction on Biomarkers of Longevity, Metabolic Adaptation, and Oxidative Stress in Overweight Individuals: A Randomized Controlled Trial." *Jama* 295 (13): 1539–1548. https://doi.org/10.1001/jama.295.13.1539.

Imam Ghazali Institute. "Tafsīr of Surah al-Kahf, Verses 11–26: The Story of the Sleepers Of The Cave." https://www.imamghazali.org/blog/tafsir-surah-kahf-11-26-sleepers-cave.

Ingram, Donald K., Min Zhu, Jacek Mamczarz, Sige Zou, Mark A. Lane, George S. Roth, and Rafael DeCabo. 2006. "Calorie Restriction Mimetics: An Emerging Research Field." *Aging Cell* 5 (2): 97–108. https://doi.org/10.1111/j.1474-9726.2006.00202.x.

IslamQ&A. 2012. "The Difference between *mu'jizah*, *karaamah* (Two Types of miracles) and Witchcraft." *Islam Question and Answer*. https://islamqa.info/en/answers/124838/the-difference-between-mujizah-karaamah-two-types-of-miracles-and-witchcraft.

IslamQ&A. 2013. "How Long Did Noah (Peace be Upon Him) Live?". Islam Question and Answer. Islam Question and Answer. https://islamqa.info/en/answers/105695/how-long-did-noah-peace-be-upon-him-live. Accessed 24 March 2024.

Islamweb 2017. "التوفيق بين حديث: ما أنزل الله داء إلا أنزل له دواء.. وحديث: غير داء واحد الهرم" ["Reconciling the *ḥadīth*: 'Allah has not sent down a disease without sending down a cure for it' and the *ḥadīth*: 'Other than one disease, old age'"]. June 21. https://www.islamweb.net/ar/fatwa/355309.

Kassem, Moustapha. 2006. "Stem Cells: Potential Therapy for Age-Related Diseases." *Annals of the New York Academy of Sciences* 1067 (1): 436–442.

Kirkland, J.L., and T. Tchkonia. 2020. "Senolytic Drugs: From Discovery to Translation." *Journal of Internal Medicine* 288 (5): 518–536. https://doi.org/10.1111/joim.13141.

Koestenbaum, Peter. 1972. "The Vitality of Death." *Omega: Journal of Death and Dying* 2 (4): 253–271. https://doi.org/10.2190/BXMU-81HF-FVLL-0MGQ.

Li, Jinyang, Salina Yuan, Robert J. Norgard, Fangxue Yan, Yu H. Sun, Il-Kyu Kim, Allyson J. Merrell, Yogev Sela, Yanqing Jiang, and Natarajan V. Bhanu. 2021. "Epigenetic and Transcriptional Control of the Epidermal Growth Factor Receptor Regulates the Tumor Immune Microenvironment in Pancreatic Cancer." *Cancer Discovery* 11 (3): 736–753. https://doi.org/10.1158/2159-8290.CD-20-0519.

Li, Kailong, Yuxuan Liu, Hui Cao, Yuannyu Zhang, Zhimin Gu, Xin Liu, Andy Yu, Pranita Kaphle, Kathryn E Dickerson, and Min Ni. 2020. "Interrogation of Enhancer Function by Enhancer-Targeting CRISPR Epigenetic Editing." *Nature Communications* 11 (1): 485. https://doi.org/10.1038/s41467-020-14362-5.

Liang, Yaru, Chang Liu, Maoyang Lu, Qiongye Dong, Zimu Wang, Zhuoran Wang, Wenxiang Xiong, Nannan Zhang, Jiawei Zhou, and Qingfei Liu. 2018. "Calorie Restriction is the Most Reasonable Anti-Ageing Intervention: A Meta-Analysis of Survival Curves." *Scientific Reports* 8 (1): 5779. https://doi.org/10.1038/s41598-018-24146-z.

Lindvall, Olle, Roger A. Barker, Oliver Brüstle, Ole Isacson, and Clive N. Svendsen. 2012. "Clinical Translation of Stem Cells in Neurodegenerative Disorders." *Cell Stem Cell* 10 (2): 151–155. https://doi.org/10.1016/j.stem.2012.01.009.

Lombard, Jessica. 2023. "Ageing and Death: A Focus on How to Transcend Diseases for Transhumanist Movements." *FORUM: University of Edinburgh Postgraduate Journal of Culture & The Arts* 34 (1). https://doi.org/10.2218/forum.1.9154.

Lorrimar, Victoria. 2018. "Human Flourishing, Joy, and the Prospect of Radical Life Extension." *The Expository Times* 129 (12): 554–561. https://doi.org/10.1177/0014524618767222.

Macrotrends. (n.d.). "World Life Expectancy 1950–2024." *Macrotrends*. https://www.macrotrends.net/global-metrics/countries/WLD/world/life-expectancy. Accessed 24 March 2024.

Magalhães, João Pedro de. 2012. "Programmatic Features of Aging Originating in Development: Aging Mechanisms beyond Molecular Damage?" *The FASEB Journal* 26 (12): 4821. https://doi.org/10.1096/fj.12-210872.

Masci, David. 2013. "To Count Our Days: The Scientific and Ethical Dimensions of Radical Life Extension." Pew Research Center, August 6. https://www.pewresearch.org/religion/2013/08/06/to-count-our-days-the-scientific-and-ethical-dimensions-of-radical-life-extension/.

Mazini, Loubna, Luc Rochette, Brahim Admou, Said Amal, and Gabriel Malka. 2020. "Hopes and Limits of Adipose-Derived Stem Cells (ADSCs) and Mesenchymal Stem Cells (MSCs) in Wound Healing." *International Journal of Molecular Sciences* 21 (4): 1306. https://doi.org/10.3390/ijms21041306.

Mohammed, Hossameldeen, and Ray Jureidini. 2022. "Umma and the Nation-State: Dilemmas in Refuge Ethics." *Journal of International Humanitarian Action* 7 (1): 17. https://doi.org/10.1186/s41018-022-00124-z/.

Most, Jasper, Valeria Tosti, Leanne M. Redman, and Luigi Fontana. 2017. "Calorie Restriction in Humans: An Update." *Ageing Research Reviews* 39: 36–45. https://doi.org/10.1016/j.arr.2016.08.005.

Musa, Aisha Y. 2009. “A Thousand Years, Less Fifty: Toward a Quranic View of Extreme Longevity.” In *Religion and the Implications of Radical Life Extension*, edited by Derek F. Maher and Calvin Mercer, 123–131. New York: Palgrave Macmillan.

Nguyen, Hung, Sydney Zarriello, Alexandreya Coats, Cannon Nelson, Chase Kingsbury, Anna Gorsky, Mira Rajani, Elliot G. Neal, and Cesar V. Borlongan. 2019. “Stem Cell Therapy for Neurological Disorders: A Focus on Aging.” *Neurobiology of Disease* 126: 85–104. https://doi.org/10.1016/j.nbd.2018.09.011.

Niclis, Jonathan C., Christopher Turner, Jennifer Durnall, Stuart McDougal, Jessica A. Kauhausen, Bryan Leaw, Mirella Dottori, Clare L. Parish, and Lachlan H. Thompson. 2017. “Long-Distance Axonal Growth and Protracted Functional Maturation of Neurons Derived from Human Induced Pluripotent Stem Cells after Intracerebral Transplantation.” *Stem Cells Translational Medicine* 6 (6): 1547–1556. https://doi.org/10.1002/sctm.16-0198.

NIH. 2022. “NIH Clinical Research Trials and You: The Basics.” National Institutes of Health (NIH). https://www.nih.gov/health-information/nih-clinical-research-trials-you/basics. Accessed 19 March 2024.

Ortiz-Ospina, Esteban. 2017. “‘Life Expectancy’—What Does This Actually Mean?” *Our World in Data*. https://ourworldindata.org/life-expectancy-how-is-it-calculated-and-how-should-it-be-interpreted. Accessed 4 July 2024.

Pandey, Prasoon, and Neelam Balekar. 2018. “Target-Specific Delivery: An Insight.” In *Drug Targeting and Stimuli Sensitive Drug Delivery Systems*, edited by Alexandru Mihai Grumezescu, 117–154. Kidlington: William Andrew, Elsevier.

Papadopoli, David, Karine Boulay, Lawrence Kazak, Michael Pollak, Frédérick A. Mallette, Ivan Topisirovic, and Laura Hulea. 2019. “mTOR as a Central Regulator of Lifespan and Aging.” *F1000Research* 8. https://doi.org/10.12688/f1000research.17196.1.

Parker, Robert E., and Vianett Garcia Achaval. 2017. “The Myth of Increasing Life Expectancy and Its Social Policy Implications.” *Journal of Sociological Research* 8 (2): 17–31. https://doi.org/10.5296/jsr.v8i2.11138.

Petr, Michael, Frida Matiyevskaya, Brenna Osborne, Magnus Berglind, Simon Reves, Bin Zhang, Michael Ben Ezra, Lina Maria Carmona Marin, Muhammad Farraz Syahzha, and Marta Cortés Mediavilla. 2024. “Pharmacological Interventions in Human Aging.” *Ageing Research Reviews* 95: 102213. https://doi.org/10.1016/j.arr.2024.102213.

Pew Research Center. 2011. “The Future of The Global Muslim Population: Main Factors Driving Population Growth.” Pew Research Center. https://www.pewresearch.org/religion/2011/01/27/future-of-the-global-muslim-population-main-factors/#ftn9_rtn.

Rafi, Mohammad A. 2011. "Gene and Stem Cell Therapy: Alone or in Combination?" *BioImpacts: BI* 1 (4): 213–218. https://doi.org/10.5681/bi.2011.030.

Raper, Steven E., Narendra Chirmule, Frank S. Lee, Nelson A. Wivel, Adam Bagg, Guang-ping Gao, James M. Wilson, and Mark L. Batshaw. 2003. "Fatal Systemic Inflammatory Response Syndrome in a Ornithine Transcarbamylase Deficient Patient following Adenoviral Gene Transfer." *Molecular Genetics and Metabolism* 80 (1–2): 148–158. https://doi.org/10.1016/j.ymgme.2003.08.016.

Rattan, Suresh I. 1998. "Is Gene Therapy for Aging Possible?" *Indian Journal of Experimental Biology* 36 (3): 233–236.

Razzak, Junaid A., Uzma Rahim Khan, Iqbal Azam, Muazzam Nasrullah, Omrana Pasha, M. Malik, and A. Ghaffar. 2011. "Health Disparities between Muslim and Non-Muslim Countries." *EMHJ: Eastern Mediterranean Health Journal* 17 (9): 654–664. https://doi.org/10.26719/2011.17.9.654.

Robine, Jean-Marie, Michel Allard, François R. Herrmann, and Bernard Jeune. 2019. "The Real Facts Supporting Jeanne Calment as the Oldest Ever Human." *The Journals of Gerontology: Series A* 74 (Supplement_1): S13–S20. https://doi.org/10.1093/gerona/glz198.

Scanlon, Jennamarie. 2020. "Gene Editing in Healthcare as it Relates to Oncological Practices." Honors College Thesis. https://digitalcommons.pace.edu/honorscollege_theses/340/.

Scholarly Community Encyclopedia. 2022. "Civil Registration." *Scholarly Community Encyclopedia*. https://encyclopedia.pub/entry/33277.

Schuler, Emily. 2021. "ICD-11 and an Argument about 'Old Age.'" *The Oxford Institute of Population Ageing*. June 30. https://www.ageing.ox.ac.uk/blog/ICD-11-and-an-argument-about-old%20age%20.

Sedick, Shaykh Irshaad. 2023. "Did People Live for Hundreds of Years?" *Seekers Guidance*: The Global Islamic Academy, March 3. https://seekersguidance.org/answers/islamic-belief/did-people-live-for-hundreds-of-years/.

Shamshirgaran, Yasaman, Jun Liu, Huseyin Sumer, Paul J. Verma, and Amir Taheri-Ghahfarokhi. 2022. "Tools for Efficient Genome Editing; ZFN, TALEN, and CRISPR." *Applications of Genome Modulation and Editing*: 29–46. https://doi.org/10.1007/978-1-0716-2301-5_2

Shetty, Ashok K., Maheedhar Kodali, Raghavendra Upadhya, and Leelavathi N. Madhu. 2018. "Emerging Anti-Aging Strategies: Scientific Basis and Efficacy." *Aging and Disease* 9 (6): 1165–1184. https://doi.org/10.14336/AD.2018.1026.

Solís-Paredes, Mario, Pilar Eguía-Aguilar, Fernando Chico-Ponce de León, Stanislaw Sadowinski-Pine, Mario Pérezpeña-Diazconti, and Francisco Arenas-Huertero. 2014. "Epigenetic Modifications in Cell Lines of Human Astrocytoma Differentially Regulate Expression of Apoptotic Genes."

Child's Nervous System 30 (1): 123–129. https://doi.org/10.1007/s00381-013-2258-6.

Somerville, Margaret A. 2009. *The Ethical Imagination: Journeys of the Human Spirit.* Montreal: McGill-Queen's Press.

Statistics, National Research Council Committee on National. 2009. "The US Vital Statistics System: A National Perspective." Vital Statistics: Summary of a Workshop.

Sulaiman, Mohd Ebrahim. 2020. "Seven Stages of Life in Islam." *IslamiCity.* https://www.islamicity.org/7887/seven-stages-of-life-in-islam/.

Suzuki, Satoru, Hideyuki Yamatoya, Masashi Sakai, Akito Kataoka, Masayoshi Furushiro, and Satoshi Kudo. 2001. "Oral Administration of Soybean Lecithin Transphosphatidylated Phosphatidylserine Improves Memory Impairment in Aged Rats." *The Journal of Nutrition* 131 (11): 2951–2956. https://doi.org/10.1093/jn/131.11.2951.

Tacutu, Robi, Thomas Craig, Arie Budovsky, Daniel Wuttke, Gilad Lehmann, Dmitri Taranukha, Joana Costa, Vadim E Fraifeld, and João Pedro de Magalhães. 2012. "Human Ageing Genomic Resources: Integrated Databases and Tools for the Biology and Genetics of Ageing." *Nucleic Acids Research* 41 (D1): D1027–1033. https://doi.org/10.1093/nar/gks1155.

Al-Tahānwī, Ashraf ʿAlī. 1996. *al-Intibāhāt al-Mufīda fī Ḥall al-Ishtibāhāt al-Jadīda.* Translated by Nūr al-Bashar Muḥammad Nūr al-Ḥaqq. Karachi: Maktaba Dār al-ʿUlūm al-Karatshī.

Tenchov, Rumiana, Janet M Sasso, Xinmei Wang, and Qiongqiong Angela Zhou. 2024. "Antiaging Strategies and Remedies: A Landscape of Research Progress and Promise." *ACS Chemical Neuroscience* 15 (3): 408–446. https://doi.org/10.1021/acschemneuro.3c00532.

Turner, Bryan S. 2009. *Can We Live Forever? A Sociological and Moral Inquiry.* London: Anthem Press.

United Nations. 1991. *Handbook of Vital Statistics Systems and Methods.* New York: United Nations.

Verdaguer, Ester, Felix Junyent, Jaume Folch, Carlos Beas-Zarate, Carme Auladell, Mercè Pallàs, and Antoni Camins. 2012. "Aging Biology: A New Frontier for Drug Discovery." *Expert Opinion on Drug Discovery* 7 (3): 217–229. https://doi.org/10.1517/17460441.2012.660144.

Vijg, Jan, and Aubrey D.N.J. de Grey. 2014. "Innovating Aging: Promises and Pitfalls on the Road to Life Extension." *Gerontology* 60 (4): 373–380. https://doi.org/10.1159/000357670.

Wang, Lixin, Xu Zuo, Zhuoer Ouyang, Ping Qiao, and Fang Wang. 2021. "A Systematic Review of Antiaging Effects of 23 Traditional Chinese Medicines." *Evidence-Based Complementary and Alternative Medicine* 2021: 1–13. https://doi.org/10.1155/2021/5591573.

Warby, Paul. 2021. "A Critical Enquiry into the Metaphysics for Mind Uploading." University of Pretoria, South Africa.

Yu, Jingyu, Tianwen Li, and Jianhong Zhu. 2023. "Gene Therapy Strategies Targeting Aging-Related Diseases." *Aging and Disease* 14 (2): 398–417. https://doi.org/10.14336/AD.2022.00725.

Al-Zarqānī, Imām Sayyidī Muḥammad. 1990. *Sharḥ al-Zarqānī ʿala Muwaṭṭaʾ al-Imām Mālik*. Beirut: Dār al-Fikr.

PART III

Practical Perspectives from the Field

CHAPTER 9

The Importance of Teaching Muslim Students About Scientific and Religious Constructions of the Human Being

Usama Javed Mirza

This chapter begins by providing an overview of some of the challenges and limitations in contemporary science education, as it relates to giving students a holistic impression of what human beings are. It outlines how some prevalent curriculum design features can contribute to students implicitly being conveyed a biomedical reductionist "model" of human beings. This is to say, they are given the uncritical impression that human beings can be conceptualized entirely as being made of, and hence explained mechanistically in terms of, atomic and sub-atomic particles and their mutual interaction. With this come associated scientific paradigms about the origins of human beings as having evolved from other life forms on Earth, and the related implications such as positivistic

U. J. Mirza (✉)
Faculty of Education, University of Cambridge, Cambridge, UK
e-mail: ujm20@cam.ac.uk

© The Author(s), under exclusive license to Springer Nature Switzerland AG 2026

A. I. Padela (ed.), *Bridging Bioscience and Islam*, Islam & Science,
https://doi.org/10.1007/978-3-032-09996-9_9

views about whether humans have "free will" and reducing human experiences of and beliefs in the sacred as being evolutionary mechanisms to create the impetus for the will to live and reproduce.

The chapter then suggests why this may be incompatible—and hence create cognitive dissonance—with Muslim students studying biomedical disciplines, and who within their Islamic worldviews conceptualize human beings as being created by Allah and as having souls, and the meaning of life having a divine purpose with strong bearing on their afterlives. After building this case, the chapter provides an overview of various contemporary initiatives in Muslim contexts that aim to recenter Islamic worldviews within the study of science, and suggests what further work is needed to build on this momentum.

Some Current Challenges and Limitations of Modern Science Education

Ayaan sits in his science class, listening intently as his teacher explains how the human brain controls thoughts, emotions, and decision-making. The lesson is fascinating—neural pathways, chemical signals, and electrical impulses working together to create consciousness. But as he takes notes, a question lingers in his mind: *What about the soul?*

At home, Ayaan has grown up learning that humans are more than just biological machines—that the soul plays a role in consciousness, free will, and moral responsibility. Yet, in the classroom, any mention of the soul is absent. It is not debated, it is not questioned—it is simply not part of the conversation. When he hesitantly asks whether science considers the possibility of the soul influencing human thought, the teacher dismisses it as a philosophical or religious matter, unrelated to scientific inquiry. Ayaan is left with an unsettling contradiction: in one part of his life, the soul is central to understanding human nature, while in another, it is treated as irrelevant or non-existent.

This experience of cognitive dissonance is common among Muslim students studying science, as it is for those of many other faiths as well. Though the educational systems may vary from country to country across the world, and from region to region within a country, scholars of science education argue that there are certain norms and patterns in the way that learning is structured and in the way that science is taught that are globally consistent (Puttick 2017; Santos 2016). This is not to say that there are no exceptions—there certainly are. But from a bird's-eye view of global science education, where students from the USA to South Africa to Japan study in mass public or private schooling systems and then pursue "higher studies" at universities is part of the norms of modernity (Pereira de Sousa et al. 2005; Baker 2012).[1]

To help explain global patterns in science education, it is useful to draw upon what Boaventura de Sousa Santos, a Portuguese sociologist and legal scholar, calls the "Monoculture of Modern Science" (Santos 2016). This concept refers to the widespread assumption that there is only one valid way to produce scientific knowledge—one that is rooted in a specific US–Western European centric historical and cultural tradition. For instance, in medical research, the dominant scientific approach often prioritizes biomedical models, such as in psychology. Though there is growing interest and space for incorporating the role of spirituality in such research and in psychosocial therapy, such approaches are not yet mainstream (Culliford 2011). This is often at the expense of other knowledge traditions that have rich legacies of inquiry and discovery that have in fact strongly contributed to the knowledge base of monocultural modern science. For example, in the Muslim world, traditional medical practices, such as those found in Islamic medicine, which blends spiritual, ethical, and physical understandings of health, have been integral in shaping medical thought for centuries (Awaad et al. 2021; Iqbal 2002).

Santos suggests that, while monocultural modern science has made pivotal contributions to society, it has also become the dominant framework for understanding the world. He argues, along with other scholars in the fields of sociology and history, that this dominance of a US–Western European intellectual tradition is not simply on the basis of the merits of the framework versus other knowledge systems. As Dr. Aasim Padela also notes in the Introduction to this book, according to scholars, there has been a suppression or erasure of diverse knowledge systems that occurred as part of European colonial ventures in many parts of the world. This is what Santos terms as "epistemicide" (Santos 2016). For example, such

[1] For example, the progression of the study of sciences (and all other subjects) for all students into "grades" that progress on an annual basis and lead from primary to secondary education is a norm in modernity. As an example of historical alternatives, in pre-colonial Islamicate societies, the norm was for students to study under individual instructors and to progress as quickly as their aptitude allowed (Makdisi 1981). The instructors, not the madrassas, would provide the students then with degrees.

scholars may argue that an epistemicide occurred in the Indian subcontinent following the First War of Independence against the British East India Company in 1857. In the aftermath, the British authorities initiated a widespread dismantling of madrassas (Islamic institutions of learning),[2] leading to the destruction of Muslim scholarship. Additionally, institutional practices marginalized the use of Urdu or Persian in academic and public life, while Muslim scholars faced targeted violence and suppression (Kolsky 2010; Mangan 2012). The legal system also reinforced these divisions, with separate criminal codes for British settlers and the indigenous population, further entrenching the inequalities of colonial rule (Kolsky 2010).

Furthermore building on the historical narrative that Dr. Aasim Padela introduces earlier in this book, historians of colonialism, sociologists, and academics in the field of education note that key aspects of epistemicide often included the imposition of standardized, centrally controlled mass schooling systems by European colonial powers. Whether examining the experiences of First Nations people in North America in the 1800s or those of the Indonesians and Malaysians, such scholars argue that the establishment of mass schooling was a critical tool used to reshape the cultural sensibilities, ways of thinking, practices, and beliefs of indigenous populations (Baker 2012). The goal was to instill admiration for the colonizers and foster a sense of cultural inferiority in relation to the indigenous peoples' own heritage. This process aimed not only at changing local educational structures but also at altering the broader social fabric by eroding traditional values and reinforcing the dominance of the colonial power's culture and worldview (Baker 2012).

With this discourse taken into account, it can be posited that the Monoculture of Modern Science is strongly interlinked with contemporary science education. To speak to the topic and central concern of this book, in monocultural education, science is frequently presented as

[2] This history is highly relevant to the contemporary separation of religion and science curricula in many Muslim countries that were formerly colonized by European powers. For centuries, madrassas and mosques served as centers of learning and research in the Muslim world, where subjects like the natural sciences, mathematics, medicine, and logic were taught alongside Islamic jurisprudence and theology, all within an Islamic worldview (Iqbal 2002; Brentjes 2024). In this context, all knowledge was understood as a subset of religious knowledge, with the belief that all human knowledge is a gift from Allah (Iqbal 2002). These educational institutions were often funded through religious trusts known as *waqf*s (Brentjes 2024).

an entirely materialist endeavor, focusing on physical explanations while leaving little room for discussions about how different cultures and traditions have approached questions of consciousness, ethics, or the nature of reality.

This is not to argue or give the impression that science education systems, book publishers, curriculum designers, and teachers are still today intentionally perpetuating this monocultural colonial legacy. Rather, within the field of sociology of education, this can be thought of as a systemic design issue. Seminal theorists in this field, such as Michael Apple (1971) and William Pinar (2019) have posited the idea that "curriculum" can be thought of as much more than simply the stated learning objectives and syllabus that is taught in an educational setting. Instead, they posit that "curriculum" also includes the implicit messages, values, and assumptions that are communicated (often unintentionally and subconsciously as a social norm) through the way subjects are taught, the perspectives that are emphasized, and the knowledge that is either included or omitted. This broader understanding is often referred to as the hidden curriculum—the unspoken lessons that shape students' worldviews, attitudes, and sense of belonging in a discipline (Apple 1971). In the context of science education, the hidden curriculum can manifest in various ways. For instance, the framing of the human mind as being *entirely in totality* explainable and reducible to the physical biochemical processes of the brain—with little to no discussion of alternative conceptualizations—may reinforce a particular way of thinking about reality, while excluding perspectives that integrate metaphysical or spiritual dimensions. These implicit messages can shape students' understanding of science, influencing not only what they learn but also how they perceive the relationship between science, culture, and identity (Mirza 2024, 2025; Gandolfi 2021).

A key design element in monocultural science education that contributes to such a hidden curriculum is the norm of compartmentalization of knowledge across taught subjects in schools (Jacobs 1989; Mathison and Freeman 1998). Scholars who study the effects of this compartmentalization argue that science education tends to focus almost exclusively on the empirical theories and findings of the discipline, while relegating its history, philosophy, and broader cultural context to the humanities, which are often considered less relevant or secondary to scientific practice (Apple 2018; Galili 2012; Höttecke and Silva 2011).

This can contribute to a hidden curriculum of scientism, where disconnecting non-scientific fields from science reinforces the belief that they are inferior. Scientism—the belief that natural science methods are superior to all other disciplines and the only reliable means of understanding the world—rests on unquestioned assumptions like empiricism, positivism, reductionism, and determinism (Ridder et al. 2018). As discussed in Nazif Muhtaroğlu and Macksood Aftab's Chapter 4, on "Scientific Advances in Understanding Human Nature: An Islamic Occasionalist Perspective," these ideas elevate the natural sciences to an epistemic pedestal, positioning them as superior to other forms of inquiry (Feyerabend 1993; Stenmark 2017). Scholars in philosophy (de Regt 2023) and science education (Gasparatou 2017; Mirza 2024) argue that this belief is reinforced by science curricula that fail to critically examine science's underlying assumptions, limitations, and failures. When philosophical ideas like positivism and reductionism are implicitly conveyed without explicit critique, science is framed as a monolithic worldview (Mirza 2025). This uncritical engagement can lead students to accept the biomedical model of human beings as a complete metaphysical account rather than one perspective among many.

Effects of the Monoculture of Modern Science on Muslim Student Worldviews

This then brings us to the question: What do students undergo when they encounter the hidden curricula from the monocultural science education they may be receiving? In too many cases, this may work to diminish the religious faith of students (Haider et al. 2021; Youssef 2018) and to create a sense of cultural inferiority (Tan 2017). Broadly speaking, the challenges that students experience can be placed into four different categories:

1. Issues of Epistemic Authority
2. Issues of Creed
3. Issues of Identity
4. Socio-religio-scientific issues

Issues of Epistemic Authority Muslim students, the world over, often find themselves facing challenges deciding which knowledge source,

Table 9.1 Examples of issues of epistemic authority

Does the positionality of the arguer matter?	Does authority stem from moral character or from academic prestige or the merits of the argument regardless of who is making them? (Ihram 2009)
If it works, is it wholly true?	Does the success of modern science indicate that religious knowledge about nature and reality is secondary to scientific knowledge about the same? (Edis 2007)
How to relate scientific findings with the Qur'an?	Should the Qur'an be used to confirm scientific findings? Should scientific findings be used to interpret the Qur'an? (Ahmad and Iksan 2023)

Islamic or secular scientific, should be relied on to solve a problem. Table 9.1 illustrates some of these common conundrums.

In response, a particular form of Islamic apologetics has emerged that seeks to align Islam with the frameworks and conclusions of modern science. This approach has been critiqued for inadvertently reinforcing scientism rather than upholding Islamic epistemology, which regards the Qur'an as the divine word of Allah, inviting reflection and the discovery of divine signs in the world. For instance, in the nineteenth century, under British colonial rule in the Indian subcontinent, Syed Ahmed Khan—a Muslim intellectual and educational reformer—attempted to interpret miracles through the lens of modern science (Iqbal 2002). More recently, Adnan Oktar (writing under the pen name "Harun Yahya") has faced similar criticism for adopting scientific arguments in ways that may misrepresent both Islam and science (Edis 2007; Gutron and Dupret 2016; Malik 2023b). However, as rigorous Islamic epistemology is seldom included in school curricula—whether in Muslim faith-based or secular institutions—these knowledge gaps can leave Muslim students vulnerable to developing uncertainties in their understanding of science and religion (Qureshi 2021). This then can contribute to students uncritically buying into this sort of an apologetics-based defensive mentality.

Issues of Creed Stemming from these issues of epistemic authority then also emerge issues of creed, as illustrated in Table 9.2. The general nature of such conundrums stem from extrapolating paradigms in science into the realm of metaphysical truth, and then holding those as a criteria against which to assess the truth of the Islamic worldview of the universe. So, for example, within the secular scientific worldview, the decisions made by a human being are reducible to the motions of electrons and chemicals in the brain.[3] When Newtonian mechanics was in vogue during the nineteenth century, the view that in principle the motion of all the particles in the brain could be mathematically mapped according to the laws of motion provided support to the philosophical view that humans have no free will—their brains are simply complex machines operating according to the predictable laws of physics. In other words, the brain is "deterministic." There was no need to hypothesize a soul and the unseen, and so religion was increasingly viewed as an unnecessary superstition. Later, when quantum mechanics was developed, the secular argument was that the particles in the brain now function according to the laws of probability. So, though the trajectory of any individual particle in the brain could not be predicted for certain, this meant that the brain is "indeterministic"—that "free will" is the external effect of the probabilistic behavior of the sub-atomic particles in the brain that constitute its cells, tissues, and electro-chemical impulses.

Muslim students may encounter such arguments suggesting that religious belief functions as a "god of the gaps" approach—where divine or spiritual explanations are invoked only in areas where scientific understanding is incomplete. This susceptibility can be influenced by a lack of

Table 9.2 Examples of issues of creed

Creationism	Does the common modern biological assumption that humans have common ancestry with other life forms conflict with religious beliefs about God creating Adam and Eve? (Malik 2023c)
Miracles in the modern world	Do miracles happen in today's world? (Billingsley et al. 2021)
Science and free will	Do humans have free will or are their actions the result of the random interactions of sub-atomic particles in their brains? (Yunus et al. 2024)

[3] As discussed in Chapter 5, these theological stances bear upon how Muslim students should assess whether human beings have free will and, as well as neuroscientific claims that their agency can be reduced to the overall effects of billions of electrochemical interactions in their physical brains, as noted in Chapter 4.

Table 9.3 Examples of issues of identity

Racism	Does studying primarily the accomplishments of White, Western scientists indicate their civilizational superiority? (Park et al. 2023)
Personal vs. Professional identity	Should the Islamic beliefs of Muslim professionals in biomedical fields contribute to their professional research and practice contributions, or should they be left to the personal domain? (Yunus et al. 2024)
Biological sex and gender roles	Is gender a spectrum or is there only masculine and feminine? (Tabatabaie 2015)
Scientific legacy of the Muslim world	Is the Muslim world in decline in terms of scientific output and education? (Iqbal 2002)

exposure to Islamic perspectives on divine action and the concept of free will (Qureshi 2021; Yunus et al. 2024).

Issues of Identity A related concern is how hidden curricula, embedded in monocultural school curricula and reinforced through media and social discourse, may influence perceptions of identity among Muslim students (see Table 9.3).

For example, Dr. Aasim Padela in the Introduction to this book already elaborates on how Muslim biomedical professionals in Western contexts may feel that they need to compartmentalize their religious beliefs and their professional identities. This sense of identity compartmentalization, or "double consciousness" as it is termed in the field of sociology (Du Bois 2006; Islam 2020; Meer 2007), may begin before entering the workplace—very much as a consequence of the absence of conversation about Islamic perspectives while studying biomedical topics in school and university. In my own doctoral research (publications forthcoming) within an Islamic faith-based high school in England that follows the secular English National Curriculum, I have found evidence that some students may sometimes compartmentalize their moral reasoning when posed with moral dilemmas that invite them to think about both scientific knowledge as well as their religious beliefs. Almost like a "moral double consciousness," by which I mean that some students appear to mentally and subconsciously compartmentalize morality into Islamic morality and secular morality. This is as opposed to having one consistent moral framework through which they ethically evaluate dilemmas they are posed with. Table 9.4 outlines some of these examples from the data. This data was collected from focus group activities in which groups of 4–5 students from Year 9 (ages 13–14) were presented with moral dilemmas, as part of participatory ethnographic research at the school, with me embedded in the school community as a volunteer teacher over a span of almost a complete academic year.

Socio-Religio-Scientific Issues These concerns of epistemic authority, creed, and identity then feed into how Muslims respond to socio-religio-scientific issues (this is a term I have coined) (Mirza 2024).[4] These are social issues that are sensitive to religious beliefs and practices of the concerned communities, and are also affected by both the medical and scientific practices and beliefs of those communities. Table 9.5 gives some examples of such issues.[5] As another example from Table 9.5, public mental health is a socio-religio-scientific issue for many Muslims as they face a dilemma: Does the use of secular psychological counselling services for mental wellness clash with belief in God's healing (Iqbal and Skinner 2021)? This concern, for example, contributes to the hesitancy of many Pakistani Muslims in reaching out to secular mental health services in the midst of an ongoing mental health crisis (Humair 2021).

The kinds of questions posed by all the tables in this section are the same kinds of specific questions that students may have on certain topics, but unfortunately they do not get access to the answers. Not the least this is because in mass schooling systems across the West and in many Muslim-majority countries, teachers hesitate (often because they feel they do have the training and background knowledge to address these questions) and often school policies are in practice against talking about "controversial" issues—science and religion falling squarely into that category (Mirza 2024, 2025; Farooqui and Kaushik 2022). Indeed, fear of Islamophobia may also cause hesitation on the part of teachers and students to broker conversation on such topics, even when there may be no such regulatory pressure from the school administration and policies (Tahseen et al. 2018; Farooqui and Kaushik 2022). Furthermore, students from

[4] Though more conventionally the term "socio-scientific" issues is used in the field of science education to refer to discussions relating to social issues stemming from scientific research, I bring in the term "religio" to qualify that a subset of these kinds of social issues have to do with religious belief and sensibilities of the general public as well, and that such social issues cannot be understood holistically without an appreciation of these religious beliefs.

[5] For more on artificial intelligence, see Chapter 4. Further, Chapters 7 and 8 also illustrate the dilemmas posed by genetic engineering and body alteration.

Table 9.4 Examples of secular/religious compartmentalization in moral reasoning

No	*Statement*	*Focus group discussion topic*	*Rationale for inclusion*
1	Student N: "I get that. But like, is this like from Islamic point of view? Like, not everyone in the world is Islamic I think in Islamic countries, they should enforce that. But we live in a Christian country. Yeah, we can't go around and saying that, Oh, this freedom of speech. But you know, I think everywhere in the world. It should be because yeah, because everyone has their own beliefs and everything Islamic countries, they should abide by it. But I think I don't think everyone in the world should because everyone's entitled to their own."	Ethics of surrogacy	First, there seems to be a misconception that Islam obligates non-Muslims to follow Islamic laws in non-Muslim majority country contexts. But granted that this is the perspective that Student N holds, then she is drawing upon moral values that are outside her religious framework to make the "live and let live" argument. Hence, there is a moral double consciousness possibly at play—since the student is making inconsistent use of two separate moral frameworks
2	Student A: "It's not unethical. Because, like, what, in a religious point of view, especially in Islam, like one of the reasons why suicide is not permissible is because you can't pick when you're going to die or when you're going to live."	Ethics of cryostasis of human life for medical preservation	The phrase "in a religious point of view, especially Islam..." in a group of participants that are all Muslim, and where the purpose is to express one's own personal ethical stance, indicates that there is a sense of moral double consciousness at play
3	Student K: "I think she shouldn't be allowed to. Because who knows? Like in the future, somebody might not want to revive them if they find them and they might just keep it that way. And yeah, on the religious aspect...."	Ethics of cryostasis of human life for medical preservation	The student first presents her moral reasoning. Then as a secondary support she moves toward religious argumentation ("yeah, on the religious aspect..."), in a confessional Muslim group setting

Note: Names of the participating students have been anonymized

Table 9.5 Examples of socio-religio-scientific issues

Mental health	Should a person suffering from depression or other psychological conditions go to a psychologist for help or to a religious expert or both? (Iqbal and Skinner 2021)
Climate change	Is it justified to harm the environment in the name of economic progress? (Mohamed 2013; Nasr 1990)
Artificial Intelligence	Is it human arrogance or in some way unethical to create machines that are made in the "image" of humans? (Malik 2023a)
Knowledge for the sake of knowledge about the natural world?	Should Muslims do research in nuclear physics, genetics, and other such fields that could potentially lead to knowledge that could be used for harm? (Keeler 2019)

minority Muslim sects may feel an additional sense of hesitation to discuss their queries, given that it may offend the sensibilities of the majority, for example, a Shīʿī minority student in a Sunnī majority class context in Pakistan (Syed and Ali 2021). As a consequence, students are too often left to reflect for themselves with very limited knowledge. Certainly, very few students are likely to know what school of theological thought within Islam they ascribe to, and so are likely to get quite lost and intimidated in any debates they encounter in the perspectives of, say, Ashʿarī versus Māturīdī positions on whether humans have common ancestry with apes is something plausible to consider from within an Islamic worldview (Malik 2023c).

Opportunities and Challenges to Developing Educational Approaches for Muslim Students

The previous sections have emphasized the need for Muslim students to learn about religious and scientific constructions of being. This section considers what it will take to actually make that happen, from an educational design standpoint.

In this spirit of bouncing back, Muslim discourse about "revival" of the pre-colonial Islamic Science Tradition (IST) (Iqbal 2002) has grown, even though Muslim thinkers still have diverging opinions on how to go about this revival. One stance, as Dr. Aasim Padela notes in the Introduction to this book, has been toward the "Islamization of knowledge" movement, led by thinkers like Seyyed Hossein Nasr (2008), Muzaffar Iqbal (2002), Ismail Al Faruqi (1989), and Naquib al-Attas (2023). This has entailed efforts to re-conceptualize the teaching of the humanities, social sciences, and natural sciences through a framework embedded in an Islamic epistemology and ontology. In these efforts, emphasis has been put on adapting primary sources from what is debatably called the "Golden Age." This movement has also involved attempts to create institutes of learning that counter the dominance of the Monoculture of Modern Science education, but here there is divergence. Muzaffar Iqbal (2003) believes that the way forward is to bring together experts versed in traditional Islam learning as well as in modern science to develop a robust intellectual framework and then pilot it in an institution that is removed from the monocultural mass-education public and private education systems (Mignolo 2007). However, Syed Naquib al-Attas, as evidenced by his leading a public Malay University project on the Islamization of education in the 1980s, believes that it is possible to bring change through collaboration with governments (Merdeka Award, n.d.).

In practice, unfortunately, these approaches can get tied up with a problematic nostalgia of a Golden Age. Certainly, the Muslim world from the seventh to eighteenth centuries made breakthroughs in areas from medicine to astronomy (Iqbal 2002). However, given the gap of centuries of colonial rule, there is no lived experience of that tradition nor yet a rich body of decolonizing research[6] unearthing the practices and institutions of scientific inquiry that were prevalent during the so-called Golden Age (Iqbal 2003; Lapidus 1992). The academic vacuum around knowledge of the tradition invites space for Muslim reformers to unintentionally or intentionally fill it in with historical fallacies—what Muzaffar Iqbal terms "precursorism."

Nevertheless, in recent times, there has been momentum building toward this front, though practical approaches to studying "science and religion" within an Islamic paradigm that can address the needs of the multicultural schooling context and those of Muslim school faith-based

[6] Decoloniality is an advocacy position that is proposed as the counter-movement against coloniality, involving praxis in all spheres of human experience that resists the logic, metaphysics, ontology, and matrix of power of colonialism (Maldonado-Torres 2016).

contexts, which are quite nascent. For example, though there is very little empirical literature exploring the particular case of encounters between science and Islam in the context of primary and secondary education, the trend has been a rapid increase in research output on this topic since the early 2000s. This is indicated by Zainun Mustafa's (2021) bibliometric analysis of Elsevier's "SciVerse Scopus" abstracts and citations database,[7] using the key terms "Science," "Islam," and "Education." Her analysis is summarized in Fig. 9.1.

More recently, beyond Islamization of knowledge and related theological movements as elaborated on in the Introduction to this book, there have been three other broad approaches advocated by Muslims thinkers at the university level to specifically tackle challenges Muslims may face from modern science to matters of creed. These, along with their challenges, are illustrated in Table 9.6.

Niyozov and Memon, who are leading contemporary Islamic education thinkers, argue that educational approaches within an Islamic worldview in this modern era must:

(i) blend the theory coming from Islamic traditions;
(ii) the empirical findings of the educational experiences of Muslim students in faith-based and secular schools; and
(iii) the best ideas coming from the theory and practice of non-Muslim scholarship on education (Niyozov and Memon 2011: 25).

Given that in today's modernity the challenges that Muslim students face are strongly influenced by the legacies of colonial epistemicide, to fulfill requirements of (iii) it is important to take inspiration and lessons from decolonial theory, as well as from other educational social

[7] These documents were predominantly in English, with significantly lesser numbers in other Global North languages and some in Arabic, Turkish, and Persian. Mustafa outlines the fact that due to the digital divide it is likely that there are other works that were written in languages like Arabic that were not compiled into the database and so did not come up in the search. Notably prominent thinkers in the field like al-Attas, Hussein Nasr, and Ismail Farooqi also did not emerge from the search; they wrote a lot of their seminal work as books in the 1970s, when likely their work as Global South thinkers may not have made it to Elsevier's database—especially since the database favors publication indexed journal articles. Additionally, Mustafa (2021) mentions that some documents in the database could not be retrieved or had incomplete citation information, though they seemed to be pertinent on the basis of their titles.

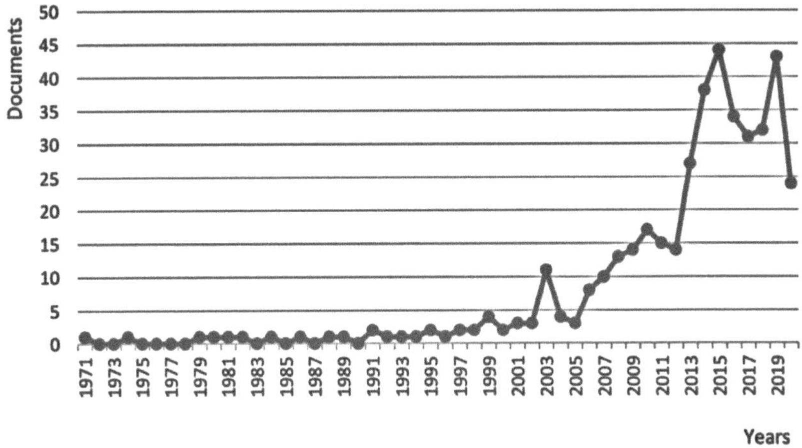

Fig. 9.1 Trend in publication of articles on science, Islam, and education (*Note* The figure shows the number of publications that could be found since the 1970s on Elsevier's SciVerse Scopus database on Islam, Science, and Education. It is evident that literature on this topic has only really started picking up momentum over the last 20 years. Mustafa outlines that from 1970–2000 most of the literature was on medical training for Muslim doctors. Then, from 2000–2010, there emerged philosophical literature on topics like evolution and religion in education, and, in 2010 onwards, there was an increased proportion of literature on teaching through technology as well. *Source* Reproduced from Mustafa [2021: 53])

justice bodies of knowledge like critical pedagogy. Additionally, much learning can and should be drawn upon from the general literature on science and religion education. This is where critically applying decolonial and Islamic philosophy to scrutinize and take the many things that are useful from such bodies of work can be helpful, while filtering out those aspects that operate within a Eurocentric and/or secular paradigm. Here, it will be of especial importance to consider the Nature of Science[8] education literature—that explores how to teach students critically about science's underlying philosophies, and its historical, social, political, and economic implications (e.g., Erduran and Dagher 2014; Abd-El-Khalick 2011; McComas et al. 1998). It is fortunate that much work has already

[8] "Nature of Science" is capitalized as it is a specific subfield in science education.

Table 9.6 Three approaches to countering colonial hegemony in science education for Muslims

Approach	*Example*	*Challenge*
Teaching the internal limitations of theories themselves	Identifying gaps in fossil records to make the case that the evidence of humans descending from apes is "inconclusive" (Elshakry 2013; Qunaybi 2023)	This approach runs into the challenge that it leaves open the possibility that the picture of reality being suggested by the theory (that is potentially challenging the religious worldview) may still in fact be true—maybe over time the evidence will grow stronger even if now it can still be dismissed as "inconclusive."
Teaching about the nature of science	This is so that students may appreciate that even seemingly certain scientific worldviews may be upended by paradigm shifts (Ahmed, n.d.)	Such lines of argument may also run into challenges such as the level of philosophical sophistication required to understand nature of science arguments may be considered beyond the intellectual training and capacity of school students and the general public
Teaching students about theology	By students appreciating the theology they align to in terms of the (1) the nature of God, (2) the nature of creation, and (3) the relationship between God and creation, they can then better assess whether a worldview like Adamic evolution is in conflict or confluence with their religious beliefs (Malik 2023c)	This approach runs into the challenge that most students may not be aligned with any particular theological sub-school of thought within Islam

Note Reproduced from Mirza (2024)

been started on developing multicultural approaches to Nature of Science education, that shift the center of gravity from Western historical and philosophical discourses to meaningfully engage with the Global South and other marginalized communities' perspectives (Gandolfi 2024; Lodge 2021; Boisselle 2016; Mirza 2025). Such an approach is part of the

Islamic Science Tradition (in all its plurality and diversity), where Muslims assimilated the knowledge from non-Muslim lands and took what was useful and framed it within an Islamic framework. For example, Islamic philosophy draws upon the work of ancient Greek and Persian thinkers, and scientific research such as in astronomy, mathematics, and medicine began for Muslims by building upon the knowledge accumulated from these lands (Saliba 2007).

With that said, at the school level, there are several prominent educational approaches with tangible resources for Muslim educators that have been developed trying to keep in mind (i), (ii), and (iii) above. Table 9.7 highlights some of these along with their characteristics.

In the previous section, it was outlined that Muslim students face these kinds of challenges to their faith: issues of epistemic authority, creed, identity, and regarding socio-religio-scientific issues—detailed below. Though a detailed comparative analysis is beyond the scope of this chapter, it is worth noting that all the initiatives in Table 9.7 emphasize tackling head on issues of epistemic authority and of creed. The level of complexity of such discussions vary as various initiatives target students in different age brackets.

Issues of Epistemic Authority and Creed Coming first to epistemology, for example, the 5D thinking approach emphasizes integrating into all science lessons reflections on the big idea that the physical world is intelligible because Allah designed it with order and beauty, making it possible for humans to acquire knowledge through reason, observation, and reflection. Exploring Bioscience and Islam also discusses epistemology, and in addition to making implicit connections in science lessons, it also incorporates a whole explicit module on Sunnī Islamic epistemology within its first teaching module before moving to explore topic wise the big questions in biomedical science and Islam. Similarly, issues of creed are integrated into all of the above initiatives, tailored to the level of schooling of the students as well. For example, Holistic Science Teaching trains teachers on how to critically discuss topics like evolution and the big bang in a way that informs students of the diversity of perspectives regarding these theories within Islamic and scientific worldviews, and Concentric Circles emphasizes that everyday phenomena like rain are in fact miraculous signs of Allah (*ʾāyat*) that we can experience in everyday life by reflecting on the intelligent design of nature, as the Qurʾan invites us to.

Table 9.7 Teaching approaches specific to science and Islam that have tangible resources for educators

Name	*Key thinkers*	*Relevant sources*	*Student age level*	*Detail*
5D Thinking	Necati Aydin, Alparslan Acikgenc, Nadine Kamal, Yamina Bouguenaya	"5d Thinking Org" (n.d.)	6–10 years	The 5D Thinking approach is based on the work of Ottoman scholar Said Nursi, and has been adapted to help teachers of Abrahamic faiths (especially Islam) connect their study of science with a sense of wonder at Allah's creations and with religious morality. The developers of this approach have created publicly accessible lesson plans to help educators appreciate how they can translate theory into practice

(continued)

Table 9.7 (continued)

Name	*Key thinkers*	*Relevant sources*	*Student age level*	*Detail*
Concentric Circles	Elma Ruth Harder and Muzaffar Iqbal	Harder (2006)	5–6 years	This is a book that provides concrete pedagogical examples of how Muslim educators working with students in primary school can help students connect their learning and play with an Islamic worldview and values
Holistic Science Teaching	Nidhal Guessoum, Athar Osama	Pakistan Innovation Foundation (n.d.)	8–13 years	This is an initiative to train schoolteachers in teaching science "holistically" by integrating teaching of the history, philosophy, and Islamic religious connections to science. Currently it is being piloted in several Global South Muslim-majority countries

(continued)

Table 9.7 (continued)

Name	*Key thinkers*	*Relevant sources*	*Student age level*	*Detail*
Islamic Scientific Critical Consciousness	Usama Javed Mirza	Mirza (2024)	13–18 years	This a theoretical framework and initiative being piloted with Muslim and public Muslim-majority schools in the UK to: (i) empower high school students with the ability to critically recognize colonial and anti-Islamic messages/ideologies they may be receiving from their school curricula; and (ii) to teach them how to counter these problematic messages/ideologies through taking social and personal action

(continued)

Table 9.7 (continued)

Name	*Key thinkers*	*Relevant sources*	*Student age level*	*Detail*
Exploring Bioscience and Islam	Omar Qureshi, Aasim Padela, Raudah Yunus	Chapters in this book	Secondary, collegiate and medical school years	This approach supplements traditional secondary school education with a series of activity-based discussion seminars, each co-led by Muslim theologians and scientists, on Islamic ontology, epistemology, and ethics through exploring application on case study problems relating to the field of bioscience

Source Adapted from Mirza (2024)

Issues of Identity and Socio-Religio-Scientific Issues A key distinction among these initiatives lies in their varying levels of emphasis on identity-related themes and socio-religio-scientific issues. Sakina/Concentric Circles and Holistic Science Teaching highlight the contributions of Muslim scientists to global knowledge, aiming to provide students with historical role models and a sense of cultural connection. Islamic Scientific Critical Consciousness (ISCC), on the other hand, incorporates a broader critical perspective by examining the historical suppression of diverse knowledge systems. It approaches science education through a decolonial social justice lens, encouraging students to reflect on how historical and contemporary structures influence knowledge production.

Similarly, in discussions on socio-religio-scientific issues, Exploring Bioscience and Islam focuses on the philosophical and theological dimensions of topics such as genetic engineering and the development of General AI. ISCC, in contrast, places additional emphasis on encouraging students to consider how they might engage with and influence these issues in their personal and social contexts.

What follows is a reflection on why some of these variations in design considerations between the educational initiatives are there. In categorizing such contemporary approaches to teaching about Islam and science, there are several key philosophical, political, and pragmatic considerations that education initiatives need to make a decision on, given the ongoing structural legacies in monocultural education from colonial epistemicide.

Inside Versus Outside the System All schools are embedded in ecosystems that are influenced by the ongoing structures that have been inherited from colonial epistemicide. This influences the "sustainability" of initiatives that aim to break free from the norm of mass schooling in a neoliberal world order. In today's monocultural education systems at the high-school level in many parts of the world, standardized testing is important as these tests then act as an entryway for youth to get access to higher education and to better employment opportunities. In this context, there is much literature discussing the phenomenon of "isomorphism"—the structural factors and incentives at play that, over time, cause institutions to become like each other rather than to radically break free from existing models and norms (Sakib 2020). However, it is not just market forces; it is also in many parts of the world with continued hegemonic control over Islamic education—given the "counter-terror" regime (Tahir 2022; Tan 2017). There are checks and balances and bureaucracy that have been set up by design in order to make it hard to set up and sustain radically different models. Furthermore, as elaborated in Chapter 1, it can also be difficult to speak openly about certain topics (such as Islam and evolution, or Islamic perspectives on medical treatment in the professional workplace) as it may invite a strong backlash. Finally, the struggle is made more difficult by the lack of educational resources on science and Islam that are available at the secondary school level, and the lack of teachers with the kind of training and comfort level to engage in such kinds of teaching (Mirza 2024).

Given these kinds of challenges, most educators choose to have initiatives that work within the realities of the current system. For example, 5D thinking has been promoted in already established schools (which may be faith-based or multicultural) in various parts of the world—ranging from Canada to Pakistan. The 5DT team's model is to provide teacher professional development and curriculum resources to interested schools, who then adapt the framework as appropriate to their local context. The 5D thinking model is versatile enough that it can be adapted by educators of other Abrahamic faiths as well (Table 9.8).

In contrast, Concentric Circles is an example of an educational approach that necessitates separation from the norms of secular mass public schooling, and the emphasis is on the design of all aspects of the school experience, not just specifically the science curriculum. For example, Sakinah Circles is a full-time Islamic faith-based school in Edmonton, Canada following the Concentric Circles philosophy ("Sakinah Circle: Integrating Faith and Learning" n.d.). As an example of Concentric Circle educational philosophy in action, Fig. 9.2 illustrates how the theme of "water" is integrated with Islamic concepts across all

Table 9.8 The essentials of 5 dimensional thinking

Analytical thinking	*Analogical thinking*	*Critical thinking*	*Meditative thinking*	*Moral thinking*
Studying the universe through science after filtering the embedded secular ideology	Using analogy to compare and contrast human-made objects with natural objects and events for better understanding	Engaging in critical thinking through questioning to find out the ultimate reality behind the natural objects and events	Reflecting on the interconnectivity, complexity and benefits of natural objects and events to know about their Maker	Reading the universe as the most meaningful book and deriving moral lessons from scientific studies of natural objects and events

Note The figure elaborates on the five dimensions of 5D thinking. It is intended that whenever a student studies a scientific concept, they should think critically about how that concept relates to these dimensions. These dimensions invite reflection about scientific phenomena that signify that there is a Creator. This 5D framework aims to present the realization that there is a Creator Who can be recognized through His creations will then serve as an impetus to apply scientific knowledge ethically and in harmony with nature. Notably these can be conceptualized without specific reference to Islamic concepts, and hence there is the possibility of adoption by Christian and Jewish educators as well. However, the 5D team does have curriculum support materials that are specifically tailored to the needs of Muslim educators as well

Source Reproduced from 5d Thinking Org (n.d.)

the subjects taught at the Sakinah Circles school. For example, water is conceptualized as an *ʾāyat* of Allah—as being a sign of his wisdom, mercy, and power. In the Fine Arts subject, this is integrated with learning about the use of water in Islamic architecture, and in science, it is linked with the theories of the Islamic Golden Age scientists al-Biruni and Ibn Sina.

Models that work within the system, and those that offer alternative possibilities, both have something to offer in terms of learning for other educational institutions, and the viability of one model or another may indeed come down to the factors of the local context.

Positional Neutrality or Being Upfront About One's Commitments Another key consideration for teachers and schools is whether to be upfront, and if so to what extent, about their personal religious commitments when teaching about Islam and science. Some initiatives in Islam and science do not delve deeply into the colonial entanglements and legacies—sticking instead to emphasizing the ethical dimensions of the practice of science from an Islamic perspective and teaching students to reflect on the beauty, significance, and intelligent design in nature, for example, the 5D thinking framework takes this approach. Whereas others make a very explicit philosophical commitment to both tackling upfront these colonial entanglements, for example, Holistic Science Teaching, and to naming the socio-cultural power dynamics between the individuals engaged in dialogue, for example, Islamic Scientific Critical Consciousness (ISCC) (Mirza 2024). Table 9.9 illustrates part of a lesson outline developed by me as an illustration of the ISCC teaching approach (Mirza 2024). Students are invited to critically reflect from within an Islamic framework, by making use of the Islamic idea of "*fitna*" (trial/tribulation from Allah that tests the faith of the believer), and about how the science curriculum materials they study may privilege certain social identities and may marginalize others.

It is not simply a matter of to what extent the local context makes it safe to have such conversations about the positionality of the teacher and the participants, that is, which Muslim sect are they from such as Sunnī and Shīʿa, which sub-school of theological thought do they subscribe to (if any), and what insights and limitations do their own lived experiences as having gone through different educational systems and having lived in potentially different socio-cultural contexts provide. Rather, it is also a matter of differing pedagogical views about whether it is better to let students assess evidence (presented by the teacher) from themselves without the added psychological pressure of knowing what the

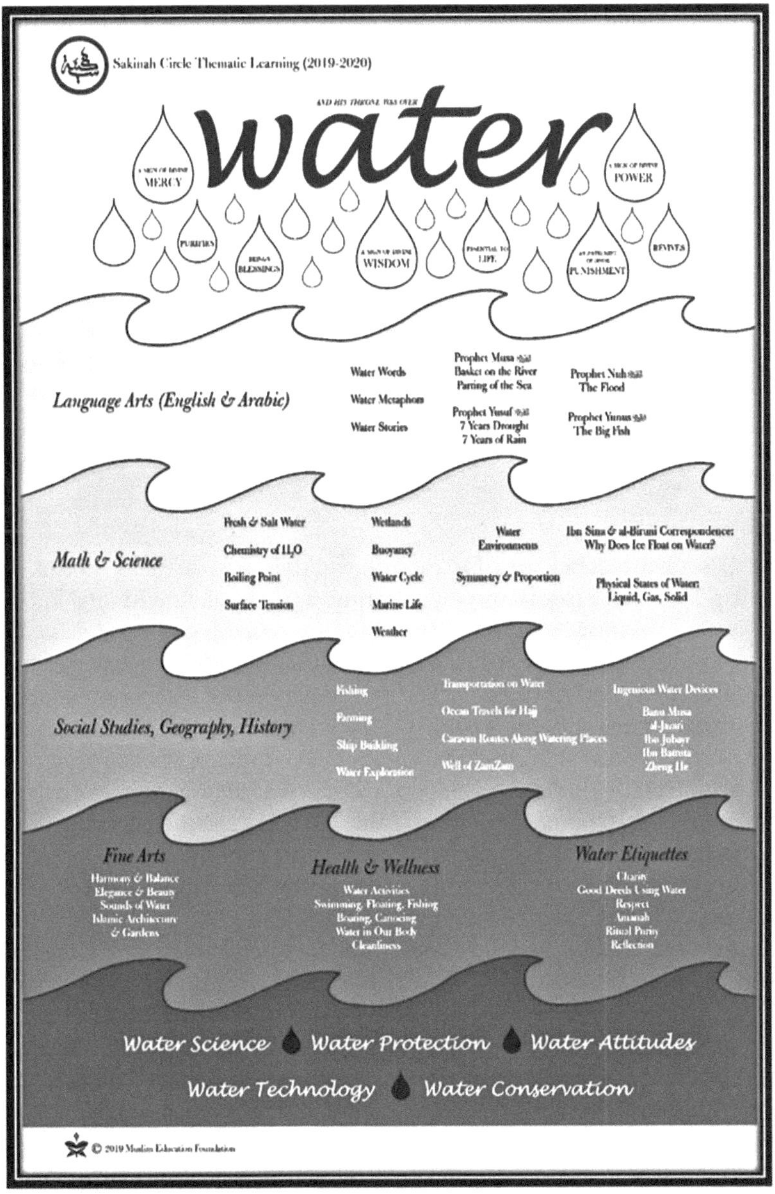

Fig. 9.2 Example of interconnected theme of "water" in Sakina Circle curriculum *Source* Reproduced from Sakinah Circle Curriculum (n.d.)

Table 9.9 Reflection questions in Islamic Scientific Critical Consciousness (ISCC) lesson outline

Lesson(s) outline	*Questions to promote ISCC (these can be private self-reflection or in group discussion)*
Ask students to flip through their science textbooks and table the gender, religion, and race of the famous scientists mentioned	a. Why are some groups (women, non-White ethnicities and so on) under-represented? b. What do the Quran, Sunnah, and Muslim scholarship say about the *fitna* of success and challenges of life in this world for individuals, and for communities as a whole? How can that relate to the erasure and emphasis of certain historical narratives being conveyed in these science textbooks?

Source Reproduced from Mirza (2024)

teacher's own stances are. Or rather, whether that actually makes the implicit bias more profound by creating a garb of "neutrality" about the sources of knowledge that are being given privilege in the class. For example, in an Islamic Scientific Critical Consciousness approach, teachers should be transparent to the extent that is safe in the local context about their own beliefs and the extent of their confidence about the knowledge on the topics being discussed, and invite students to challenge them if a certain point of view or kind of evidence source is being privileged over another. However, in contrast, the Exploring Bioscience and Islam approach emphasizes dialogue among the students themselves based on the lecture materials and literature to read provided, without the teachers necessarily sharing their personal views.

Interdisciplinary, Separate, or Add-On in Subjects and Curricula Another key concern, that stems from the aforementioned isomorphism and the lack of teaching resources as well as to scientism, is to what extent learning about "science and Islam" is compartmentalized or integrated across the school curriculum. For example, should such conversations be an "add-on" to the traditional monocultural science curriculum, or should it be a "module" in religious education or Islamic Studies? Or should it be part of neither, and instead be dealt through a "theory of knowledge" separate course? Or should such discussions be thoughtfully integrated keeping in mind a larger thought out architecture for integrating the study of the natural sciences with the humanities and social sciences?

For example, Holistic Science Teaching, 5D Thinking, and Islamic Scientific Critical Consciousness all are versatile in the sense that they aim to build the capacity of teachers to dialogue about Islam and science in whatever school system they are already embedded in. Exploring Bioscience and Islam is similarly open to a diverse array of students, regardless of what kind of school model they are enrolled at. However, Concentric Circles necessitates working in an Islamic faith-based school, with everything from the structure of the day to the integration of knowledge across subjects being designed in a way that is aligned with its visionary philosophy.

Social Action or Personal Reflection Another key consideration is what is the actionable knowledge, skills, and dispositions that are intended to be inculcated through the teaching of Islam and science? Is the aim simply to increase the faith of Muslim students by helping them to critically analyze the *ʾāyat* of Allah in nature through a proper consideration of Islamic epistemology and scientific findings? Most such science and Islam initiatives indeed take this route. However, as I advocate for in Islamic Scientific Critical Consciousness (ISCC), it is important to consider social action learning outcomes as well. What will students learn about how to mobilize people in their communities to counter sensibilities that diminish the space to think about Islam and science collectively? Such as when it comes to many socio-religio-scientific issues?

Table 9.10 illustrates an example of a social action project from the Islamic Scientific Critical Consciousness approach. Here the idea is to teach students how to mobilize their communities to tackle socio-religio-scientific issues, such as distrust and lack of knowledge of mental health services. However, such teaching ambitions need to be balanced against considerations about the safety of teaching social action in cultural contexts where this may cause harm to students and teachers, for example, from societal backlash in the form of hate crimes.

From the discussion in this section, we can see that there has been a resurgent interest in developing approaches to engaging science and Islam in school classrooms. Many of these initiatives are still in their nascent stages, in the processes of gaining traction, piloting their pedagogical models, and developing more robust curriculum materials. These initiatives all have negotiated differently the challenges that emerge from

Table 9.10 Example of a social action project in the ISCC approach

Topic	*Detail*	*Rationale*
Awareness project on mental health	Destigmatizing discussions about, and experience of, mental health conditions by running campaigns in collaboration with mental health experts and religious scholars	This counters the *Fitna* of Knowledge and Being by helping communities appreciate that seeking support for mental well-being is Islamic, and creating awareness about the work of contemporary Islamic scholars in the field of psychology, so as to encourage trust that counselling services can be coherent with one's own religious beliefs, rather than necessitating a secular world view

Source Reproduced from Mirza (2024)

enabling a critical discussion of science from within an Islamic worldview. A curriculum can easily be developed, along with teaching materials, if there is will and resources, but sustaining its meaningful implementation in the long run is a much more uphill task. This is because teachers, students, and schools are embedded in larger socio-cultural ecosystems that gravitate toward maintaining the same status quo (such as pressure to take state mandated standardized tests that determine a school's performance rating). It will not be easy for a Muslim school in almost any part of the world to swim against the tide and sustain an Islamic model of education that may be different from both the surrounding Muslim schools and the secular ones. Furthermore, there will likely be no one-size-fits-all solution that would work for all schools, students, and teachers across the world.

Conclusion

In conclusion, this chapter has highlighted the importance of teaching Muslim students about both religious and scientific perspectives on human existence. By examining how mass education has shaped knowledge systems, the discussion underscored the value of introducing students to diverse approaches to science that align with their faith. It also addressed the challenges posed by monocultural science education, emphasizing the need for a more integrated understanding of Islam's relationship with science. Finally, the chapter considered both practical and theoretical approaches to developing educational strategies that support Muslim students and educators in multicultural and faith-based schooling contexts, both in the West and in Muslim-majority countries.

Perhaps concluding with this encouraging quote from Said Nursi, whose ideas are the basis for the 5D thinking model is apt, as this chapter outlines that there is much work to be done and there are many different approaches being tried:

> Intention is a "spirit" that causes ordinary deeds and customs to become "living" acts of worship. Its "soul" is sincerity or acting solely for God's sake. Salvation is found only in sincere actions done in His cause. Sincerity multiplies an action's worth, and through it one can reach Paradise with a little action and in a short life. (Nursi 2007: 113)

References

"5d Thinking Org." n.d. https://5dthinking.org/. Accessed June 14, 2024.

Abbasi, Iskander. 2020. "Islam, Muslims, and the Coloniality of Being: Reframing the Debate on Race and Religion in Modernity." *Journal for the Study of Religion* 33 (2): 1–31.

Abd-El-Khalick, Fouad. 2011. "Nature of Science in Science Education: Toward a Coherent Framework for Synergistic Research and Development." In *Second International Handbook of Science Education*, edited by Barry J. Fraser, Kenneth G. Tobin, and Campbell J. McRobbie, 1041–1060. Dordrecht: Springer.

Adamson, Peter. 2024. "Al-Kindi." In *The Stanford Encyclopedia of Philosophy*, edited by Edward N. Zalta and Uri Nodelman, Summer 2024. Metaphysics Research Lab, Stanford University. https://plato.stanford.edu/archives/sum2024/entries/al-kindi/.

Aftab, Macksood. 2013. "Ghazali, Islamophobia and the Myth of Islamic Decline." *The Muslim Observer TMO* (blog). December 12, 2013. https://www.muslimobserver.com/ghazali-islamophobia-and-the-myth-of-islamic-decline/.

Ahmad, Fawarni Hj, and Zanaton H. Iksan. 2023. "Science Teachers' Perspectives on the Integration of the Quran in Science Learning." *Islāmiyyāt* 45 (1): 253–262. https://doi.org/10.17576/islamiyyat-2023-4501-21.

Ahmed, Farah. 2019. "The Potential of Halaqah to Be a Transformative Islamic Dialogic Pedagogy." In *The Routledge International Handbook of Research on Dialogic* Education, 1st ed., edited by Neil Mercer, Rupert Wegerif, and Louis Major, 647–659. Abingdon: Routledge.

Ahmed, Subhoor. n.d. "Darwinian Delusions." https://www.youtube.com/c/DarwinianDelusions. Accessed August 10, 2023.

Alexander, Denis R. 2007. "Models for Relating Science and Religion." *The Faraday Papers*. https://www.faraday.cam.ac.uk/wp-content/uploads/resources/Faraday%20Papers/Faraday%20Paper%203%20Alexander_EN.pdf.

Altinyelken, Hülya Kosar. 2021. "Critical Thinking and Non-Formal Islamic Education: Perspectives from Young Muslims in the Netherlands." *Contemporary Islam* 15 (3): 267–285. https://doi.org/10.1007/s11562-021-00470-6.

American Institute of Physics. 2015. "P.A.M. Dirac—Session III." January 21, 2015. https://www.aip.org/history-programs/niels-bohr-library/oral-histories/4575-3.

Apple, Michael W. 1971. "The Hidden Curriculum and the Nature of Conflict." *Interchange* 2: 27–40 (1971). https://doi.org/10.1007/BF02287080.

Apple, Michael W. 2018. *Ideology and Curriculum*, 4th ed. New York: Routledge.

Arjmand, Reza. 2017. "Islam and Education in the Modern Era: Social, Cultural, Political Changes and Responses from Islamic Education." In *Handbook of Islamic Education*, edited by Holger Daun and Reza Arjmand, 1–17. Cham: Springer International Publishing.

Al-Attas, Syed Muhammad Naquib. 2023. *The Concept of Education in Islam: A Framework for an Islamic Philosophy of Education*. Doha: Dar Ul Thaqafah.

Awaad, Rania, Danah Elsayed, Sara Ali, and Aneeqa Abid. 2021. "Islamic Psychology: A Portrait of its Historical Origins and Contributions." In *Applying Islamic Principles to Clinical Mental Health Care: Introducing Traditional Islamically Integrated Psychotherapy*, edited by Hooman Keshavarzi, Fahad Khan, Bilal Ali, and Rania Awaad, 69–95. New York: Routledge.

Aydin, Necati. 2019. *Said Nursi and Science in Islam: Character Building through Nursi's Mana-i Harfi*. London: Routledge.

Bakar, Mukhlis Abu, ed. 2019. *Rethinking Madrasah Education in a Globalised World*. London: Routledge.

Baker, Michael. 2012. "Modernity/Coloniality and Eurocentric Education: Towards a Post-Occidental Self-Understanding of the Present." *Policy Futures in Education* 10 (1): 4–22. https://doi.org/10.2304/pfie.2012.10.1.4.

Barbour, Ian G. 1990. *Religion in an Age of Science*, 1st ed. The Gifford Lectures, vol. 1, 1989–1991 [i.e. 1990]. San Francisco, CA: Harper & Row.

Bashford, Alison, and Philippa Levine. 2010. *The Oxford Handbook of the History of Eugenics*. Oxford: Oxford University Press.

Bazian, Hatem. 2015. "The Islamophobia Industry and the Demonization of Palestine: Implications for American Studies." *American Quarterly* 67 (4): 1057–1066.

Bewley, Aisha Abdurrahman. 2004. *Muslim Women: A Biographical Dictionary*. London: Ta-Ha.

Billingsley, Berry, Keith S. Taber, and Mehdi Nassaji. 2021. "Secondary School Students' Perceptions of Scientific and Religious Positions on Miracles." *Science & Christian Belief* 33 (2): 99–120.

Billingsley, Berry, Keith Taber, Fran Riga, and Helen Newdick. 2013. "Secondary School Students' Epistemic Insight into the Relationships Between Science and Religion—A Preliminary Enquiry." *Research in Science Education* 43 (4): 1715–1732. https://doi.org/10.1007/s11165-012-9317-y.

Boisselle, Laila N. 2016. "Decolonizing Science and Science Education in a Post-colonial Space (Trinidad, a Developing Caribbean Nation, Illustrates)." *SAGE Open* 6 (1). https://doi.org/10.1177/2158244016635257.

Boli, John, Francisco O. Ramirez, and John W. Meyer. 1985. "Explaining the Origins and Expansion of Mass Education." *Comparative Education Review* 29 (2): 145–170.

Brentjes, Sonja. 2024. *The Sciences in Islamicate Societies in Context: Patronage, Education, Narratives*. Abingdon: Routledge.

"Cambridge IGCSE Subjects." n.d. https://www.cambridgeinternational.org/programmes-and-qualifications/cambridge-upper-secondary/cambridge-igcse/subjects/. Accessed May 27, 2024.

Castro-Gómez, Santiago. 2021. *Zero-Point Hubris: Science, Race, and Enlightenment in Eighteenth-Century Latin America*. Translated by George Ciccariello-Maher and Don T. Deere. Lanham, MD: Rowman & Littlefield.

Culliford, Larry. 2011. *The Psychology of Spirituality: An Introduction*. London: Jessica Kingsley Publishers.

Damper, R.I. 2022. "Science and Religion in Conflict, Part 2: Barbour's Four Models Revisited." *Foundations of Science* (September). https://doi.org/10.1007/s10699-022-09871-z.

De Cordier, Bruno. 2009. "The 'Humanitarian Frontline,' Development and Relief, and Religion: What Context, Which Threats and Which Opportunities?" *Third World Quarterly* 30 (4): 663–684. https://doi.org/10.1080/01436590902867086.

De Regt, Herman. 2023. "School, Science, Scientism: A Utopian View of a University College Curriculum." In *Educational Utopias: Liber Amicorum Prof. Alkeline van Lenning*, edited by Gerwin van der Laan, Tessa Leesen,

Michiel Bot, Ellen Dreezens, Vikas Lakhani, Martijn J. Loos, Anna Shekiladze, Geno Spoormans, and Tom Willems. Tilburg: Open Press Tilburg University. https://openpresstiu.pubpub.org/educational-utopias.
Du Bois, W.E.B. 2006. "Double-Consciousness and the Veil." In *Social Class and Stratification: Classic Statements and Theoretical Debates*, edited by Rhona F. Levine, 203–209. Lanham, MD: Rowman & Littlefield.
Dussel, Enrique. 2008. "Meditaciones anti-cartesianas: sobre el origen del anti-discurso filosófico de la Modernidad Anti-Cartesian Meditations: About the Origin of the Philosophical Anti-Discourse of Modernity Meditações anti-cartesianas: sobre a origem do anti-discurso filosófico da modernidade." *Tabula Rasa* 9 (July–December): 153–197. http://www.scielo.org.co/pdf/tara/n9/n9a10.pdf.
Edis, Taner. 2007. *An Illusion of Harmony: Science and Religion in Islam*. New York: Prometheus Books.
Edis, Taner. 2009. "Modern Science and Conservative Islam: An Uneasy Relationship." *Science & Education* 18 (6): 885–903. https://doi.org/10.1007/s11191-008-9165-3.
Edis, Taner. 2023. "Islam's Encounter with Modern Science: A Mismatch Made in Heaven." *Elements in Islam and Science* (November). https://doi.org/10.1017/9781009257473.
Edis, Taner, and Saouma BouJaoude. 2014. "Rejecting Materialism: Responses to Modern Science in the Muslim Middle East." In *International Handbook of Research in History, Philosophy and Science Teaching*, edited by Michael R. Matthews, 1663–1690. Dordrecht: Springer Netherlands.
Elshakry, Marwa. 2013. *Reading Darwin in Arabic, 1860–1950*. Chicago, IL: University of Chicago Press.
Erduran, Sibel, and Zoubeida R. Dagher. 2014. "Reconceptualizing the Nature of Science for Science Education." In *Reconceptualizing Nature of Science for Science Education: Scientific Knowledge, Practices and Other Family Categories*, 1–18. Dordrecht: Springer.
Fanon, Frantz. 2007. *The Wretched of the Earth*. New York: Grove Atlantic.
Farooqi, Jannat Fatima, and Archana Kaushik. 2022. "Growing Up as a Muslim Youth in an Age of Islamophobia: A Systematic Review of Literature." *Contemporary Islam* 16: 65–88. https://doi.org/10.1007/s11562-022-00482-w.
Farooqi, Sadaf. 2016. "Fellow." The Academy of Medical Sciences. August 1. https://web.archive.org/web/20160801012720/http://www.acmedsci.ac.uk/fellows/fellows-directory/ordinary-fellows/professor-sadaf-farooqi/.
Farrar, Max, Simon Robinson, Yasmin Valli, and Paul Wetherly. Eds. 2012. *Islam in the West: Key Issues in Multiculturalism*. London: Palgrave Macmillan.

Faruqi, I. 1989. *Islamization of Knowledge: General Principles and Work Plan*, edited by ʿAbdulḤamīd AbūSulaymān and International Institute of Islamic Thought, 2nd ed. Herndon, VA: International Institute of Islamic Thought.

Feyerabend, Paul. 1993. *Against Method: Outline of an Anarchistic Theory of Knowledge*. London: Verso.

Fischer, Claude S. 1996. *Inequality by Design: Cracking the Bell Curve Myth*. Princeton, NJ: Princeton University Press.

Fuess, Albrecht. 2007. "Islamic Religious Education in Western Europe: Models of Integration and the German Approach1." *Journal of Muslim Minority Affairs* 27 (2): 215–239. https://doi.org/10.1080/13602000701536166.

Galili, Igal. 2012. "Promotion of Cultural Content Knowledge Through the Use of the History and Philosophy of Science." *Science & Education* 21 (9): 1283–1316. https://doi.org/10.1007/s11191-011-9376-x.

Gandolfi, Haira Emanuela. 2021. "Decolonising the Science Curriculum in England: Bringing Decolonial Science and Technology Studies to Secondary Education." *The Curriculum Journal* 32 (3): 510–532. https://doi.org/10.1002/curj.97.

Gandolfi, Haira Emanuela. 2024. "(Re)Considering Nature of Science Education in the Face of Socio-Scientific Challenges and Injustices: Insights from a Critical-Decolonial Perspective." *Science & Education*. https://doi.org/10.1007/s11191-024-00536-w.

Gasparatou, Renia. 2017. "Scientism and Scientific Thinking." *Science & Education* 26 (3): 1–14. https://doi.org/10.1007/s11191-017-9931-1.

Ġazzālī, Abū-Ḥāmid Muḥammad Ibn-Muḥammad al-. 2005. *Imām Al-Ghazālī's Deliverance from Error and the Beginning of Guidance*. Kuala Lumpur: The Other Press.

Ghazzālī. 1977. *The Jewels of the Qurʾan: Al-Ghazali's Theory: A Translation, with an Introduction and Annotation, of al-Ghazali's Kitab Jawahir al-Qurʾan*. Kuala Lumpur: The Other Press.

Golding, David, and Kyle Kopsick. 2019. "The Colonial Legacy in Cambridge Assessment Literature Syllabi." *Curriculum Perspectives* 39 (1): 7–17. https://doi.org/10.1007/s41297-018-00062-0.

Griffin, Rosarii. 2006. *Education in the Muslim World: Different Perspectives*. Oxford: Symposium Books.

Grosfoguel, Ramón. 2013. "Epistemic Racism/Sexism, Westernized Universities and the Four Genocides/Epistemicides of the Long 16th Century." December, 31–58.

Guessoum, Nidhal, and Stefano Bigliardi. 2023. "Islam and Science: Past, Present, and Future Debates." *Elements in Islam and Science*, March. https://doi.org/10.1017/9781009266550.

Gutron, Clémentine, and Baudouin Dupret. 2016. "Islamic Positivism and Scientific Truth." *Human Studies* 39 (4): 621. https://doi.org/10.1007/s10746-016-9402-8.

Haider, Sarah, Muhammad Syed, Hasan Tauha, and Jason Mickevich. 2021. "Apostate Report." *Ex-Muslims of North America*. https://exmuslims.org/apostate-report/.

Haq, Syed Nomanul. 1994. *Names, Natures and Things: The Alchemist Jābir ibn Hayyān and his Kitāb al-Ahjār (Book of Stones)*. Dordrecht: Springer Netherlands.

Haq, S Nomanul. 2003. "Science, Scientism, and the Liberal Arts." *Islam & Science* 1 (2): 267.

Harder, Elma Ruth. 2006. *Concentric Circles: Nurturing Awe and Wonder in Early Learning: A Foundational Approach*. Sherwood Park, AL: Al-Qalam.

Häuberer, Julia. 2011. "The Founding Concepts of Social Capital—Bourdieu's Theory of Capital and Coleman's Rational-Choice Approach to Social Capital." In *Social Capital Theory: Towards a Methodological Foundation*, edited by Julia Häuberer, 35–51. Wiesbaden: VS Verlag für Sozialwissenschaften.

Haught, John F. 1995. *Science and Religion: From Conflict to Conversation*. New York: Paulist Press.

Herrnstein, Richard J., and Charles Murray. 1994. *The Bell Curve: Intelligence and Class Structure in American Life*. New York: Free Press Paperbacks.

Hoodbhoy, Pervez Amirali. 2007. "Science and the Islamic World—The Quest for Rapprochement." *Physics Today* 60 (8): 49–55. https://doi.org/10.1063/1.2774098.

Höttecke, Dietmar. 2001. *Die Natur der Naturwissenschaften historisch verstehen: fachdidaktische und wissenschaftshistorische Untersuchungen*. Zweite unveränderte Auflage. Studien zum Physiklernen 16. Berlin: Logos-Verlag.

Höttecke, Dietmar, and Cibelle Celestino Silva. 2011. "Why Implementing History and Philosophy in School Science Education Is a Challenge: An Analysis of Obstacles." *Science & Education* 20 (3): 293–316. https://doi.org/10.1007/s11191-010-9285-4.

Humair, Yusuf. 2021. "Culture and Mental Health in Pakistan." In *The Routledge International Handbook of Race, Ethnicity and Culture in Mental Health*, edited by Roy Moodley and Eunjung Lee. Abingdon: Routledge.

Ihram, Silma. 2009. "Muslim Youth and the Mufti: Youth Discourses on Identity and Religious Leadership under Media Scrutiny." Master's thesis, Western Sydney University. http://handle.uws.edu.au:8081/1959.7/496471.

Iqbal, Mohammad. 2013. *The Reconstruction of Religious Thought in Islam*. Stanford, CA: Stanford University Press.

Iqbal, Muzaffar. 2002. *Islam and Science*. Ashgate Science and Religion Series. Burlington, VT: Ashgate.

Iqbal, Muzaffar. 2003. "Challenges to Islam and Muslims: What Is to Be Done?" *Islamic Studies* 42 (4): 595–637.

Iqbal, Muzaffar. 2022. "Why Muslims, Especially Muslim Teachers, Should Get Vaccinated against Precursorism." January 20. https://mi.cis-ca.org/why-muslims-especially-muslim-teachers-should-get-vaccinated-against-the-virus-called-precursorism.

Iqbal, Naved, and Rasjid Skinner. 2021. "Islamic Psychology: Emergence and Current Challenges." *Archive for the Psychology of Religion* 43 (1): 65–77. https://doi.org/10.1177/0084672420983496.

Islam, Inaash. 2020. "Muslim American Double Consciousness." *Du Bois Review: Social Science Research on Race* 17 (2): 429–448. https://doi.org/10.1017/S1742058X20000235.

Jacobs, Heidi Hayes. 1989. "Interdisciplinary Curriculum: Design and Implementation." Association for Supervision and Curriculum Development, 1250 N. Eric, Institute of Education Sciences. https://eric.ed.gov/?id=ED316506.

Jacoby, Tim. 2023. "Islam, Violence and the 'Four Dogmas of Orientalism.'" *Race & Class* 65 (2): 75–97. https://doi.org/10.1177/03063968231164586.

Keeler, Ahmed Paul. 2019. *Rethinking Islam and the West: A New Narrative for the Age of Crises.* Equilibra Press.

Khan, Muhammad Aftab, G. Hussein Rassool, and Shaikh Mabud. 2020. *Sexuality Education from an Islamic Perspective.* Newcastle upon Tyne: Cambridge Scholars Publishing.

Kohlstedt, Sally Gregory. 1995. "Women in the History of Science: An Ambiguous Place." *Osiris* 10 (January): 39–58. https://doi.org/10.1086/368742.

Kolsky, Elizabeth. 2010. *Colonial Justice in British India: White Violence and the Rule of Law.* Cambridge: Cambridge University Press.

Kuropatnicki, Andrzej K., Małgorzata Kłósek, and Marek Kucharzewski. 2018. "Honey as Medicine: Historical Perspectives." *Journal of Apicultural Research* 57 (1): 113–118. https://doi.org/10.1080/00218839.2017.1411182.

Lapidus, Ira M. 1992. "The Golden Age: The Political Concepts of Islam." *The ANNALS of the American Academy of Political and Social Science* 524 (1): 13–25. https://doi.org/10.1177/0002716292524001002.

Lodge, Wilton. 2021. "Confronting Repressive Ideologies with Critical Pedagogy in Science Classrooms." *Cultural Studies of Science Education* 16:609–620. https://doi.org/10.1007/s11422-021-10047-7.

Lyotard, Jean-François. 1984. *The Postmodern Condition: A Report on Knowledge.* Translated by Geoff Bennington and Brian Massumi. Minneapolis, MN: University of Minnesota Press.

McComas, William F., Hiya Almazroa, and Michael P. Clough. 1998. "The Nature of Science in Science Education: An Introduction." *Science & Education* 7: 511–532. https://doi.org/10.1023/A:1008642510402.

Makdisi, George. 1981. *The Rise of Colleges: Institutions of Learning in Islam and the West*. Edinburgh: Edinburgh University Press.

Maldonado-Torres, Nelson. 2007. "On the Coloniality of Being: Contributions to the Development of a concept." *Cultural Studies* 21 (2–3): 240–270. https://doi.org/10.1080/09502380601162548.

Maldonado-Torres, Nelson. 2016. "Outline of Ten Theses on Coloniality and Decoloniality." Frantz Fanon Foundation. https://fondation-frantzfanon.com/outline-of-ten-theses-on-coloniality-and-decoloniality/.

Malik, Shoaib Ahmed. 2023a. "Artificial Intelligence and Islamic Thought: Two Distinctive Challenges." *Journal of Islamic and Muslim Studies* 8 (2): 108–115. https://doi.org/10.2979/jims.00020.

Malik, Shoaib Ahmed. 2023b. "A Historical Introduction to Islam, Science, and Evolution: The Book Symposium on 'Islam and Evolution: Al-Ghazālī and the Modern Evolutionary Paradigm.'" *Theology and Science* 21 (4): 564–587. https://doi.org/10.1080/14746700.2023.2255945.

Malik, Shoaib Ahmed. 2023c. "Challenges and Opportunities in Teaching Interdisciplinary Courses on Islam and Evolution: A Theology-Centric Perspective." *Religions* 14 (1): 95. https://doi.org/10.3390/rel14010095.

Malik, Shoaib Ahmed. 2023d. "Defending 'Islam and Evolution: Al-Ghazālī and the Modern Evolutionary Paradigm': Abrahamic Dialogues and Interdisciplinary Insights." *Theology and Science* 21 (4): 745–780. https://doi.org/10.1080/14746700.2023.2255955.

Mangan, J., ed. 2012. *The Imperial Curriculum: Racial Images and Education in the British Colonial Experience*. London: Routledge.

Maqbool, Aleem. 2011. "Fury over Doctor's Book on Sex Education for Muslims." *BBC News*, January 10, sec. South Asia. https://www.bbc.com/news/world-south-asia-12117519.

Markel, Howard. 2021. *The Secret of Life: Rosalind Franklin, James Watson, Francis Crick, and the Discovery of DNA's Double Helix*. New York: W.W. Norton & Co.

Massad, Joseph. 1993. "Palestinians and the Limits of Racialized Discourse." *Social Text* 34: 94–114. https://doi.org/10.2307/466356.

Massad, Joseph. 2005. "The Persistence of the Palestinian Question." *Cultural Critique* 59: 1–23.

Massad, Joseph A. 2008. *Desiring Arabs*. Chicago, IL: University of Chicago Press.

Mathison, Sandra, and Melissa Freeman. 1998. "The Logic of Interdisciplinary Studies. Report Series 2.33." Eric, Institute of Education Sciences. https://eric.ed.gov/?id=ED418434.

Meer, Nasar. 2007. "Citizenship and Double Consciousness: Muslims and Multiculturalism in Britain." PhD dissertation, Department of Sociology, University of Bristol. https://research-information.bris.ac.uk/ws/portalfiles/portal/34502800/486084.pdf.

Memon, Nadeem A. 2016. "Diverse Communities, Divergent Aspirations? Islamic Schooling in the West." In *Philosophies of Islamic Education: Historical Perspectives and Emerging Discourses*, edited by Nadeem A. Memon and Mujadad Zaman. New York: Routledge.

Merdeka Award. n.d. "Tan Sri Professor Dr Syed Muhammad Naquib al-Attas." https://www.merdekaaward.my/the-award/past-recipients/outstanding-scholastic-achievement/tan-sri-professor-dr-syed-muhammad-naquib-al-attas. Accessed August 10, 2023.

Mignolo, Walter D. 2007. "Delinking: The Rhetoric of Modernity, the Logic of Coloniality and the Grammar of de-Coloniality." *Cultural Studies* 21 (2–3): 449–514. https://doi.org/10.1080/09502380601162647.

Mirza, Usama Javed. 2024. "Islamic Scientific Critical Consciousness as A Theoretical Framework for Muslim Science Educators." *London Review of Education* 22 (1): 9. https://doi.org/10.14324/LRE.22.1.09.

Mirza, Usama Javed. 2025. "Regime Change: The Need for Scientific Critical Consciousness." *International Journal of Critical Diversity Studies* 7 (1): 86–110. https://doi.org/10.13169/intecritdivestud.7.1.0086.

Mohamed, Najma. 2013. "Islamic Education, Eco-Ethics and Community." *Studies in Philosophy and Education* 33 (3): 315–328. https://doi.org/10.1007/s11217-013-9387-y.

Moore, Alison M.D. 2023. "Modern European Sexological and Orientalist Assimilations of Medieval Islamicate *ʿilm al-Bah* to Erotology." *History of the Human Sciences* 36 (5): 15–41. https://doi.org/10.1177/09526951211056152.

Mustafa, Zainun. 2021. "Islam, Science and Education: Delving into the Progress, Collaboration and Biases." *Journal of Islamic Thought and Civilization* 11 (2). https://doi.org/10.32350/jitc.11.2.03.

Nadwi, Mohammad Akram. 2021. *Al-Muḥaddithāt: The Women Scholars in Islam*. Jeddah: Dar al-Minhaj.

Nasr, Seyyed Hossein. 1990. "Islam and the Environmental Crisis." *Islamic Quarterly* 34 (4): 217–234.

Nasr, Seyyed Hossein. 2008. "Islam and Science." In *The Oxford Handbook of Religion and Science*, edited by Philip Clayton. Oxford: Oxford University Press. https://doi.org/10.1093/oxfordhb/9780199543656.003.0006.

Nicolaides, Demetris. 2020. *In Search of a Theory of Everything: The Philosophy Behind Physics*. New York: Oxford University Press.

Niyozov, Sarfaroz. 2016. "Religious Pluralism and Islamic Education: Addressing Mutual Challenges." In *Philosophies of Islamic Education: Historical Perspectives and Emerging Discourses*, edited by Nadeem A. Memon and Mujadad Zaman. New York: Routledge.

Niyozov, Sarfaroz, and Nadeem Memon. 2011. "Islamic Education and Islamization: Evolution of Themes, Continuities and New Directions." *Journal of Muslim Minority Affairs* 31 (1): 5–30. https://doi.org/10.1080/13602004.2011.556886.

Noor, Farish A., Yoginder Sikand, and Martin van Bruinessen. *The Madrasa in Asia: Political Activism and Transnational Linkages*. 2008. Amsterdam: Amsterdam University Press.

Nord, Warren A. 1999. "Science, Religion and Education." *Religion & Education* 26 (2): 55–66. https://doi.org/10.1080/15507394.1999.11000904.

Nursi, Bediüzzaman Said. 2007. The Risale-i Nur Collection. http://archive.org/details/the-risale-i-nur-collection.

Pakistan Innovation Foundation. n.d. "Holistic Science Teaching Workshop & Fellowship." http://pif.org.pk/scienceteaching/. Accessed September 9, 2023.

Paraskeva, João M. 2016. "EPISTEMICIDES: Toward an Itinerant Curriculum Theory." *Counterpoints* 491: 261–289. https://www.jstor.org/stable/45157416.

Park, Jaddon, and Sarfaroz Niyozov. 2008. "Madrasa Education in South Asia and Southeast Asia: Current Issues and Debates." *Asia Pacific Journal of Education* 28 (4): 323–351. https://doi.org/10.1080/02188790802475372.

Park, Wonyong, Alison Cullinane, Haira Gandolfi, Sahar Alameh, and Günkut Mesci. 2023. "Innovations, Challenges and Future Directions in Nature of Science Research: Reflections from Early Career Academics." *Research in Science Education* 54: 27–48. https://doi.org/10.1007/s11165-023-10102-z.

Pereira de Sousa, Cynthia, Denice Barbara Catani, António Nóvoa, and Frank Simon. 2005. School and Modernity: Knowledge, Institutions and Practices. Introduction." *Paedagogica Historica: International Journal of the History of Education* 41 (1–2): 1–8. https://doi.org/10.1080/0030923042000335439.

Pinar, William F. 2019. *What Is Curriculum Theory?* New York: Routledge.

Polkinghorne, John C. 2009. *One World: The Interaction of Science and Theology*. Templeton Foundation Press.

Puttick, Steven. 2017. "'You'll See That Everywhere': Institutional Isomorphism in Secondary School Subject Departments.' *School Leadership & Management* 37 (1–2): 61–79. https://doi.org/10.1080/13632434.2017.1293633.

Qunaybi, Iyad. 2023. الدكتور إياد قنيبي—القناة الرسمية—YouTube. https://www.youtube.com/c/eyadqunaibi/videos.

Qureshi, Omar. 2021. "Science Curriculum from an Islamic Worldview." In *Curriculum Renewal for Islamic Education: Critical Perspectives on Teaching Islam in Primary and Secondary Schools*, edited by Nadeem Memon, Mariam Alhashmi, and Mohamad Abdalla, 113–124. New York: Routledge.

Rahman, Tariq. 2001. "Pluralism and Intolerance in Pakistani Society: Attitudes of Pakistani Students towards the Religious 'Other.'" In *New Perspectives on Pakistan: Visions for the Future*, edited by Saeed Shafqat. Karachi: Oxford University Press.

Ridder, Jeroen de, Rene van Woudenberg, and Rik Peels (eds.). 2018. *Scientism: Prospects and Problems.* New York: Oxford University Press.

Rothman, Abdallah, and Adrian Coyle. 2018. "Toward a Framework for Islamic Psychology and Psychotherapy: An Islamic Model of the Soul." *Journal of Religion and Health* 57 (October). https://doi.org/10.1007/s10943-018-0651-x.

Said, Edward W. 1979. *Orientalism.* New York: Vintage Books.

Sakib, Nurul Huda. 2020. "Institutional Isomorphism." In *Global Encyclopedia of Public Administration, Public Policy, and Governance*, edited by Ali Farazmand, 1–7. Cham: Springer International Publishing. https://doi.org/10.1007/978-3-319-31816-5_3932-1.

"Sakinah Circle Curriculum." n.d. https://sakinahcircle.com/curriculum-document. Accessed June 14, 2024.

"Sakinah Circle: Integrating Faith and Learning." n.d. https://sakinahcircle.com/. Accessed June 14, 2024.

Saliba, George. 1995. *A History of Arabic Astronomy: Planetary Theories During the Golden Age of Islam.* New York: New York University Press.

Saliba, George. 2007. *Islamic Science and the Making of the European Renaissance.* Cambridge, MA: The MIT Press.

Santos, Boaventura de Sousa. 2016. *Epistemologies of the South: Justice against Epistemicide.* London: Routledge.

"Sayedaty Ep. 6: Sutayta Al-Mahāmali." n.d. Yaqeen Institute for Islamic Research. https://yaqeeninstitute.org/watch/series/sayedaty-ep-6-sutayta-al-mahamali. Accessed May 27, 2024.

Shakeel, M. Danish. 2018. "Islamic Schooling in the Cultural West: A Systematic Review of the Issues Concerning School Choice." *Religions* 9 (12): 392. https://doi.org/10.3390/rel9120392.

Stenmark, Mikael. 2017. *Scientism: Science, Ethics and Religion.* London: Routledge.

Syed, Jawad, and Faiza Ali. 2021. "A Pyramid of Hate Perspective on Religious Bias, Discrimination and Violence." *Journal of Business Ethics* 172: 43–58. https://doi.org/10.1007/s10551-020-04505-5.

Tabatabaie, Alireza. 2015. "Childhood and Adolescent Sexuality, Islam, and Problematics of Sex Education: A Call for Re-Examination." *Sex Education* 15 (3): 276–288. https://doi.org/10.1080/14681811.2015.1005836.

Taber, Keith S. 2013. "Conceptual Frameworks, Metaphysical Commitments and Worldviews: The Challenge of Reflecting the Relationships Between Science and Religion in Science Education." In *Science Education for Diversity: Theory and Practice*, edited by Nasser Mansour and Rupert Wegerif, 151–177. Dordrecht: Springer Netherlands. https://doi.org/10.1007/978-94-007-4563-6_8.

Tahir, Izza. 2022. "Decolonizing Madrassa Reform in Pakistan." *Current Issues in Comparative Education* 24 (1). https://doi.org/10.52214/cice.v24i1.8853.

Tahseen, Madiha, Sawssan R. Ahmed, and Sameera Ahmed. 2018. "Muslim Youth in the Face of Islamophobia: Risk and Resilience." In *Islamophobia and Psychiatry: Recognition, Prevention, and Treatment*, edited by H. Steven Moffic, John Peteet, Ahmed Zakaria Hankir, and Rania Awaad, 307–319. Cham: Springer.

Tamdgidi, Mohammad H. 2012. "Beyond Islamophobia and Islamophilia as Western Epistemic Racisms: Revisiting Runnymede Trust's Definition in a World-History Context." *Islamophobia Studies Journal* 1 (1): 54–81. https://doi.org/10.13169/islastudj.1.1.0054.

Tan, Charlene. 2017. "Colonialism, Postcolonialism, Islam, and Education." In *Handbook of Islamic Education*, edited by Holger Daun and Reza Arjmand, 1–12. Cham: Springer International Publishing. https://doi.org/10.1007/978-3-319-53620-0_22-1.

Tan, Charlene, Paul Anderson, and Yasir Suleiman. 2011. *Reforms in Islamic Education: Report of a Conference Held at the Prince Alwaleed Bin Talal Centre of Islamic Studies*. University of Cambridge, April 9–10, 2011.

Tbakhi, Abdelghani, and Samir S. Amr. 2007. "Ibn Al-Haytham: Father of Modern Optics." *Annals of Saudi Medicine* 27 (6): 464–467. https://doi.org/10.5144/0256-4947.2007.464.

Turner, Howard R. 2010. *Science in Medieval Islam: An Illustrated Introduction*. Austin, TX: University of Texas Press.

Valk, John. 2009. "Knowing Self and Others: Worldview Study at Renaissance College." *Journal of Adult Theological Education* 6 (1): 69–80. https://doi.org/10.1558/jate.v6i1.69.

Wahid, Muhammad Abdul. 2023. "From Orientalism to Neo-Orientalism: Medial Representations of Islam and the Muslim World." *Textual Practice* (December): 1–20. https://doi.org/10.1080/0950236X.2023.2288112.

Winter, Tim, ed. 2008. *The Cambridge Companion to Classical Islamic Theology*. Cambridge Companions to Religion. Cambridge: Cambridge University Press.

Yasri, Pratchayapong, Shagufta Arthur, Mike U. Smith, and Rebecca Mancy. 2013. "Relating Science and Religion: An Ontology of Taxonomies and Development of a Research Tool for Identifying Individual Views." *Science & Education* 22 (10): 2679–2707. https://doi.org/10.1007/s11191-013-9623-4.

Youssef, Chouhoud. 2018. "What Causes Muslims to Doubt Islam? A Quantitative Analysis." Yaqeen Institute for Islamic Research. 2018. https://yaqeeninstitute.org/read/paper/what-causes-muslims-to-doubt-islam-a-quantitative-analysis.

Yunus, Rauhad M., Aasim I. Padela, Shaza Khan, and Mohd Yusof Mohamad. 2024. "Tackling 'Big' Questions at the Bioscience–Islam Interface: Reporting on the Efficacy of an Educational Programme Designed for Muslim High School Students." *International Journal of Science Education*. 1–25. https://doi.org/10.1080/09500693.2024.2367774.

Zack, Naomi. 2002. *Philosophy of Science and Race*. London: Taylor & Francis Group. http://ebookcentral.proquest.com/lib/cam/detail.action?docID=1679724.

Zine, Jasmin. 2002. "Muslim Women and the Politics of Representation." *American Journal of Islam and Society* 19 (4): 1–22. https://www.ajis.org/index.php/ajiss/article/view/3034.

CHAPTER 10

Addressing "Big Questions" About the Human Being Through Bioscience and Islam: Findings from an Educational Intervention Across Two Countries

Raudah M. Yunus and Aasim I. Padela

Introduction

For millennia, humans have wondered about themselves and the universe, and have attempted to understand the mechanics and purposes behind natural phenomena. Contemplation of life and its travails gives rise to

R. M. Yunus · A. I. Padela (✉)
Medical College of Wisconsin, Milwaukee, WI, USA
e-mail: apadela@mcw.edu; draasim@medicineandislam.org

R. M. Yunus
e-mail: raudah.yunus@gmail.com

R. M. Yunus
Universiti Teknologi MARA, Selangor, Malaysia

A. I. Padela
The Initiative On Islam and Medicine, Glendale Heights, IL, USA

© The Author(s), under exclusive license to Springer Nature Switzerland AG 2026

A. I. Padela (ed.), *Bridging Bioscience and Islam*, Islam & Science,
https://doi.org/10.1007/978-3-032-09996-9_10

questions like "How did life begin?" "Who am I?" "Where am I going?" and "Are humans alone in this universe?" These existential inquiries reflect the intrinsic thirst human beings have for knowledge, truth, and meaning-making. While such "big" questions have traditionally been the domain of religion and philosophy, modern science as a positivistic-inclined, empirically driven, and naturalistically oriented knowledge system has also laid claim to providing answers to these fundamental questions. Certainly, one cannot deny that information garnered from studies in biology, cosmology, computer science, physics, and other fields has revolutionized human thinking about the nature of the universe and the place of humans within it (Ward 2008). Classic examples include Darwin's theory of evolution, or more accurately certain postulates commonly held in evolutionary biology (e.g., common ancestry), that have challenged the religious narrative of human creation by God, and the potential existence of extraterrestrial intelligent life that has upended the long-established notion of human uniqueness in the Western tradition.[1]

Within both popular narratives and academic research in the West,[2] the idea of conflict between the two domains of knowledge is pervasive. Some trace this dynamic to a post-Enlightenment European heritage that was marshaled to global society alongside colonialism, and others further link the schism between science and religion to the emergence of modern nation-states arising in response to the socio-political crises perpetrated by religiously oriented governments (Harrison 2006; Chaney 2016; Knepper 2003). Regardless of its specific intellectual historical roots, popular opinions, particularly within industrialized nations at the leading edge of scientific research, tend to see the domain of religion as private and individual, whereas scientific knowledge is more broadly relevant to guiding

[1] These two ideas and their relationship with Islamic theology are picked up in Chapters 3 and 6 respectively.

[2] By this term, we refer to the "intellectual" West which overlaps with, but does not fully map onto, the geographic west. Principally we refer secular, rationalistic, and liberal European intellectual heritages which have worked their way into political, economic, and social systems.

public life and policy making (Knepper 2003; Montuschi 2017). Moreover, when the two bodies of knowledge do meet, many perceive the claims emerging from each worldview[3] to be in conflict.

Within the academic literature, nuanced taxonomies depict a wide spectrum of relations between the two domains, ranging from inherent conflict to coalescence and complementarity (Yasri et al. 2013). For instance, scientist-philosophers Mahner and Bunge contend that science and religion are integrally incompatible with respect to their doctrinal, metaphysical, methodological, and attitudinal perspectives (Mahner and Bunge 1996). Meanwhile, prominent theistic scientists like Sir Francis Collins, the former director of the National Institutes of Health in the USA, and Nidhal Guessoum, an Algerian astrophysicist, writer, and educator in the field of science and Islam, maintain the consonance and compatibility of the two systems (Guessoum 2010; Collins 2006).

Sir John Polkinghorne, a physicist and theologian, provides a larger typology involving four possible ways of relating science and religion (Polkinghorne 2009). The first relation is of conflict as they provide contradictory responses to the same questions. The second view aligns with natural theology, where science is seen as a tool to explain the nature of God and His works. A third perspective considers science and religion to share certain features but focus on different subject matters (e.g., physical and objective versus spiritual and subjective). This view is called Modes-of-Thought. The fourth category, termed the One World view, considers that "each works in its own realm in search of its own truth, and ultimately these combine to form a richer understanding of reality" (Polkinghorne 2009; Yasri et al. 2013). Physicist, theologian, and priest Ian Barbour constructs a similar taxonomy but with slightly different terminologies and emphasis; he proposes four ways of assessing religion and science relations: (i) conflict, (ii) independence, (iii) dialogue, and (iv) integration (Barbour 1990). On the first view (conflict), the metaphysical and epistemological distinctions between science and religion make it necessary to choose one over the other; either science or religion is the only true source of knowledge. On the second view (independence), the domains of science and religion are completely separate, the data streams are independent and autonomous, and one must recognize

[3] The term "worldview" here refers to the concept described earlier by Usama Javed Mirza in Chapter 9. Science and religion are each a spectrum of worldviews and neither of them is monolithic.

their differences in methodology and languages. The third perspective, (dialogue), considers the findings of science and religion to be mutually supportive and complementary. The final perspective, the integrative view, suggests that science and religion can be directly merged to construct a uniform explanation of reality (Barbour 2002). Other taxonomies have been sketched out but detailed discussions on science–religion relationships are beyond the remit of this chapter (Haught 1995; Nord 1999; Alexander 2007). It is critical to remind readers the point highlighted by Mirza in Chapter 9 that most of the above-mentioned frameworks emerge within a particular intellectual milieu and are not necessarily reflective of Islamic thought. Indeed, Muslim thinkers are largely absent in the dominant discourse about science and religion. This is the case, despite the fact that classical Islamic theologians and legists have devoted many treatises to describing the relationship between rational, empirical, and revelatory perspectives on natural phenomena.[4] The absence of contemporary Muslim interlocutors could reflect the fact that these schema bifurcate science and religion into separate categories whereas investigations within the natural sciences are regarded as a subset of religious inquiry in Islamic thought. Hence, the divide may be difficult to bridge.

Within science education, students' religious commitments and cultural values should be considered by teachers as they design curricula (Southerland and Scharmann 2013). This is because the efficacy of science education is influenced by students' ability to grapple with scientific data, which in turn, is shaped by their prior understandings of the world, beliefs about scientific knowledge, reactions to conflicting evidence, and emotions surroundings specific topics of inquiry (Southerland et al. 2011; Southerland and Scharmann 2013). For instance, topics like the existence of free will or the soul may evoke prejudgment, the off-hand dismissal of religious or scientific perspectives, or preconceived notions of inevitable conflict, given that learners spend more time outside of the classroom being shaped by social and cultural frameworks intrinsic to their familial and ethnic communities (Southerland and Scharmann 2013; Barnes et al. 2017). Classroom discussions that are not nuanced and dialogic in nature

[4] In the past the debates revolved around religion and philosophy, which represented rational thought that truncated revelatory knowledge. A great example of this discourse within the Islamic tradition is the *Incoherence of the Philosophers* by Imam al-Ghazālī in the eleventh century and the response to that text by Ibn Rushd in the twelfth century, in his *Incoherence of the Incoherence*.

may place stress upon students' identities and lead to disengagement and knowledge and spiritual crises (Roth 1997). Illustratively the disengagement of some Black Americans from science education is attributed to the "whiteness" they perceive within extant science curricula (Morton et al. 2023). In a similar vein, the Māoris of New Zealand and First Nation people in Canada have refused to engage with Western-style schools because they view the science education to disregard their beliefs and values (Smith 1995; Cardinal 1969; Sinclair 2004). Such resistance should not be surprising as research finds that some students view the science classroom as personally threatening (Smith and Scharmann 2008), and that some wrestle with a sense of self-betrayal should they accept views contrary to their religious or cultural beliefs (Aikenhead and Jegede 1999). These challenges are further compounded when teachers do not have the proper time or training to enter into potentially charged conversations. Some may avoid probing controversial topics due to their own discomfort or lack of knowledge regarding students' diverse cultural and belief systems that relate to the topic at hand (Dean 2005).

These challenges, while most prominently detailed in Christian contexts, are also present in the Muslim community and educational settings. Certainly, the way science is taught in predominantly secular, Western societies and educational systems spurred the establishment of Islamic schools in non-Muslim-majority nations (Qureshi 2021). As members of a faith minority, Muslim youth face unique challenges due to many factors. For one, national curricular standards sometimes heighten perceived tensions, for instance on topics like origins of life, while attempts to explore alternative notions and frameworks are marginalized or unrecognized. Second, Muslim students rarely, if at all, receive instruction in epistemology in public/secular or parochial/religious settings, leaving them ill-equipped to navigate through competing religious and scientific claims (Rafe'Zou'bi 2015). Indeed, a member of our project team, theologian-educator Omar Qureshi, who has worked as a science teacher as well as principal in both parochial and public schools states that in Islamic schools, science teachers tend to dismiss issues that arise from perceived tensions between religion and science or relegate the discussion to religion class instructors (Qureshi 2021). While these schools generally have little awareness of specific theological and metaphysical tensions with science, Muslim teachers are largely not equipped to deal with such intricacies (Dean 2005). Third, few venues exist within or outside of school for students to engage in civil discourse at the intersection of religion and science. Fourth, regular encounters with religious discrimination and Islamophobic sentiments may further discourage Muslim students in secular schools from expressing religious beliefs in the classroom (Abu Khalaf et al. 2023; Amjad 2018).

On the other hand, science education within Muslim-majority countries (MMCs) is challenged due to different factors. To begin with, many MMCs were colonized for centuries, thus inheriting public educational systems that privilege secularity and marginalized religious discourse within education. New Islamic schools thus emerged—and older ones were reformed—to provide alternatives to these colonial heritage mass public schooling systems (Tan 2017). Second, given how scientific technologies have transformed high-income nations in the West, MMC leadership is often highly supportive of science, technology, engineering, and mathematics (STEM) education, in their efforts to catch up with the West in terms of technology. As a result, they borrow curricula from the West and eschew religiously inflected curricula that may question scientific evidence (Hashim 1999). Despite the scientific achievements seen in countries like United Arab Emirates (UAE) and Qatar, science education in most MMCs lags behind their Western counterparts. For instance, reviews of science teaching and curricula in some Arab nations reveal outdated textbooks and an overemphasis on memorization at the expense of inquiry-type activities and critical and problem-solving skills (Guessoum 2018; BouJaoude 2003). Relatedly, sensitive topics like evolution are either absent, censored, or partially represented within biology syllabi (Osman, Razali, and Bahri 2018; Lay et al. 2018). For example, the topic is banned in Saudi Arabia, Oman, Algeria, and Morocco, whereas in Malaysia, only selected components of evolution are taught (Osman et al. 2018; Alassiri 2020).

Against this backdrop, curricular initiatives that center the relationship between science and theology for Muslim students are under development. Examples include: (a) a three-volume textbook titled *Natural Science from the Worldview of the Quran* (2017) edited by the late Malaysian scholar Mohd. Kamal Hassan (Hassan 2017); (b) a textbook series titled "Science," published by a group of Muslim educators in California (Ahmadi et al. 2020); (c) the 5D Thinking project, a pedagogical approach based on the work of Syed Nursi, aiming at assisting teachers in connecting science education with the sense of wonder at Allah's creation (5DThinking n.d.; Mirza 2024); (d) a book titled *Concentric*

Circles, by two Islamic scholars and authors Elma Harder and Muzaffar Iqbal, that outlines methods of connecting learning and play with an Islamic worldview among primary school students (Harder and Iqbal 2006); and (e) Holistic Science Teaching, an initiative that trains school teachers to teach science "holistically" by integrating history, philosophy, and Islamic studies with science (Pakistan Innovation Foundation 2024). These preliminary efforts have been welcomed by Muslim educators worldwide, though critics underscore that existing curricula have not adequately covered Islamic epistemology nor addressed the philosophical commitments of modern science (Haseeb 2022). Consequently, these initiatives might still be insufficient in preparing students for holistically and effectively contending with perceived science–religion conflicts.

To address these curricular gaps, our project designed, implemented, and evaluated a seminar-style curriculum addressing fundamental questions about the human being through Islamic and bioscientific perspectives, targeting Muslim pre-medical students (we tailored the curriculum to medical students later on). Our curriculum aimed to build students' conceptual literacy at the bioscience–Islam interface, develop their capacities to critically assess religious and scientific evidence and, in so doing, reflect on whether there are conflicting truth claims, and promote attitudinal change away from perceived conflict toward the possibility of harmony between Islam and bioscience.

With those overarching goals in mind, this chapter discusses the efficacy of our curricular intervention by presenting findings from our formal program evaluation. More specifically, we report on student knowledge gain, changes in their interest, intention, and preparedness to learn, engage with, and pursue further learning and opportunities at the intersection of bioscience and theology, and changes in views on science–religion compatibility. Additionally, in order to assess benefits of the program for specific student groups, we identified socio-demographic predictors for knowledge gain and other behavioral outcomes, and compared students from Malaysia (representing a Muslim-majority context) to those from the United States (representing a Muslim-minority context).

Methodology

Authors' Positionality

In the spirit of self-reflexivity, we acknowledge our standpoint as American and Malaysian Muslim scholar-researchers with postgraduate training in health education and healthcare-related research, as well as advanced education in Sunnī theological, epistemological, and ethico-legal frameworks. Overall, our underlying philosophical assumptions align with the epistemological schema elaborated by Omar Qureshi in Chapter 2. The interpretive framework we adopt herein inclines toward post-positivism.[5] While we are critical of the nature of modern science being empirically driven and naturalistically oriented, we do not reject its utility for evaluations such as this one. Therefore, in this chapter, we adopt an empirical approach because we view this inquiry to involve a series of logically related steps that warrant rigorous methods of analysis.

Curriculum Design and Pedagogical Elements

A team of bioscientists, theologians, and educators co-designed an approximately 20-hour curriculum covering five "Big Questions" about the human being: its origin, its nature, its physical malleability and future, its fatedness, and its uniqueness. These questions allowed for juxtaposing theological alongside bioscientific understandings and, in this way, made the instruction relevant to Muslim students interested in pursuing the health sciences and medicine. In total, there were seven seminars; aside from five seminars dedicated to the five questions, there was an introductory session on epistemology and a closing review session.

Several innovative elements marked our educational design. Our instructional team comprised individuals with a level of "dual training in Islamic studies and the biosciences, as well as experiences in teaching at this intersection. For instance, Omar Qureshi holds a PhD in Cultural and Educational Policy, possesses traditional teaching licenses in multiple Islamic disciplines, and has served as a high school science teacher as well as Muslim school administrator. Similarly, Shoaib Malik holds doctorates in engineering and theology and widely teaches on religion and

[5] We use the term post-positivism—rather than positivism—to indicate that we do not believe in the necessity of cause and effect relationships; rather, we recognize that all apparent cause and effect is based on probabilities and could be interrupted.

science (see Table 10.1). The interdisciplinary literacy and familiarity with teaching at the interface of Islam and bioscience provided team members with critical insights and lived experiences to draw upon as they designed and taught the seminars. Together the team reviewed the extant literature on Islam and the five fundamental questions to identify key concepts relevant to the religious and bioscientific narratives, and subsequently referred to the various existing typologies relating science and religion to discern areas of contention, compatibility, and uncertainty. Critically, leveraging the bridging scholarship of the team, a bioscientist–theologian dyad mapped out lesson plans and co-taught each seminar. In this way, they exemplified engagements across the so-called religion and science divide. Our group was supported by two pedagogical experts with extensive experience in designing curricula for Muslim students. The pedagogists observed the curriculum development process and provided feedback on appropriate ways to simplify the contents so as to tailor teaching to the cognitive levels of targeted audiences, and provided pointers on delivering the seminars effectively and interactively (see the seminar learning objectives and outlines in the Appendix).

Table 10.1 presents the seminar topics and instructors and Fig. 10.1 illustrates the relationship between the seminar topics.

Additionally leaning on constructive learning pedagogies, rather than trying to solve intractable problems and provide definitive answers, our instruction focused on relational issues; how to assess, align/integrate, or privilege/marginalize truth claims. Thus, we sought to prevent learners from becoming fixed on absolute truth-finding and instead have them focus on critical inquiry, thereby training students to develop the ability to make more reasonable judgments and to embrace uncertainty and plurality. Given the early career stage of the learners, we utilized popular film and cultural media clips, and relatively easy-to-understand Qur'anic passages, to distill important concepts. Finally, the seminars utilized a multi-modality teaching approach—referring to the application of multiple instructional methods such as texts, image, audio, and technology-enriched tools—to cater to varied learning styles and maintain student engagement. Thus, each seminar involved didactics, facilitated group activities, and media reflections.

The seminar series took place in the Chicago suburbs at a Muslim community center over two weekends in January and February 2023 with two seminars a day for the first three days and the review session and awards ceremony taking place on the last day. In Malaysia, the program

Table 10.1 Seminar topics and instructors

Seminar number	*Seminar topics*	*Instructors*
1	What do we know and how do we know it? Reasoning exercises in religion and science	Omar Qureshi, PhD Holds a bachelor's degree in microbiology and master's in Education in Curriculum and Instruction, and a PhD in Cultural and Educational Policy Studies. Also has advanced training in Islamic theology, ethics, and law through seminary studies Hamza Karamali, MSc Holds master's degrees in Computer Engineering, as well as an equivalent graduate degree in Islamic Law and Legal Theory through seminary studies
2	What is human origin? Human evolution and god's special creation	Shoaib Ahmed Malik, PhD, PhD Holds dual expertise in Science and Religion with a PhD in Chemical Engineering and a PhD in Theology Omar Qureshi, PhD—see above
3	What is human nature? Neuroscience, artificial intelligence, and Human nature	Macksood Aftab, DO, MS Holds a medical degree with expertise in neuroradiology through clinical training, and a master's degree in History of Science Hamza Karamali, MSc—see above
4	Human enhancement and transhumanism: are we changing/should we change our being?	Aasim I. Padela, MD, MSc Holds graduate degrees in Medicine and Healthcare Research and has advanced training in bioethics and emergency medicine through post-graduate training. Also has undergraduate degrees in Biomedical Engineering and Classical Arabic. Has training in Islamic theology and law through visiting fellowships at the Oxford Centre for Islamic Studies and the International Institute for Islamic Thought, as well as seminarian tutorials Hamza Karamali MSc—see above

(continued)

Table 10.1 (continued)

Seminar number	*Seminar topics*	*Instructors*
5	Are we programmed or Do we have free will? Discussions on fate and genetic determinism	Abdurezak Hashi, PhD Trained in Islamic Studies and completed a PhD in Theology and Comparative Religion Mohd Yusof Muhammad, PhD Holds undergraduate and postgraduate degrees in Biomedical Sciences, Human Anatomy and Tissue Engineering, and a postgraduate Diploma in Islamic Studies
6	Are humans unique in this universe? Science and theological inquiry about extra-terrestrial intelligent life	Ahsan Arozullah, MD, MPH Holds a master's degree in public health and a doctorate in medicine Bilal Ali Ansari Studied Computer Engineering at an undergraduate level, and holds a bachelor's degree in the Applied Behavioural Sciences. Holds graduate level degrees in the traditional Islamic sciences through seminar studies
7	Knowledge Cafés and Harvest: panel presentations & moderated discussions	Aasim I. Padela, MD, MSc—see above Raudah M. Yunus, MBBCH, MPH, DrPH Holds a bachelor's degree in medicine, and a master's and doctorate degree in Public Health from the University of Malaya. Received training in Classical Arabic Language and Literature

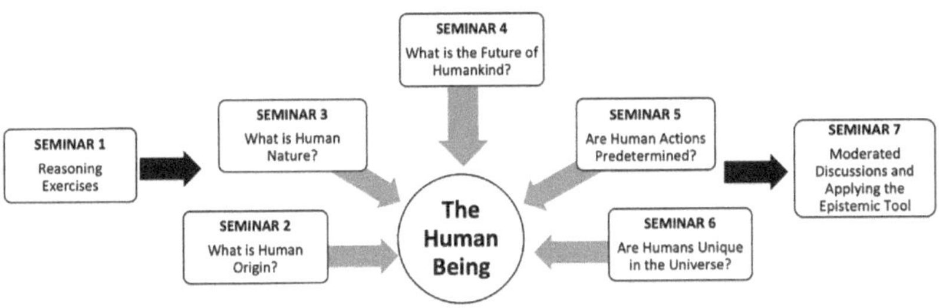

Fig. 10.1 Seminar topics and the inter-relationships between them

was held at the Centre for Foundation Studies (CFS) of the International Islamic University Malaysia (IIUM) over four consecutive days in July 2023 with two seminars a day for the first three days and the review session on the last day. These locations and institutions were selected based on the project collaborators and partner organizations.

Participant Recruitment and Data Collection

In the USA, students were recruited through flyer advertisements at five private Islamic high schools[6] in the Chicago and Milwaukee regions, personal referral from school administrators, and via social media posts. A short promotional video along with a brochure provided greater details about the program. Interested individuals were required to submit a short essay or a video explaining their motivation for the seminars, and a recommendation letter. Students who completed the application were registered for the seminar series and provided logistical details and an informational letter describing the program components, participation requirements, and incentives ($100 gift card for each weekend). Eligibility criteria included self-identifying as Muslim, being in the 11th or 12th grade, and an expressed interest in pursuing further studies in biosciences or medicine.

In Malaysia, recruitment was performed through CFS at IIUM. The program was advertised to CFS students using flyers and a promotional video, and faculty members held an open forum to discuss program details with students. Those interested self-registered using an online form, or alternatively contacted designated CFS members to assist them in enrolling. Participant incentives for the Malaysian cohort included embossed T-shirts and flash drives. To further incentivize engagement during the seminars, select students were gifted signed copies of books related to Islam and biomedicine after each seminar in both the USA and Malaysia.

Although the two student cohorts are different in terms of geography and societal context, they are similar in several ways. For one, they are of similar age groups and educational backgrounds; students in 11th and

[6] These five schools were Salam School, the Muslim Community Center (MCC) Academy, College Preparatory School of American (CPSA), Universal School, and Islamic Foundation School.

12th grade in the USA have completed at least ten years of general education, while entrance into CFS also requires ten years of education as Malaysian students must complete the *Sijil Pelajaran Malaysia* (SPM; equivalent to the General Certificate of Education—Ordinary Level). Moreover, both groups of students are English literate as CFS students must pass an English competency exam in order to enroll. Second, both groups are religiously oriented in that the US students have elected to attend private Islamic schools where coursework in general Islamic studies (Qur'an, Muslim history, and Arabic) is mandatory. CFS students, on the other hand, are mostly Islamic high school graduates who then enrolled in one- or two-year foundational studies before they can get admitted into undergraduate programs. At IIUM, all CFS students must sit for Arabic Placement Tests (APT) upon registration. Third, in both locations, the medium of instruction was English. These similarities allow for comparing the relative efficacy of the curriculum across the two cohorts. The differences in the two cohorts in terms of their being in a Muslim-minority versus a Muslim-majority context, as well as general societal perspectives on the role of Islam were hypothesis-generating. In other words, we expected that the students in Malaysia would tend to see bioscience and Islam as less conflictual than American students and then have less "attitudinal" change given that the sociocultural milieu was more Islamically oriented and Muslim-friendly. We also expected that these students would have more background religious knowledge than Muslim high school students and thus perhaps have lesser knowledge gain.

Upon registration, participants from both cohorts were asked to complete a web-based questionnaire several days prior to the program. The post-seminar questionnaire was administered online immediately after the last seminar. Pre- and post-intervention questionnaires were linked but anonymized through a student-created unique identifier code. We chose this method of evaluation due to reasons of practicality and because pre- and post-test designs have been shown to be feasible and effective in educational settings to measure student growth and improve curriculum efficacy (Hornbuckle 2022). With regards to qualitative data, students were asked to either: (i) write an essay reflecting on their learning experiences as well as on how the seminars have affected them; or (ii) create two infographics that illustrate the new knowledge they gained. Each essay was approximately 1000–1500 words in length and were submitted three weeks after the conclusion of the seminars.

Measures and Tools

The survey instrument assessed participant knowledge, interest, preparedness, attitude, and intention as aligned with the Kirkpatrick Evaluation Model.[7] The Kirkpatrick Model is a widely recognized method of evaluating the outcomes of learning programs by measuring them against four levels of criteria: reaction, learning, behavior, and results. In this study, we focused on two domains: learning and behavior. The learning domain (level) entails knowledge, attitude, skills, and confidence and these correspond to our measured outcomes. The questionnaire comprised 30 knowledge and skills items, 10 items related to on attitude/belief toward Islam–science compatibility, and 4 items on preparedness which is analogous to confidence in the Kirkpatrick Model. The behavior domain (level) corresponds to intention as we measured it, because intention can be regarded as the most immediate antecedent to action. We included items on interest because even though it is not explicitly stated in the Kirkpatrick Model, various studies corroborate the importance and role of interest in the pathway toward learning and behavioral change (Crutzen and Ruiter 2013; Crutzen et al. 2014; Thoman et al. 2011).

The items were compiled into a questionnaire that subsequently underwent expert panel review by the team to remove redundant items and ensure the question stem precision and response category alignment. Specific survey domains and items are noted below.

Outcome Variables

(a) **(a) Knowledge** was measured using 30 questions formulated by the bioscientist-theologian dyads based on the key learning objectives of the seminars. Each seminar, save for the review session, was assessed with five questions with true/false or multiple-choice response categories. The learning outcomes for each seminar are provided in Table 10.1. Examples of items in this scale are:

1. A proposition is a statement which can be true or false.

[7] The Kirkpatrick Evaluation Model is not without criticism or challenge. Ideally, we should have used an evaluation model that adopts an Islamic approach. But literature on this topic is relatively scarce and existing Islamic models do not easily fit our purpose (e.g., built for elementary level, variables measured are different, etc.). Using the Kirkpatrick Model enables comparison with other mainstream studies given its wide usage.

2. The similarity between the Islamic and naturalistic epistemology is that both view cause-and-effect relations as necessary.
3. Neuroscientific reductionism is compatible with the Islamic understanding of the human soul.

(b) **Interest** was assessed by a series of 14 statements exploring students' keenness to learn more about specific seminar topics, as well as the intersection of bioscience and Islamic theology in general. Items were modeled after the educational and career interest scale in science, technology, and mathematics for high school students developed by Oh and colleagues (Oh et al. 2013). Responses were categorized according to a five-point Likert-type scale from "not interested at all" (1) to "extremely interested" (5). Examples of items in the interest scale are: Indicate your interest in...

1. Learning how to integrate knowledge from bioscience and Islam in your future career.
2. Learning skills that will prepare me for a career at the interface of bioscience and Islam.
3. Pursuing research topics at the intersection of bioscience and Islam.

(c) **Preparedness** was assessed by questions on students' levels of readiness to learn, pursue relevant courses, and pursue research opportunities and relevant careers at the bioscience–Islam interface. Responses were recorded along a five-point Likert-type scale from "not prepared at all" (1) to "extremely prepared" (5). Examples of items measuring preparedness were: Indicate your level of preparedness to do the following:

1. Learn about the intersection of bioscience and Islam.
2. Pursue courses at the intersection of bioscience and Islam.
3. Pursue a career at the intersection of bioscience and Islam.

(d) **Intention** Six statements gauged students' plan or desire to take the following actions: (i) obtain further training; (ii) study with experts/scholars; (iii) seek advanced interdisciplinary degrees; (iv) conduct research; (v) inform others about relevant information; and (vi) assure that their career focuses on research and education, at the intersection of bioscience and Islamic theology. These items were adapted from the Career Intentions in Science scale and a scale that measures intention to pursue a high-tech career developed by Saifuddin (Saifuddin 2015). Responses were rated along a five-point Likert-type scale from "no intention at all" (1) to "already taking steps" (5). Examples of items in this domain were: Indicate your intention to do the following:

1. Obtain further training at the intersection of bioscience and Islam.
2. Study with experts/scholars working at the intersection of bioscience and Islam.
3. Seek advanced interdisciplinary degrees at the intersections of bioscience and Islam.

(e) **View on Science–Religion Compatibility** The term view here refers to the degree to which science (bioscience) and religion (Islam) are perceived as compatible by the student and measured as a spectrum through ten statements. These statements were identified through a literature review and expert consultation, and subsequently rephrased to suit a Muslim lexicon. Some of the items were derived from the Science and Religion Compass Questionnaire developed by Theos and the Faraday Institute (Waite and Spencer 2023), while some others were adapted from published studies in a similar field (Yasri et al. 2013; Tenneson et al. 2015; Longest and Smith 2011). Responses were recorded along a five-point Likert-type agreement scale from "strongly agree" (1) to "strongly disagree" (5). Examples of the scale items are:

1. There is irreconcilable conflict between bioscience and Islam.
2. Bioscience cannot complement Islamic views on the human being.
3. Science is the only way to obtain reliable knowledge.

See the complete list of items for all the outcome variables in the Appendix.

Independent Variables

Explanatory measures included participants' socio-demographic characteristics such as age, sex, and ethnicity. We also assessed religiosity based on a modified version of the five-item Duke University Religious

Index (DUREL-5) (Koenig and Büssing 2010), where congregational attendance frequencies and other questions were adapted to Muslim religious practice and lexicon. Notably, we used DUREL because it is deemed an efficient tool widely adopted in numerous empirical studies and has been validated in various languages and across multiple settings (Toscanelli et al. 2022). There are three subscales under DUREL: Organizational Religious Activity (ORA); Non-Organizational Religious Activity (NORA); and Intrinsic Religiosity (IR). Examples of this scale items are:

1. How often do you attend congregational religious services? (ORA)
2. How often do you spend time in private religious activities, specifically voluntary extra prayers or reading the Qur'an outside of prayer? (NORA)
3. In my life, I experience the presence of the Divine (Allah) (IR)
4. My Islamic beliefs are what really lie behind my whole approach to life (IR)
5. I try hard to carry Islam over into all my other dealings in life (IR)

Data Transformation and Analysis

Participants with ID mismatch (pre- and post-) were excluded from analysis, and whole-case analysis for each survey measure was used. With respect to the outcome domains, the 30 knowledge items were summed into an index score, with "1" assigned to each correct answer and 0 if otherwise. The possible score range was between 0 and 30.

For interest, intention, and preparedness, the Likert-type responses were treated as ordinal scales. Each outcome had the score of its individual items summed up to yield an index score. For interest, three items[8] were summed up to give a possible score range between 3 and 15. With regard to intention, the six items yielded a score range between 6 and 30; and for preparedness, scores from the four items were calculated in a similar fashion, giving a possible range between 4 and 20.

With respect to attitude/belief, there were originally ten items and the agreement scale had a score from 1–5, giving a possible total score

[8] See the Appendix, which contains the complete list of items. The first eleven items are not included in measuring students' interest because they are limited to individual topics. Only the last three items were used for scoring interest levels.

range between 10 and 50. Consistent with best practices, we performed exploratory factor analysis (EFA) to uncover the underlying factor structure of this measure and to examine its internal reliability. First, we analyzed the bivariate correlation matrix of all items to detect multicollinearity. Next, we performed reliability statistics to obtain the overall Cronbach's α and identify items that could potentially be removed through item-rest correlations. Thereafter, we applied principal axis factoring and varimax rotation to determine the number of factors and the factor loading for each item.

These psychometric analyses revealed high reliability of the measure with an overall Cronbach's α of 0.83. Notably items 9 and 10[9] had the lowest item-rest correlation value. Removal of any item except for item 9, lowered Cronbach's α.[10] The solution derived from this analysis yielded three unique factors for the scale accounting for 62.5% of the variation in the data. Nonetheless, the last two items, Items 9 and 10 (which loaded onto Factor 3) showed loading factors less than 0.45 (0.43, 0.36 respectively); they did not load into any dimension significantly. Hence, the EFA was repeated without these two items. Results of this new analysis confirmed a satisfactory two-factor solution where the Kaiser–Meyer–Olkin (KMO)[11] value was 0.85 and the Barlett's test[12] was significant ($p < 0.01$). The two factors yielded for the scale accounted for 61.4% of the variation in the data. The first four items loaded predominately upon the second factor, with its theme reflecting more closely the notion of religion and science compatibility. Hence, this subscale was carried forward as summed score-based continuous outcome (see EFA results in Tables 10.3 and 10.4).

[9] See the Appendix in which the items are listed.

[10] Generally an alpha value of 0.70 or more is considered acceptable, and the closer it is to 1.0, the better.

[11] The Kaiser–Meyer–Olkin (KMO) Measure of Sampling Adequacy indicates the proportion of variance in your variables that might be caused by underlying factors. Values closer to 1.0 mean that a factor analysis is appropriate for the data.

[12] The Bartlet's test of sphericity is used to test the hypothesis that the correlation matrix is an identity matrix. If the p-value is significant (<0.05), that means the correlation matrix is not an identity matrix. This indicates that the data is well suited for factor analysis.

As regards to religiosity, items from each of the three DUREL subscales were tabulated into scores and treated as continuous variables. Additionally, each subscale was treated as a separate outcome to avoid collinearity and canceling out effects (Koenig and Büssing 2010). Socio-demographic characteristics of participants were tabulated using descriptive statistics.

Following the completion of all the seminars, the two datasets (one from the USA and one from Malaysia) were merged and variables were recoded as noted above. Comparison between pre- and post-intervention scores was computed using paired-sample t-tests and its non-parametric counterpart, Wilcoxon signed rank test when data distribution was not normal. We ran Generalized Linear Models (GLiM) for multivariable analyses to determine independent predictors for change in outcome variables. To guide us in selecting predictor variables that should be included in the GLiM models, we first performed bivariate analyses to identify socio-demographic variables associated with outcome variables. Statistical significance—or more accurately termed "practical significance"—at this stage was set at 0.25 due to the relatively small sample size and our attempt to be inclusive of many possible relationships between outcomes and predictors. For every outcome variable that showed substantial pre- and post- intervention change, we built two GLiM models: one to identify socio-demographic predictors; and another to identify behavioral/knowledge-related predictors. We report β coefficients, standard errors, and 95 percent confidence intervals from the models below. At the multivariate stage, the significance level was set at 0.05 as is convention.

Our mixed-methods approach follows an explanatory sequential design (Toyon 2021) (essays written and submitted after surveys were completed); we undertook this approach as the essay contents would supplement our survey findings, by explaining further what the numerical changes meant and why changes were not as expected, if any. Essays were analyzed using a framework approach to thematic analysis (Gale et al. 2013). First, researchers familiarized themselves with the essay contents by reading them a few times and deliberating over their main points. Next, a preliminary codebook was developed, describing a list of domains, codes, and subcodes, and how each construct was defined. Subsequently, two researchers independently coded the same 12 (18% of the total) randomly selected essays in order to assure inter-rater coding reliability. This was followed by a series of meetings where they compared similarities and differences in their coding results, highlighted relevant codes not captured by the codebook, and discussed unusual findings. The codebook went through several iterations until the team was satisfied that it captured necessary information systematically and without redundancy. Throughout this process, we focused on themes and codes that matched with the outcome variables measured in the survey. Therefore, relevant quotes that demonstrated students' knowledge, interest, preparedness, intention, and view on the bioscience–Islam relationship were identified and mapped to outcome domains and reviewed in a team-based fashion as a validity-check (Smith and Firth 2011). To complete data triangulation, we looked for patterns or codes that emerged across both datasets and examined the convergence or agreement between them to strengthen the overall findings.

Results

Participation and Participant Characteristics

Of the 57 American students who completed the application process and registered for the program, 51 completed the pre-seminar survey. Over the two weekends, there was attrition with 50, 44, 39, and 37 students participating respectively. Thus, the attendance rate varied between 88 percent (50/57) and 65% (37/57), with an overall attrition rate of 26% (13 dropouts between the first and the last seminar). The majority of participants were female (57%), of South Asian background (65%), and 11th graders (67%). Forty students completed both pre- and post-seminar surveys and were included in the analyses.

In Malaysia, 198 students registered, and 187 completed the pre-seminar survey. Over the four-day program, attendance ranged between 165 and 180 participants, with an attrition rate of 5.6 percent (ten dropouts between the first and the last day). The majority of participants were Malay (97.6%) and female (71.4%). A total of 168 students completed both surveys and were included in the final analysis.

With respect to essays, we received 68 submissions, of which 21 were from the US cohort and 47 from the Malaysian group. Table 10.2 shows the socio-demographic characteristics of seminar participants.

Table 10.2 Socio-demographic characteristics of student participants (n = 208)

Variable	*All* N *(%) or Mean (SD)*	*By Cohort*	
		Malaysia	*USA*
Age in years	18.5 (0.08)	19.0 (0.02)	16.4 (0.11)
Biological sex	65 (31.3)	48 (28.6)	17 (42.5)
Male	143 (68.7)	120 (71.4)	23 (57.5)
Female			
Ethnic background	165 (69.3)	165 (98.2)	26 (65.0)
Malay	26 (12.5)	3 (1.8)	10 (25.0)
South Asian	10 (4.8)		2 (5.0)
Arab	2 (1.0)		2 (5.0)
African American	5 (2.4)		
Others			
Cohort/Country	168 (80.8)	NA	NA
Malaysia	40 (19.2)		
USA			
Religiosity (DUREL)	4.40 (0.09)	4.33 (0.10)	4.73 (0.21)
ORA	4.31 (0.09)	4.38 (0.11)	4.00 (0.21)
NORA	13.1 (0.12)	13.15 (0.14)	13.13 (0.23)
IR			

*No significant difference in levels of religiosity between the two cohorts, Malaysian students were significantly older than American students (p < 0.01)

The Impact of Seminars Upon Student Knowledge Interest, Intention, Preparedness, and Opinions/Beliefs

Students' Knowledge Gain Was Substantial In aggregate, there was significant knowledge acquisition. The overall knowledge score increased from 12.1 at baseline to 18.1 at conclusion of the program (p < 0.01). In other words, the average score increased from 40.3 to 60.3%. This trend of increasing knowledge was similar across items from each of the six seminars. While both cohorts demonstrated meaningful knowledge gain, comparisons between the two cohorts revealed notable differences. The American students had higher baseline knowledge than Malaysian students (14.7 points versus 11.5 points, p < 0.01), and accordingly the relative increase in scores was greater among Malaysian students than their American peers (6.47 points versus 3.95 points respectively, p < 0.01). Nevertheless, the final average score across the cohorts was nearly identical (18 points versus 18.6 points, $p = 0.48$) (Tables 10.3 and 10.4).

Table 10.3 Changes in knowledge score among program participants (n = 208)

Seminar number	*All*		*Malaysia*		*USA*	
	Pre-	*Post-*	*Pre-*	*Post-*	*Pre-*	*Post-*
1	1.57 (1.07)	2.48 (0.82)*	1.48 (1.04)	2.43 (0.80)*	1.97 (1.11)	2.67 (0.89)*
2	2.86 (1.16)	4.05 (0.99)*	2.79 (1.15)	4.11 (0.94)*	3.13 (1.22)	3.82 (1.19)*
3	2.08 (0.98)	2.89 (0.97)*	2.03 (1.01)	2.89 (0.92)*	2.31 (0.79)	2.92 (1.18)*
4	2.37 (1.29)	3.60 (1.07)*	2.22 (1.21)	3.55 (1.02)*	3.02 (1.44)	3.79 (1.23)*
5	2.03 (1.27)	2.94 (1.28)*	1.91 (1.24)	2.97 (1.24)*	2.53 (1.27)	2.82 (1.43)
6	1.20 (1.09)	2.14 (1.14)*	1.08 (0.99)	2.02 (1.09)*	1.72 (1.33)	2.61 (1.27)*
Total score	12.13 (4.38)	18.13 (4.00)*	11.53 (3.93)	18.00 (3.64)*	14.69 (5.26)	18.64 (5.29)*

*Indicates a statistically significant change ($p < 0.05$)

Table 10.4 Changes in interest, intention, preparedness, and opinion among participants (n = 208)

Domain	*All mean (SD)*		*By Cohort Mean (SD)*			
			Malaysia		*USA*	
	Pre-	*Post-*	*Pre-*	*Post-*	*Pre-*	*Post-*
Interest	11.42 (0.18)	11.80 (0.19)*	11.42 (0.20)	12.05 (0.20)*	11.43 (0.39)	10.78 (0.42)
Intention	19.58 (0.31)	20.89 (0.31)*	19.79 (4.29)	21.46 (4.25)*	18.73 (4.20)	18.50 (4.33)
Preparedness	10.44 (0.29)	13.40 (0.27)*	10.37 (4.22)	13.55 (3.94)*	10.75 (3.68)	12.75 (3.73)*
Opinion/belief	16.3 (3.09)	15.6 (3.37)*	16.1 (3.16)	15.4 (3.38)*	17.1 (2.68)	16.6 (3.20)

*$p < 0.05$

Findings from the reflective essays underscored the knowledge gain, as participants expressed improved literacy and comprehension of the seminar contents. Illustrative excerpts are provided below:

> this seminar (series) has greatly influenced and improved my overall understanding and knowledge about the questions I have constantly had in mind but was always scared to ask.
> (Participant 2, Chicago)
> I got so much new knowledge about Islam and bioscience. The speakers were able to solve the riddle in my mind as they explained to us about the relation between Islam and bioscience.
> (Participant 7, IIUM)
> I also found it astonishing how I could easily answer most of the post-seminar questions confidently compared to the moment I did with the pre-ones.
> (Participant 35, IIUM)

Students' Interest, Intention, and Preparedness to Engage with the Islam–Bioscience Interface Increased In the aggregate, levels of interest, intention, and preparedness to learn, inform others, engage with, and pursue opportunities at the bioscience–theology interface showed a statistically significant increase. The interest score rose from 11.4 to 11.8 ($p = 0.03$), the intention score increased from 19.6 to 20.9 ($p < 0.01$), and the preparedness score rose from 10.4 to 13.4 ($p < 0.01$). While the score changes appear small, they were statistically significant, aligned with one another, and reflect positive behavioral change.

Sub-analyses revealed no significant differences between the Malaysian and American cohorts in terms of baseline interest (11.4 versus 11.4, $p = 0.99$), intention (19.8 versus 18.8, $p = 0.18$), and preparedness levels (10.4 versus 10.8, $p = 0.60$). At the conclusion of the seminars, however, American students substantially improved in preparedness levels only, while Malaysian students improved in all the three domains of interest, intention, and preparedness.

Excerpts from students' essays reinforced these findings:

> After completing the seminar series, my perspective on science and Islam changed... they have made me more interested in further pursuing a career in science.
> (Participant 12, Chicago)

the seminar [series] has had a profound impact… It has sparked passion in me to examine how science and faith interact.
(Participant 101, IIUM)
I aim to apply the insights gained from this seminar to bridge the gap between science and religious communities.
(Participant 3, IIUM)
This seminar has instilled in me a strong determination to pursue and immerse myself further in the fields of theology and science.
(Participant 5, IIUM)
It [seminar series] increased my confidence and has made me more excited to study about both [domains] and their relationships.
(Participant 11, Chicago)

Students' Views on the Relationship Between Bioscience and Islam Are Diverse Prior to the seminars, American students had a more favorable opinion on science–religion compatibility than their Malaysian counterparts (17.1 points versus 16.1 points, $p = 0.03$). Post-intervention, overall, students' perception shifted toward greater science–religion *incompatibility* (from 16.3 to 15.6, $p < 0.01$). This overall decline was mainly driven by the Malaysian cohort as there was no significant change among American participants, whereas Malaysian students had significant score decline (from 16.1 to 15.4, $p < 0.01$).

When examining essay contents surrounding this subject, our findings were mixed. Some essays exhibited a tendency toward perceiving greater compatibility between bioscience and Islam after the program, while others focused on the tensions raised during the seminars. Interestingly, some students expressed contradictory views; while they admitted "tensions" exist, a harmonious relationship was still possible. Table 10.5 illustrates the different types of views held by students regarding the relationships between bioscience and Islam.

Multivariable Models Showed that Lower Baseline Preparedness and Knowledge Predicted Larger Gains Overall, our multivariable models showed that lower levels of preparedness at the baseline predicted knowledge gain when controlled for other factors (B: −0.03, $p = 0.01$), while no other socio-demographic characteristic contributed to knowledge acquisition. Analysis by cohort corroborated this finding for the Malaysian group (B: −0.03, $p = 0.05$). For the American students, higher interest in bioscience and Islam at baseline—and not preparedness level—predicted greater knowledge acquisition (B: 0.68, $p = 0.02$).

Table 10.5 Essay quotes depicting participants' views on bioscience–Islam relationships

Types of view	*Quotes*
Bioscience and Islam are in harmony	"There is frequently a belief that science and religion cannot coexist and that one develops at the expense of the other. This is a false statement, it fails to capture the rich and complex nature of the interaction between science and religion. This seminar[s] gave a clearer understanding of how science and religion can coexist and complement one another" (Participant 8, Chicago) "The most noticeable difference I can see is a shift in my thought process when it comes to Islam and science, from stating that Islam and science are incompatible to asserting that both are compatible" (Participant 11, IIUM) "…the (seminars) really helped me realize that Islam and science are more closely related than I originally thought" (Participant 13, IIUM)
There are many conflicts between these two domains	"Formerly, I thought everything about science and Islam was consistent with each other. It turns out that there are conflicts and contradictions. The difference is just the way we resolve them" (Participants 25, IIUM) "This session taught me about various scientific concepts that are at odds with Islamic viewpoints" (Participants 47, IIUM) "There are many conflicts (raised) during this seminar and also ways to resolve them and how our thinking process should be" (Participant 59, IIUM)

(continued)

Table 10.5 (continued)

Types of view	*Quotes*
Tensions exist, but a harmonious relation is possible	"Conflicts may arise when attempting to reconcile Islam and bioscience, but from my perspective, such conflicts are acceptable. Resolving conflicts between bioscience and Islam requires a thoughtful and balanced approach… there can be a harmonious relationship between bioscience and Islam." (Participant 61, IIUM) "Islam and science have different points of view… I also learnt that Islam and science are totally compatible with each other" (Participant 9, IIUM) "I think that this seminar helped me to see science and Islam in a different perspective… I will try my best to link Islam and science in a harmonious manner" (Participant 99, IIUM)

When change in preparedness was set as the outcome in regression models, we found that lower knowledge scores at baseline predicted higher increase in preparedness (B: −0.02, 95%CI: −0.05, 0.00, $p = 0.05$), when controlled for other factors. Cohort analyses supported this finding only for the American students (B: −0.07, 95%CI: −0.13, −0.01, $p = 0.02$) and not for the Malaysian group. Tables 10.6, 10.7, 10.8, 10.9, 10.10, and 10.11 presents results of these multivariable analyses.

Discussion

The core question to answer is: Was our curriculum effective? From an acceptability and feasibility standpoint, we would hazard certainly it was. None of the Islamic schools we approached declined having their students participate. Likewise, the CFS administrators were keen on having all students within their school participate, requesting us not to restrict student enrollment based on class size; they subsequently provided a large gym space for the seminar series. Additionally, students, overall, found the seminars to be informative, engaging, and valuable. Practically, we were able to recruit students through school-based advertisements and were able to implement the seminar series across four

Table 10.6 Generalized Linear Model to identify predictors of knowledge gain ($n = 191$)

Independent variable	*β*	*Standard error*	*95% Confidence interval*	*p value*
Model 1: Socio-demographic and religious domains				
Sex[a]	−0.15	0.09	−0.34, 0.04	0.13
Male				
Ethnicity[b]	0.62	0.58	−0.52, 1.75	0.29 0.77
Malay	−0.06	0.21	−0.46, 0.34	
South Asian				
Cohort	−0.45	0.59	−1.62, 0.72	0.45
Malaysia				
USA				
Religiosity (ORA)	−0.02	0.03	−0.09, 0.04	0.49
Model 2: Behavioral and belief domains				
Interest at T1	0.02	0.02	−0.03, 0.06	0.48
Preparedness at T1	−0.03	0.01	−0.06, −0.01	**0.01***
Intention at T1	0.01	0.01	−0.02, 0.04	0.43
Opinion/belief at T1	0.01	0.01	−0.03, 0.03	0.91

*Indicates statistical significance, T1: baseline/pre-seminar
[a]Female—reference category
[b]Others—reference category

Table 10.7 Generalized Linear Model to identify behavioral and belief-related predictors of knowledge gain in the Malaysian cohort ($n = 158$)

Independent variable	*β*	*Standard error*	*95% Confidence interval*	*p value*
Interest at T1	0.01	0.02	−0.05, 0.05	0.95
Preparedness at T1	−0.03	0.01	−0.05, −0.01	**0.05***
Intention at T1	0.01	0.01	−0.02, 0.04	0.66
Opinion/belief at T1	0.01	0.01	−0.02, 0.04	0.42

*Indicates statistical significance, T1: baseline

days in both settings. The instructors were able to deliver their content, didactic, media, and exercises all together, without complications and difficulty. Indeed, our team's formal after-event debriefs were replete with reflections on how the seminar series was energizing, beneficial to the instructors themselves (some listened in on sessions they were not leading), and well-coordinated.

Table 10.8 Generalized Linear Model to identify behavioral and belief-related predictors of knowledge gain in the American cohort ($n = 34$)

Independent variable	*β*	*Standard error*	*95% Confidence interval*	*p value*
Interest at T1	0.68	0.28	0.13, 1.23	**0.02***
Preparedness at T1	0.02	0.14	−0.26, 0.29	0.92
Intention at T1	−0.04	0.18	−0.39, 0.31	0.83
Opinion/belief at T1	−0.03	0.12	−0.27, 0.21	0.81

Table 10.9 Generalized Linear Model to identify predictors of improvement in preparedness ($n = 148$)

Independent variable	*β*	*Standard error*	*95% Confidence interval*	*p value*
Model 1: Socio-demographic and religious domains				
Sex	−0.09	0.11	−0.30, 0.12	0.40
Male				
[Female]				
Ethnicity	−0.17	0.17	−0.52, 0.18	0.33
Malay	−0.40	0.24	−0.87, 0.07	0.09
South Asian				
[Others]				
Religiosity (IR)	0.04	0.03	−0.02, 0.10	0.15
Model 2: Behavioral and belief domains				
Interest at T1	−0.03	0.02	−0.08, 0.02	0.28
Intention at T1	−0.01	0.01	−0.03, 0.03	0.86
Opinion/belief at T1	0.01	0.01	−0.01, 0.03	0.41
Knowledge at T1	– 0.02	0.01	– 0.05, 0.00	**0.05***

*Indicates statistical significance, T1: baseline

From a participation standpoint, while attrition for an optional enrichment program is to be expected, the attrition in the US cohort was relatively high when compared to the Malaysian cohort. We speculate that the higher rate among American students relates to the seminar series crossing two weekends across two months, and, as such, students may have had to attend to newfound pressing commitments that precluded attendance on the final weekend. Similarly, the consecutive four days in Malaysia likely facilitated participant retention among that group. These data have implications for future iterations of the curriculum as to whether

Table 10.10 Generalized Linear Model to identify predictors for improvement in preparedness in the Malaysian cohort ($n = 127$)

Independent variable	*β*	*Standard error*	*95% Confidence Interval*	*p value*
Model 1: Socio-demographic and religious domains				
Age	0.28	0.21	−0.13, 0.69	0.19
Religiosity (IR)	0.05	0.03	−0.01, 0.11	0.08
Model 2: Behavioral and belief domains				
Interest at T1	−0.02	0.03	−0.07, 0.03	0.39
Intention at T1	−0.01	0.01	−0.03, 0.03	0.91
Opinion/belief at T1	0.01	0.01	−0.01, 0.02	0.68
Knowledge at T1	−0.02	0.01	−0.04, 0.01	0.24

*Indicates statistical significance, T1: baseline

Table 10.11 Generalized Linear Model to identify behavioral and belief-related predictors for improvement in preparedness in the American cohort ($n = 31$)

Independent variable	*β*	*Standard error*	*95% Confidence Interval*	*p value*
Interest at T1	− 0.05	0.07	− 0.19, 0.10	0.53
Intention at T1	− 0.01	0.04	− 0.08, 0.07	0.93
Opinion/belief at T1	0.04	0.03	− 0.01, 0.09	0.14
Knowledge at T1	− 0.07	0.03	− 0.13, − 0.01	**0.02***

*Indicates statistical significance, T1: baseline

it is deployed as an optional after-school extra-curricular enrichment activity, or as an in-school elective, or as a required course within the school curriculum. In our view, as an enrichment activity, the intensive format across consecutive dates maybe a better approach. However, a less intensive format is preferred if the course is integrated with a longitudinal school curriculum as attrition would not be an issue.

Moving on to the specific learning outcome measures, as reported above, our curriculum was efficacious. Students' knowledge gain was substantial, and levels of interest, intention, and preparedness to learn more, engage with, and pursue further opportunities in this field increased significantly. In what follows, we comment on several notable findings and the implications thereof.

Among the American cohort, there was no change in levels of interest and intention post-seminar, unlike the trend among their Malaysian counterparts. Empirical evidence in this regard is mixed; some studies corroborate interest as a consequence of learning while others report that learning itself is the outcome of interest (Rotgans and Schmidt 2017; Ainley et al. 2002). Similarly, the empirical link between learning and intention is inconsistent. In some circumstances, knowledge, that is, learning gain, alone is inadequate to motivate intention or actual behavior (Arlinghaus and Johnston 2018; Bang et al. 2000); importantly, these two constructs are linked through multiple other mediating and moderating factors (Levine and Strube 2012; Lucero and Chen 2020). Hence, while we desired an increase in interest and intention post-learning, theory suggests that such results may not always follow. Nonetheless, we speculate that the increase in intention levels post-seminar among Malaysian participants was due to their being older and more advanced in their studies as compared to the American cohort, as such, they are more likely to have thought about future careers or next steps in their learning journey and thus become more interested in the intersectional topics. With respect to preparedness levels, both cohorts demonstrated significant improvement. This is in line with existing literature that underscores how "knowing" is associated with "feeling prepared" (Michalski et al. 2020).

Students' views on the relationship between science and religion warrant a more nuanced discussion. Although quantitative data suggested an overall score decline (more prominent among Malaysian students), indicating that students viewed the relationship as less compatible or more conflictual upon seminar completion, the essays indicated mixed and diverse standpoints. The multiplicity of views in this regard, along with the lack of collective attitudinal change away from "conflictual relationship" are in fact coherent with the seminar contents and our pedagogical approach; a distinctive feature of our curriculum was to avoid providing definitive answers. Rather, we sought to stimulate critical inquiry and curiosity, while teaching students to accept plurality and uncertainty. Highlights of discord do not necessarily mean students hold a negative attitude toward the relationship between science and religion;

rather, our findings may reflect students' preoccupation with the "conflicts" deliberately brought up by the instructors to stimulate discussion.[13] Likewise, in our opinion, the coexistence of conflict and harmony demonstrated in Table 10.5 does not imply confusion or incongruity; rather, it reflects a realistic approach toward the complex nature of the discourse and relationship between science and theology. The students did not hold uniform perspectives analogous to the ways the discourse beguiles a singular approach. However, it is important to note that only 68 students chose to write essays, while almost all ($n = 208$) completed the survey. Hence, the window into reality is much smaller from the qualitative data, and these data are subject to selection and spectrum biases. Nonetheless, the essays indicate that the shift toward lowered compatibility as seen in the declining scores could represent a change of opinion from complete harmony to a more nuanced relation between the two domains.

Adding another layer of complexity, studies show that undergoing a belief change process—in which a learner's perspective or worldview is challenged and about to change—can result in a transitional state of "liminality," or a suspended state, where "the thought process oscillates back-and-forth between old and emergent understandings" (Cousin 2006; Land et al. 2014). Our prompt assessment (conducted immediately post-seminar) could have captured this phase of liminality; a real change in belief or attitude therefore requires sufficient time and is perhaps more detectable by a repeat measurement. Overall, the picture is complicated and deserves further dedicated study through implementations beyond the present sample. In our view, the resulting nuanced and complicated conceptions of the relationships between Islam and bioscience, and the attitudes about the discourse between these domains, reflect positively as the outcome of learning a critical approach to assessing truth claims.

For the Malaysian cohort, lower preparedness at baseline predicted knowledge acquisition post-seminar whereas for the American cohort, knowledge gain was predicted by higher interest levels. We speculate that lower preparedness—in some circumstances—can signify self-awareness of one's ability and current state of knowledge, an insight that can propel

[13] Some "conflicts" were deliberately brought up by the instructors to initiate discussions. Examples include the claims of biological evolutionists versus the Qurʾanic narrative of human creation (under the topic of Human Evolution) and the claims of some neuroscientists about the non-existence of the human soul versus the Qurʾanic assertion (under the Topic of Human Nature).

greater motivation to study and learn. Indeed, Friedman and Mandel who conducted a study among college students in New York found that needs for achievement—defined as desire to achieve results and improve performance—predicted higher cumulative GPA scores (Friedman and Mandel 2011). On the other hand, interest leads to knowledge gain by improving one's cognitive and affective functioning (Schiefele et al. 1992; Schiefele and Krapp 1996) and ability to focus (Ainley et al. 2002). The distinction in factors that prognosticate knowledge gain between the two cohorts potentially indicate differences in learning styles and motivational beliefs embraced by students in different geographical and educational settings. For instance, empirical studies demonstrate that in the Western educational contexts, individual interest and passion are more valued, and these factors predict academic achievement (Li et al. 2021). However, in collectivistic societies, interest is not the most critical factor predicting students' achievements; it is balanced out by other criteria such as parental support, family expectations, and fear of failure (Li et al. 2021; Eaton and Dembo 1997).

Among the American participants, poorer knowledge at baseline predicted a better sense of preparedness post-intervention. We postulate that students with lower baseline scores have more room for improvement and hence could make a bigger leap than their well-informed peers. This resultant "score gap" could have led to a higher sense of readiness. Moreover, we found a negative correlation between baseline knowledge and score increase ($r = -0.61$, $p < 0.01$) [data not shown].[14] From another perspective, this could imply a misconstrued perception of preparedness, stemming from a common tendency among lower-performing students to overestimate their abilities, while higher-performing students underestimate theirs—a well-documented phenomenon in various empirical studies (Hacker, Bol, and Bahbahani 2008; Miller and Geraci 2011; Vadhan and Stander 1994).

Our study findings should be interpreted in light of several limitations. First, our sampling frame in the USA primarily involved private Islamic schools in two cities, therefore, it does not represent the full diversity of American Muslim high school students. Likewise, the Malaysian cohort came from a single institution. Neither cohort was fully representative of the student population in these settings and thus our findings

[14] This means the lower the baseline knowledge, the higher the score leap.

may not be wholly generalizable. Nonetheless, our positive findings are a starting point for further replication within Islamic schools/institutions with more diverse samples and in different geographic settings. Second, seminar participation was driven by interest in the bioscience–theology interface and in bioscientific and medical careers. Our students therefore are likely to be primed to finding the topics and teaching to be of value. A broader high school student population may be less engaged in learning and the impact of the curriculum may not be as positive. Therefore, our curriculum may be better suited as an elective, enrichment activity for select student groups. Third, our instruments were derived from a combination of published scales and consultation with field experts. While face and content validity were conducted, we did not perform statistical validation of the survey instrument. It is worth highlighting here that the scarcity of validated tools among the Muslim population can hamstring research in this area, and our work may set up future validation studies. Fourth, our post-intervention assessment was conducted one time and immediately after the last seminar; a repeat assessment could have yielded more rigorous results regarding knowledge retention as well as behavior change. It is also critical to recognize that novel elements of the program may reduce the replicability of our findings. Specifically, the instructors were dually trained scholars with professional identities at the interface of bioscience and Islam. Their personal narratives, enthusiasm, and characteristics may have impacted students in ways that were not measured by our surveys nor captured in student reflections. Future implementation of our curriculum may require teacher training workshops and the utilization of scripts and resources from our instructors.[15] Lastly, while our principal approach was to emphasize critical inquiry and plurality, and to avoid prescribing definitive answers, we acknowledge that affirming neutrality in real-life educational settings is not easy, given the existence and nature of human subconscious biases.

That said, our findings have several important implications for educators and for future curricular implementation. First, an open and constructive pedagogical approach that embraces critical thinking and plurality of answers (within certain boundaries) may be impactful in the long term and suitable for Muslim youth today who are constantly

[15] Teacher training workshop materials including seminar teaching videos, curriculum guides containing lesson plans and scripts, and a teacher's companion book is available at Initiative on Islam and Medicine. n.d.

challenged by a myriad of ideologies and perspectives challenging their religious worldview. These include the Monoculture of Modern Science (MMS) that perpetuates the coloniality of knowledge as elaborated by Usama Javed Mirza in Chapter 9. Second, factors that predict program effectiveness might differ from one sociocultural and geographical setting to another; educators should be cognizant of learners' backgrounds and use a variety of teaching strategies to maximize curricular benefits. Third, data collection prior to program implementation is highly useful as it helps educators understand their prospective students and plan their approach accordingly. Our data shows that educators need not excessively worry about students who appear less prepared or less knowledgeable at the baseline because these factors do not necessarily hamper their ability to benefit from an educational intervention. Fourth, while the unique dual training of our seminar instructors is a major strength, the scalability of this program could be challenged in the absence of similar instructor expertise. However, with proper training and teaching resources, instructors around the globe teach subjects they are not masters of. Indeed, we view this issue as an opportunity for training and dissemination. As such, the project has embedded teacher trainings and dissemination of teaching modules as part of its activities.[16] Indeed, for those who desire to increase fidelity, the project website and teaching modules contain videos of our instructors teaching the material which could be used for didactic purposes in the classroom, and also contain detailed scripts that instructors can utilize.

In a similar vein, some may wonder whether our curricular model can be applied to public or secular institutions, particularly due to its religious content. Truth be told, aspects of the curriculum are already being implemented in a secular, private, medical school setting as an elective series and has been received with great enthusiasm by students of various faith (and no faith) backgrounds.[17] Furthermore, it is essential to note

[16] These are available at Initiative on Islam and Medicine. n.d.

[17] This implementation, led by Dr. Aasim Padela, involves co-teaching the seminars with a philosopher-ethicist, Prof. Fabrice Jotterand, and expanding the teaching to include other religious and philosophical perspectives on the Big Questions. The elective, hosted at the Medical College of Wisconsin, is titled "Between Bioscience, Philosophy, and Theology: Big Questions about the Human Being." The course is in its third iteration and has had approximately 15% of the medical students at the school participate in any given year.

that our project goal was to foster literacy and interest in this field among Muslim students, hence, transferability was not a primary consideration.

Lastly, despite students' interest and keenness to engage with educational activities, their attitude/beliefs toward a subject matter may not necessarily change significantly or quickly. This is something to be accepted and understood by educators whose roles should focus on cultivating understanding, appreciation, and love for knowledge, rather than altering students' views at the outset. Indeed, an appreciation for the need for lifelong learning is the north star to be aimed for.

In summary, our novel curriculum, "Exploring Bioscience and Islam," was feasible to implement as an extracurricular, seminar-based enrichment program for Muslim secondary school and college students. The program was judged to be highly acceptable in terms of content by both instructors and students within Islamic school and university contexts. Moreover, it was effective in improving students' knowledge about bioscientific and theological conceptions of the human being, as well as their interest, intention, and preparedness to engage at the intersection of Islam and bioscience. While students' attitude toward religion–science compatibility varied, many of them seem to have acquired more nuanced views upon completing the seminars. Given these successes and positive findings, our program presents a model that can be replicated and built upon to engage pre-medical students in topics at the intersection of Islam and bioscience.[18]

Appendix

Domain: Interest

Item	*Pre*	*Post*	*p-value*
1. Artificial intelligence	3.69 (1.06)	3.85 (0.83)	0.41
2. Islamic theories of knowledge	4.10 (0.72)	3.68 (0.88)	**<0.01**
3. Islamic theories on the origin of universe	3.92 (1.13)	3.95 (0.93)	0.83
4. Philosophy of bioscience	3.33 (0.87)	3.23 (1.05)	0.46
5. Theories of evolution	3.26 (1.14)	3.08 (1.25)	0.25

(continued)

[18] Toward this end, our team also offers teacher training workshops as well as other assistance in implementing the program (please see Initiative on Islam and Medicine n.d., XXX).

(continued)

Item	*Pre*	*Post*	*p-value*
6. Islamic conceptions of human beings	3.67 (1.00)	3.55 (1.00)	0.29
7. Intelligent life beyond earth	3.77 (1.33)	3.73 (1.30)	0.55
8. The use of genetic/genomic therapies in medicine	3.41 (1.12)	3.50 (0.99)	0.75
9. Islamic doctrines on free will and human agency	3.49 (1.07)	3.95 (1.06)	**0.02**
10. The use of biotechnology in healthcare	3.77 (0.93)	3.53 (1.01)	0.05
11. Islamic bioethics	3.26 (0.94)	3.78 (0.80)	**0.01**
12. Learning how to integrate knowledge from bioscience and Islam in my future career	3.97 (0.99)	3.75 (1.01)	0.07
13. Learning skills that will prepare me for a career at the interface of bioscience and Islam	3.92 (0.93)	3.68 (1.19)	0.19
14. Pursuing research topics at the intersection of bioscience and Islam	3.46 (0.99)	3.35 (1.00)	0.58

The stem was: "Indicate your interest in the following topics," and the response was a Likert scale from 1 (not interested at all) to 5 (extremely interested).

Domain: Intention

Item	*Pre*	*Post*	*p-value*
1. Obtain further training at the intersection of bioscience and Islam	3.28 (0.96)	3.23 (0.80)	0.84
2. Study with experts/scholars working at the intersection of bioscience and Islam	3.33 (0.92)	3.05 (0.98)	0.14
3. Seek advanced interdisciplinary degrees at the intersections of bioscience and Islam	2.70 (0.99)	2.85 (0.89)	0.32
4. Conduct research at the intersections of bioscience and Islam	3.05 (1.06)	2.88 (1.04)	0.24
5. Inform others about the intersection of bioscience and Islam	3.68 (0.83)	3.65 (1.09)	0.86
6. Assure that my career focuses on research and education at the interface of bioscience and Islam	2.70 (0.99)	2.85 (1.23)	0.44

The stem was: "Indicate your intention to do the following," and the response was a Likert scale from 1 (no intention at all) to 5 (already doing it).

Domain: Preparedness

Item	*Pre-*	*Post-*	*p-value*
1. Learn about the intersection of bioscience and Islam	3.03 (1.09)	3.50 (0.88)	**0.02**
2. Pursue courses at the intersection of bioscience and Islam	2.73 (1.11)	3.13 (1.07)	**0.03**
3. Pursue research opportunities at the intersection of bioscience and Islam	2.68 (0.97)	3.00 (1.09)	**0.04**
4. Pursue a career at the intersection of bioscience and Islam	2.33 (1.16)	3.13 (1.14)	**<0.01**

The stem was: "Indicate your level of preparedness to do the following," and the response was a Likert scale from 1 (not prepared at all) to 5 (extremely prepared).

Items in the Belief/View Questionnaire (on Science–Religion Relationship)

No	*Item*
1	There is irreconcilable conflict between bioscience and Islam
2	Bioscience cannot complement Islamic views on the human being
3	Integrating bioscience and Islam is not possible
4	There is no meeting point between bioscience and Islam
5	The more modern science reveals about the universe, the less likely it seems that there is a Divine force behind it
6	Science is the only way to obtain reliable knowledge
7	Religious belief will hinder you from becoming a good scientist
8	The belief that God exists is not compatible with science
9	Science can sometimes damage people's sense of morality
10	Modern science has disproved the existence of the afterlife

Seminar Outlines

Seminar	*Learning objectives*	*Outline/main contents*
1. What do we know and how do we know it? Reasoning exercises in religion and science	1. Identify and explain the sources of knowledge and their relationship to each other 2. Define, provide examples, and explain the relationship between the three types of propositions (rational, revelatory, empirical) 3. Classify various propositions and provide justification	*Didactics* * Islamic and naturalistic epistemologies: overlapping epistemologies * Islamic and naturalistic epistemologies: Karl Popper and the unseen—is revelation a source of knowledge? * Integrating science and revelation: how do we resolve conflict? *Popular media discussion/ reflection elicitation tools* * Galileo—Science and the Church: https://www.youtube.com/watch?v=NMM8vx9vDiE * Karl Popper—Science is the only way to truth: https://www.youtube.com/watch?v=rVo6tTQKa70 *Facilitated group work and classroom discussions* * Galileo—Science and the Church: discuss probing questions about the conflict Galileo had with the church regarding scientific principles * Karl Popper and the Unseen: group discussion as to how Karl Popper would analyze certain prepositions versus how a Muslim would analyze those same prepositions * Where is the overlap between revelation and science? * What is a revelatory and empirical proposition?

(continued)

(continued)

Seminar	*Learning objectives*	*Outline/main contents*
2. What is human origin? Human evolution and god's special creation	1. Describe and explain the scientific account of Neo-Darwinian evolution 2. Describe and explain the theological account of non-humans and humans 3. Describe, categorize, and evaluate various Muslim responses to Neo-Darwinian evolution 4. Distinguish and evaluate the various kinds of conflicts (potential, apparent, and actual) between Islam and evolution	*Didactics* * Islamic narratives on creation * The science of evolution *Popular media discussion/ reflection elicitation tools* DNA evidence that humans & chimps share a common ancestor: endogenous retroviruses https://youtu.be/oXfDF5Ew3Gc *Facilitated classroom discussions* * After observing the skeletal structure, determine which animal each structure belongs to * After watching video: share thoughts from students about the conclusiveness or probabilistic nature of evolutionary evidences

(continued)

(continued)

Seminar	*Learning objectives*	*Outline/main contents*
3. What is human essential nature? Neuroscience, artificial intelligence, and human nature	1. Describe the Islamic concept of the soul and personal identity 2. Define consciousness and discuss how this experience cannot be reduced to physical processes 3. Explain the basic concepts in neural-correlates-of consciousness 4. Discuss what the neuroscience leaves out about consciousness 5. Describe how artificial intelligence differs from human consciousness	*Didactics* * Islamic view of human nature: the soul * What does death mean for humans? * What is the purpose of life? * Challenges to the Islamic concept of human nature * Introduction to ontology—nature of the world * The mind, the brain and neuroscience—Dr. Macksood Aftab * Consciousness, the soul and artificial intelligence * What does artificial generalized intelligence mean for humans? *Thought experiments/reflection elicitation tools* * I think therefore I am—Descartes * Mary's room and color experiment on consciousness * Turing test and Chinese room experiment on understanding & knowledge * Karamali worship experiment on spirituality and the soul *Media sources:* * Mary's room: http://www.youtube.com/watch?v=mGYmiQkah4o * How does AI work: https://youtu.be/WAiqNav2cRE * AI & Personhood: http://www.youtube.com/watch?v=39EdqUbj92U

(continued)

(continued)

Seminar	*Learning objectives*	*Outline/main contents*
4. Human enhancement and transhumanism: are we changing/should we change our being?	1. Identify key components of transhumanist ideology 2. Discern transhumanism-based narratives and elements within popular media 3. Distinguish between Islamic theological perspectives on human enhancement and its purposes and secular bio-scientific views 4. Apply scriptural and scientific concepts to evaluate uses of biomedical technologies to "change" the human body	*Didactics* * Biomedical interventions & biological futures * Revelatory insights into enhancing our being and human futures * Limits and possibilities of changing our body and soul *Popular media discussion/reflection elicitation tools* * "Transhumanism: Will humans evolve to something smarter?" A–Z of ISMs Episode 20, BBC Ideas, YouTube: https://www.youtube.com/watch?app=desktop&v=RVmuU04-X5E&feature=youtu.be * Fennic's cybernetic implants: "Fennec's Cybernetic Implants." Book of Boba Fett S01E04 Clip, 4 k, YouTube: https://www.youtube.com/watch?v=14RsOGZcfnU * Sully becomes an Avatar: transferring consciousness: https://www.youtube.com/watch?v=7ogLskyjxwc https://www.youtube.com/watch?v=e0GowLIIHck *Facilitated group work and classroom discussions* * After watching videos, reflect on the stories and re-create your own * After hearing about contemporary real "cyborgs," share their stories and reflect on them * Case review, analysis, and discussion; three cases (i) Tommy John surgery for baseball player; (ii) Lasik surgery; (iii) Brain function supplement

(continued)

(continued)

Seminar	*Learning objectives*	*Outline/main contents*
5. Are we programmed or do we have free will? Discussions on fate and genetic determinism	1. Describe the Islamic view on free will and determinism 2. Identify key components human nature related to free will and determinism 3. Identify the accountability of human related to human choice 4. Describe the features of genetic interaction to behavior with the role of epigenetics 5. Classify the difference types of genetic determinism	*Didactics* * Revelatory insights on free will and determination * Genetic determinism *Popular media discussion/ reflection elicitation tools* * Epigenetics: https://youtu.be/_aAhcNjmvhc * CRSPR animation: https://youtu.be/UKbrwPL3wXE * Big debate on free will and determinism: https://youtu.be/VlV7BCjZy8Y *Facilitated group work and classroom discussions* * Class activity using https://playgameoflife.com/ & discuss free choice and deterministic elements in the game * Discussion about the game and its relation to reality

(continued)

(continued)

Seminar	*Learning objectives*	*Outline/main contents*
6. Are humans unique in this universe? Science and theological inquiry about extra-terrestrial intelligent life	1. Identify the sources and motivations of intellectual and scientific enquiry 2. Critically analyze and position various scientific inquiries considering Islamic theology and ethics 3. Appreciate the depth and scope of scientific enquiry in the Islamic intellectual tradition 4. Apply an Islamic epistemological framework to scriptural and scientific concepts in order to prove a cohesion between all conclusive knowledge evidence	*Didactics* * Do extraterrestrial intelligent beings exist? Exploring the empirical evidence * Case study: the seven earths question: what does revelation say about ETI? Conclusions: how certain is ETI based on empirical, rational, and revelatory evidence? *Popular media discussion/ reflection elicitation tools* * What is our current state of knowledge regarding the existence of extraterrestrial life? https://www.youtube.com/watch?v=-HzjRWdXjyk *Facilitated group work and classroom discussions* * After watching video, share thoughts with groups about the conclusiveness of ETI through empirical evidence; how does this evidence inform the overall likelihood of ETI existing? * Share thoughts with one another about what revelation says about ETI; how does this revelatory evidence inform the overall likelihood of ETI existing?

(continued)

(continued)

Seminar	*Learning objectives*	*Outline/main contents*
7. Knowledge Cafés and Harvest: Panel presentations & moderated discussions	1. Apply epistemic concepts to revisit key ideas presented in the seminars 1–6 2. Discern how the knowledge gained can be used to explore "novel" issues at the intersection of Islam and bioscience 3. Participate in a humble dialogue and discussion about synergy and controversy at the interface of Islam and bioscience 4. Consider how to communicate/express your newfound insights through written or artistic methods	*Revision of the 6 topics* * Recap of the main ideas of 6 preceding seminars * Application of epistemic framework tool to the truth claims of each seminar *Case studies involving a model dialogue between theologian and bioscientist over the case, and facilitated discussions with students* 1. Case Study 1: Jinn Possession 2. Case Study 2: Polydactyly

Some Examples of the Knowledge Items/Questions

Type of question	*Item*
True or false	1. A proposition is a statement which can be true or false 2. The similarity between the Islamic and naturalistic epistemology is that both view cause-and-effect relations as necessary 3. The Qurʾan does not mention specifically the age of the Earth 4. Our experience of color and sound is an example of consciousness 5. Neuroscientific reductionism is compatible with the Islamic understanding of the human soul 6. Bodily changes are always impermissible in Islam 7. Genes and human behavior have strong and direct causal relationships 8. With the evidence we now have about ETI from empirical, rational, and revelatory sources, we can say that its existence is conclusive
Multiple choice	The statement—"There are 600 billion planets in our galaxy"—is based on: (a) Revelatory knowledge (b) Empirical knowledge (c) Rational knowledge (d) Empirical and rational knowledge

References

5DThinking. n.d. "5 Dimensional Thinking." 5DThinking. https://5dthinking.org/.

Abu Khalaf, Nadin, Ashley B. Woolweaver, Roslyn Reynoso Marmolejos, Grace A. Little, Katheryn Burnett, and Dorothy L. Espelage. 2023. "The Impact of Islamophobia on Muslim Students: A Systematic Review of the Literature." *School Psychology Review* 52 (2): 206–223. https://doi.org/10.1080/2372966X.2022.2075710.

Ahmadi, T., Huma Bokharey, Irum Sarfaraz, Asma Pal, Fatima Rodriguez, and Baasil Bokharey. 2020. *Science: Level 1 Workbook*: Independently published.

Aikenhead, Glen S., and Olugbemiro J. Jegede. 1999. "Cross—Cultural Science Education: A Cognitive Explanation of a Cultural Phenomenon." *Journal of Research in Science Teaching* 36 (3): 269–287. https://doi.org/10.1002/(SICI)1098-2736(199903)36:3<269::AID-TEA3>3.0.CO;2-T.

Ainley, Mary, Suzanne Hidi, and Dagmar Berndorff. 2002. "Interest, Learning, and the Psychological Processes that Mediate their Relationship." *Journal of Educational Psychology* 94 (3): 545–561. https://doi.org/10.1037/0022-0663.94.3.545.

Alassiri, Mohammed. 2020. "Evolution is the Disguised Friend of Islam." *Nature Human Behaviour* 4 (2): 122–122. https://doi.org/10.1038/s41562-019-0771-7.

Alexander, Denis Raven. 2007. *Models for Relating Science and Religion*. Vol. 3. Faraday Institute for Science and Religion Kanpur. https://www.faraday.cam.ac.uk/resources/faraday-papers/.

Amjad, Afshan. 2018. "Muslim Students' Experiences and Perspectives on Current Teaching Practices in Canadian Schools." *Power and Education* 10 (3): 315–332. https://doi.org/10.1177/1757743818790276.

Arlinghaus, Katherine R., and Craig A. Johnston. 2018. "Advocating for Behavior Change with Education." *American Journal of Lifestyle Medicine* 12 (2): 113–116. https://doi.org/10.1177/1559827617745479.

Bang, Hae-Kyong, Alexander E. Ellinger, John Hadjimarcou, and Patrick A. Traichal. 2000. "Consumer Concern, Knowledge, Belief, and Attitude Toward Renewable Energy: An Application of the Reasoned Action Theory." *Psychology & Marketing* 17 (6): 449–468. https://doi.org/10.1002/(SICI)1520-6793(200006)17:6%3c449::AID-MAR2%3e3.0.CO;2-8.

Barbour, Ian G. 1990. *Religion in an Age of Science*. San Francisco, CA: Harper & Row.

Barbour, Ian G. 2002. "On Typologies for Relating Science and Religion." *Zygon: Journal of Religion and Science* 37 (2): 345–360. https://doi.org/10.1111/0591-2385.00432.

Barnes, M. Elizabeth, E. Margaret Evans, Ashley Hazel, Sara E. Brownell, and Randolph M. Nesse. 2017. "Teleological Reasoning, Not Acceptance of

Evolution, Impacts Students' Ability to Learn Natural Selection." *Evolution: Education and Outreach* 10 (1): 1–12. https://doi.org/10.1186/s12052-017-0070-6.

BouJaoude, Saouma. 2003. "Science and Technology Education in the Arab World in the 21st Century." UNESDOC Digital Library. https://unesdoc.unesco.org/ark:/48223/pf0000133581.

Cardinal, Harold. 1969. *The Unjust Society: The Tragedy of Canada's Indians*. Edmonton: M.G. Hurtig.

Chaney, Eric. 2016. "Religion and the Rise and Fall of Islamic Science." Working Paper, Department of Economics, Harvard University, Cambridge, MA. https://scholar.harvard.edu/files/chaney/files/paper.pdf.

Collins, Francis S. 2006. *The Language of God: A Scientist Presents Evidence for Belief*. New York: Simon and Schuster.

Cousin, Glynis. 2006. "An Introduction to Threshold Concepts." *Planet* 17 (1): 4–5. https://doi.org/10.11120/plan.2006.00170004.

Crutzen, Rik, and Rob Ruiter. 2013. "Interest in Behavior Change Interventions: A Conceptual Model." *The European Health Psychologist* 17 (1): 6–11. https://ehps.net/ehp/index.php/contents/article/view/759/pdf_35.

Crutzen, Rik, Robert A.C.. Ruiter, and Nanne K. de Vries. 2014. "Can Interest and Enjoyment Help to Increase Use of Internet-Delivered Interventions?" *Psychology & Health* 29 (11): 1227–1244. https://doi.org/10.1080/08870446.2014.921300.

Dean, Cornelia. 2005. "Evolution Takes a Back Seat in U.S. Classes." *New York Times*, February 1. https://www.nytimes.com/2005/02/01/science/evolution-takes-a-back-seat-in-us-classes.html. Accessed on 7 January 2024.

Eaton, Martin J., and Myron H. Dembo. 1997. "Differences in the Motivational Beliefs of Asian American and Non-Asian Students." *Journal of Educational Psychology* 89 (3): 433.

Friedman, Barry A., and Rhonda G. Mandel. 2011. "Motivation Predictors of College Student Academic Performance and Retention." *Journal of College Student Retention: Research, Theory & Practice* 13 (1): 1–15. https://doi.org/10.2190/CS.13.1.

Gale, Nicola K., Gemma Heath, Elaine Cameron, Sabina Rashid, and Sabi Redwood. 2013. "Using the Framework Method for the Analysis of Qualitative Data in Multi-Disciplinary Health Research." *BMC Medical Research Methodology* 13 (117): 1–8. https://doi.org/10.1186/1471-2288-13-117.

Guessoum, Nidhal. 2010. "Science, Religion, and the Quest for Knowledge and Truth: An Islamic Perspective." *Cultural Studies of Science Education* 5 (1): 55–69. https://doi.org/10.1007/s11422-009-9208-3.

Guessoum, Nidhal. 2018. "Science and Religion Issues in Higher Education." In *Universities in Arab Countries: An Urgent Need for Change: Underpinning the Transition to a Peaceful and Prosperous Future*, edited by Adnon Badran, Elias Baydoun, and John R. Hillman, 187–197. Cham: Springer.

Hacker, Douglas J., Linda Bol, and Kamilla Bahbahani. 2008. "Explaining Calibration Accuracy in Classroom Contexts: The Effects of Incentives, Reflection, and Explanatory Style." *Metacognition and Learning* 3 (2): 101–121. https://doi.org/10.1007/s11409-008-9021-5.

Harder, Elma Ruth, and Muzaffar Iqbal. 2006. *Concentric Circles: Nurturing Awe and Wonder in Early Learning: A Foundational Approach*. Sherwood Park, Alberta: Al-Qalam Publishing.

Harrison, Peter. 2006. "'Science' and 'Religion': Constructing the Boundaries." *The Journal of Religion* 86 (1): 81–106. https://doi.org/10.1086/497085.

Haseeb, Abid. 2022. "Rethinking Islamic School Education." *Traversing Tradition*, January 26. https://traversingtradition.com/2022/01/26/rethinking-islamic-school-education/.

Hashim, Rosnani. 1999. "Islamization of the Curriculum." *American Journal of Islam and Society* 16 (2): 27–43. https://doi.org/10.35632/ajis.v16i2.2127.

Hassan, M. Kamal. 2017. *Natural Science from the Worldview of the Qur'an: An Introduction*. Kuala Lumpur: Institut Terjemahan & Buku Malaysia (ITBM).

Haught, John F. 1995. *Science and Religion: From Conflict to Conversation*. New York: Paulist Press.

Hornbuckle, Molly. 2022. "5 Reasons Pretests and Posttests in Education Matter." *YouScience*, January 25. https://www.youscience.com/pretest-and-posttest-in-education/.

Initiative on Islam and Medicine. n.d. "Fostering Connections Between Science and Theology for Medical Track Muslim Students." https://www.medicineandislam.org/fostering-connections-between-science-and-theology-2/.

Knepper, Paul. 2003. "Faith, Public Policy, and the Limits of Social Science." *Criminology & Public Policy* 2 (2): 331–352. https://doi.org/10.1111/j.1745-9133.2003.tb00130.x.

Koenig, Harold G., and Arndt Büssing. 2010. "The Duke University Religion Index (DUREL): A Five-Item Measure for Use in Epidemological Studies." *Religions* 1 (1): 78–85. https://doi.org/10.3390/rel1010078.

Land, Ray, Julie Rattray, and Peter Vivian. 2014. "Learning in the Liminal Space: A Semiotic Approach to Threshold Concepts." *Higher Education* 67 (2): 199–217. https://doi.org/10.1007/s10734-013-9705-x.

Lay, Yoon Fah, Eng Tek Ong, Crispina Gregory K. Han, and Sane Hwui Chan. 2018. "A Glimpse of Evolution Education in the Malaysian Context." In *Evolution Education around the Globe*, edited by Hasan Deniz and Lisa A. Borgerding, 357–374. Cham: Springer.

Levine, Debra Siegel, and Michael J. Strube. 2012. "Environmental Attitudes, Knowledge, Intentions and Behaviors among College Students." *The Journal of Social Psychology* 152 (3): 308–326. https://doi.org/10.1080/00224545.2011.604363.

Li, Xingyu, Miaozhe Han, Geoffrey L. Cohen, and Hazel Rose Markus. 2021. "Passion Matters but Not Equally Everywhere: Predicting Achievement from Interest, Enjoyment, and Efficacy in 59 Societies." *Proceedings of the National Academy of Sciences* 118 (11): e2016964118. https://doi.org/10.1073/pnas.2016964118.

Longest, Kyle C., and Christian Smith. 2011. "Conflicting or Compatible: Beliefs about Religion and Science Among Emerging Adults in the United States 1." *Sociological Forum* 26 (4): 846–869.

Lucero, Katie Stringer, and Pan Chen. 2020. "What Do Reinforcement and Confidence Have to Do with It? A Systematic Pathway Analysis of Knowledge, Competence, Confidence, and Intention to Change." *Journal of European CME* 9 (1): 1834759. https://doi.org/10.1080/21614083.2020.1834759.

Mahner, Martin, and Mario Bunge. 1996. "Is Religious Education Compatible with Science Education?" *Science & Education* 5: 101–123. https://doi.org/10.1007/BF00428612.

Michalski, Piotr, Michał Kasprzak, Michał Siedlaczek, and Aldona Kubica. 2020. "The Impact of Knowledge and Effectiveness of Educational Intervention on Readiness for Hospital Discharge and Adherence to Therapeutic Recommendations in Patients with Acute Coronary Syndrome." *Medical Research Journal* 5 (2): 72–78. https://doi.org/10.5603/MRJ.a2020.0023.

Miller, Tyler M., and Lisa Geraci. 2011. "Unskilled but Aware: Reinterpreting Overconfidence in Low-Performing Students." *Journal of Experimental Psychology: Learning, Memory, and Cognition* 37 (2): 502–506. https://doi.org/10.1037/a0021802.

Mirza, Usama Javed. 2024. "Islamic Scientific Critical Consciousness as a Theoretical Framework for Muslim Science Educators." *London Review of Education* 22 (1). https://doi.org/10.14324/LRE.22.1.09.

Montuschi, Eleonora. 2017. "Using Science, Making Policy: What Should We Worry About?" *European Journal for Philosophy of Science* 7: 57–78. https://doi.org/10.1007/s13194-016-0143-3.

Morton, Terrell R., Wesley Agee, Kilan C. Ashad-Bishop, Lori D. Banks, Zanethia Choice Barnett, Imari D. Bramlett, Briana Brown, Walter Gassmann, Korie Grayson, and Gail P. Hollowell et al. 2023. "Re-Envisioning the Culture of Undergraduate Biology Education to Foster Black Student Success: A Clarion Call." *CBE—Life Sciences Education* 22 (4): es5. https://doi.org/10.1187/cbe.22-09-0175.

Nord, Warren A. 1999. "Science, Religion, and Education." *Phi Delta Kappan* 81 (1): 28–33.

Oh, Youn Joo, Yueming Jia, Mhora Lorentson, and Frank LaBanca. 2013. "Development of the Educational and Career Interest Scale in Science, Technology, and Mathematics for High School Students." *Journal of Science Education and Technology* 22 (5): 780–790. https://doi.org/10.1007/s10956-012-9430-8.

Osman, Kamisah, Rezzuana Razali, and Nurnadiah Mohamed Bahri. 2018. "Biological Evolution Education in Malaysia: Where We Are Now." In *Evolution Education around the Globe*, edited by Hasan Deniz and Lisa A. Borgerding, 375–390. Cham: Springer.

Pakistan Innovation Foundation. 2024. "Holistic Science Teaching Workshop & Fellowship." Pakistan Innovation Foundation. https://pif.org.pk/scienceteaching/. Accessed on 13 April 2024.

Polkinghorne, John C. 2009. *One World: The Interaction of Science and Theology*. West Conshohocken, PA: Templeton Foundation Press.

Qureshi, Omar. 2021. "Science Curriculum from an Islamic Worldview." In *Curriculum Renewal for Islamic Education: Critical Perspectives on Teaching Islam in Primary and Secondary Schools*, edited by Nadeem Memon, Mariam Alhashmi, and Mohamad Abdalla, 113–124. New York: Routledge.

Rafe'Zou'bi, Moneef. 2015. "Science Education in the Islamic World: A Snapshot of the Role of Academies of Sciences." *Procedia-Social and Behavioral Sciences* 192: 359–363. https://doi.org/10.1016/j.sbspro.2015.06.051.

Rotgans, Jerome I., and Henk G. Schmidt. 2017. "The Relation Between Individual Interest and Knowledge Acquisition." *British Educational Research Journal* 43 (2): 350–371. https://doi.org/10.1002/berj.3268.

Roth, Wolff-Michael. 1997. "The Interaction of Students' Scientific and Religious Discourses: Two Case Studies." *International Journal of Science Education* 19 (2): 125–146. https://doi.org/10.1080/0950069970190201.

Saifuddin, Samina M. 2015. "Modeling Intention to Pursue a High Tech Career Using Social Cognitive Career Theory." PhD thesis, Carleton University. https://doi.org/10.22215/etd/2015-10939.

Schiefele, Ulrich, and Andreas Krapp. 1996. "Topic Interest and Free Recall of Expository Text." *Learning and Individual Differences* 8 (2): 141–160. https://doi.org/10.1016/S1041-6080(96)90030-8.

Schiefele, Ulrich, Andreas Krapp, and Adolf Winteler. 1992. "Interest as a Predictor of Academic Achievement: A Meta-Analysis of Research." In *The Role of Interest in Learning and Development*, edited by K. Ann Renninger, Suzanne Hidi, and Andreas Krapp, 183–212. New York: Psychology Press.

Sinclair, Raven. 2004. "Aboriginal Social Work Education in Canada: Decolonizing Pedagogy for the Seventh Generation." *First Peoples Child & Family Review* 1 (1): 49–61. https://fpcfr.com/index.php/FPCFR/article/view/10.

Smith, Graham Hingangaroa. 1995. "Falling Through the Cracks of the Constructivism Debate: The Neglect of the Māori Crisis within Science Education." *ACCESS-Critical Perspectives on Cultural and Policy Studies in Education* 13 (2): 103–121.

Smith, Joanna, and Jill Firth. 2011. "Qualitative Data Analysis: The Framework Approach." *Nurse Researcher* 18 (2): 52–62. 0.7748/nr2011.01.18.2.52.c8284.

Smith, Mike U., and Lawrence Scharmann. 2008. "A Multi-Year Program Developing an Explicit Reflective Pedagogy for Teaching Pre-Service Teachers the Nature of Science by Ostention." *Science & Education* 17: 219–248. https://doi.org/10.1007/s11191-006-9009-y.

Southerland, Sherry A., and Lawrence C. Scharmann. 2013. "Acknowledging the Religious Beliefs Students Bring into the Science Classroom: Using the Bounded Nature of science." *Theory into Practice* 52 (1): 59–65. https://doi.org/10.1080/07351690.2013.743778.

Southerland, Sherry, Alejandro Gallard, and Laurie Callihan. 2011. "Examining Teachers' Hurdles to 'Science for All.'" *International Journal of Science Education* 33 (16): 2183–2213. https://doi.org/10.1080/09500693.2010.530698.

Tan, Charlene. 2017. "Colonialism, Postcolonialism, Islam, and Education." In *Handbook of Islamic Education*, edited by Holger Daun and Reza Arjmand, 1–12. Cham: Springer International Publishing.

Tenneson, Michael, David Bundrick, and Matthew Stanford. 2015. "A New Survey Instrument and Its Findings for Relating Science and Theology." *Perspectives on Science & Christian Faith* 67 (3): 200–222.

Thoman, Dustin B., Jessi L. Smith, and Paul J. Silvia. 2011. "The Resource Replenishment Function of Interest." *Social Psychological and Personality Science* 2 (6): 592–599. https://doi.org/10.1177/1948550611402521.

Toscanelli, Cecilia, Elizabeth Shino, Sarah L. Robinson, and Amber Gayle Thalmayer. 2022. "Religiousness Worldwide: Translation of the Duke University Religion Index into 20 Languages and Validation across 27 Nations." *Measurement Instruments for the Social Sciences* 4 (1): 13. https://doi.org/10.1186/s42409-022-00041-2.

Toyon, Mohammad Abu Sayed. 2021. "Explanatory Sequential Design of Mixed Methods Research: Phases and Challenges." *International Journal of Research in Business and Social Science* 10 (5): 253–260. https://doi.org/10.20525/ijrbs.v10i5.1262.

Vadhan, Vimla, and Philip Stander. 1994. "Metacognitive Ability and Test Performance among College Students." *Journal of Psychology* 128 (3): 307–309. https://doi.org/10.1080/00223980.1994.9712733.

Waite, Hannah, and Nick Spencer. 2023. "Science and Religion Compass: Theos and The Faraday Institute." Theos and The Faraday Institute. https://www.sciencereligioncompass.org/. Accessed on 2 September.

Ward, Keith. 2008. *The Big Questions in Science and Religion*. West Conshohocken, PA: Templeton Foundation Press.

Yasri, Pratchayapong, Shagufta Arthur, Mike U. Smith, and Rebecca Mancy. 2013. "Relating Science and Religion: An Ontology of Taxonomies and Development of a Research Tool for Identifying Individual Views." *Science & Education* 22 (10): 2679–2707. https://doi.org/10.1007/s11191-013-9623-4.

Index

© The Editor(s) (if applicable) and The Author(s), under exclusive license to Springer Nature Switzerland AG 2026
A. I. Padela (ed.), *Bridging Bioscience and Islam*, Islam & Science,
https://doi.org/10.1007/978-3-032-09996-9

Q

R

S

FSC
www.fsc.org
MIX
Papier | Fördert
gute Waldnutzung
FSC® C083411

Kolibri 360 GmbH
Ferdinand-Jühlke-Straße 7
99095 Erfurt, Deutschland
produktsicherheit@kolibri360.de